T0317500

Handbook of Agricultural Entomology

Handbook of Agricultural Entomology

H. F. van Emden
Emeritus Professor of Horticulture
School of Agriculture, Policy and Development
University of Reading
Reading
UK

A John Wiley & Sons, Ltd., Publication

This edition first published 2013 © 2013 by John Wiley & Sons, Ltd.

Wiley-Blackwell is an imprint of John Wiley & Sons, formed by the merger of Wiley's global Scientific, Technical and Medical business with Blackwell Publishing.

Registered office: John Wiley & Sons, Ltd, The Atrium, Southern Gate, Chichester, West Sussex, PO19 8SQ, UK

Editorial offices: 9600 Garsington Road, Oxford, OX4 2DQ, UK
The Atrium, Southern Gate, Chichester, West Sussex, PO19 8SQ, UK
111 River Street, Hoboken, NJ 07030-5774, USA

For details of our global editorial offices, for customer services and for information about how to apply for permission to reuse the copyright material in this book please see our website at www.wiley.com/wiley-blackwell.

Library of Congress Cataloging-in-Publication Data
Van Emden, Helmut Fritz.
Handbook of agricultural entomology/H.F. Van Emden.
p. cm.
Includes bibliographical references and index.
ISBN 978-0-470-65913-7 (hardback: alk. paper)
1. Insect pests-identification. I. Title.
SB931.V25 2013
632'.7-dc23

21012025032

A catalogue record for this book is available from the British Library.

Wiley also publishes its books in a variety of electronic formats. Some content that appears in print may not be available in electronic books.

Cover image credits: Middle left: Soft scales (Coccidae) on citrus: (a) *Coccus hesperidum* and (b) olive scale (*Saissetia oleae*) (from Bayer 1968, with permission). Bottom left: *Rhyssa persuasoria* (from Mandahl-Barth 1974, with permission).
Top right: Pea moth (*Cydia nigricana*) and pod opened to show damage with caterpillars (from Bayer 1968, with permission). Middle right: Side view of head of (a) a short-nosed and (b) a long-nosed weevil (from Reitter 1908–1916). Bottom right: A spider beetle (Ptinidae) (from Reitter 1908–1916).

Cover design by Design Deluxe (e: info@designdeluxe.com)

Set in 9.5/12 pt Berkeley by Toppan Best-set Premedia Limited

1 2013

Contents

Preface

Although the Preface is almost the last part of the book to be written, at the time of writing I still don't know the title that will finish on the cover! In my mind it has always been 'Insects that matter', but this would have to be qualified so much as to become totally unwieldy. It is primarily about those insects that matter to agriculturists, horticulturists and foresters because the insects either attack their crops and livestock, are potential natural enemies of those that do, or are important pollinators. But I also include species that matter to humans in general because they attack them, transmit diseases or produce valuable or marketable substances. Also I include mites as 'honorary insects' in attacking plants in similar ways. But even that is not the end of it. Some species are included because they are eye-catching or have something about them that makes them very easy to recognise, and I also have not been able to resist adding others because they have something especially fascinating or bizarre about them! Try thinking of a title that encompasses that lot!

Why have I decided to write this book? For over 30 years from 1964 we ran an MSc course in the Technology of Crop Protection at the University of Reading. During this time I taught economic entomology to many hundreds of students from many different parts of the world. Each would find different pests of different relevance to them, but of all the books on agricultural entomology, I could only find the two books by Hill (1983 for the tropics and 1987 for temperate regions) as not restricted to a particular country or continent. In my career I have been lucky to visit each of the five major continents several times to teach and/or supervise research. This experience is coupled with my interaction with over 40 overseas research students back in Reading. I had therefore stood in most of the crops mentioned in this book, and had held many of the insects in my hand. I therefore felt I should write a textbook on 'insects that matter' with a world-wide coverage.

Of course, given the huge number of insect Families and species that are of economic importance in different parts of the world, I have had to use my experience to select those that seem to me to 'matter' most. A different author might well make a partially different selection. It is inevitable, given my location, that my account of the minor pests is probably weighted for northern Europe, I have probably also given more detailed information where I have personal knowledge of a species.

There are also many species where I have had to trawl other textbooks and the internet for information. I therefore expect the book not to be free of errors, especially as I have found inconsistent information given by different sources! Especially, information on generation times and number of generations may not be accurate for every region where the insect is found; also, world distribution and crops attacked by a pest may be incomplete. I would obviously appreciate receiving more accurate information from readers.

I have stressed in this Preface that space considerations restrict my selection of species for inclusion. I must also point out that changing crop management practices frequently result in the 'insects that matter' changing. Of what I consider the world's top ten insect pests today, none had this status when I was taught economic entomology in the early 1950s at Imperial College, and none may still have the same notoriety in 20 years' time. In those early 1950s, insecticides and today's synthetic fertilisers were not very widely used and hand labour was available for crop cultivation, including sowing and weed control; there were different crops, and the crops grown today are quite different varieties and often sown at different times. You will find quite a lot of anecdotes in this book giving examples of such changes in pest status. Something else that changes, and rather more frequently, is the Latin name given to a species and the classification of the relevant insect Order. I am not into insect taxonomy, but have tried to use Latin names that are the accepted ones today, as well as classifications that are relatively recent and widely used. Readers with some knowledge of entomology may be brought up short, as I was, by some of the recent changes in insect taxonomy; it is a strange outcome of Linnaeus's invention of binomial Latin names that these are now sometimes less useful for communication than the English common names, of which very few by comparison change with time!

This book concentrates on the entomology of the 'insects that matter'. It is not a book where you can look up how to control a particular pest problem. I focus on things like symptoms, damage caused, life histories and spot characters for recognition. I try to give sufficient information on how the different Orders are classified to enable the divisions of the Order down to relevant Families to be distinguished. I aim to equip you to be able to assign many of the insects you meet to their Family by concentrating on those characters that are relatively easy to recognise. This is sometimes only possible by ignoring a few, usually quite rare, exceptions, and therefore is in practice more useful, if not as accurate, as using definitive scientific keys. With a key you usually finish up at a name (whether or not correct). Yet you are likely to come across insects in Families I have not had space to include; I hope you will then realise that 'you won't find it here' instead of assigning it wrongly as you might with a key.

I have supplemented the text with a very large number of illustrations. You will find that many of these have arrows pointing at particular morphological characters. These are spot characters useful for recognition and I have purposely not labelled the arrows because – to be useful as well as more readily remembered – they need to be identified from the relevant description in the text. Therefore it is an unusual feature of this book that each illustration is near to its textual context rather than distantly on colour plates grouped together, and I am most grateful to Dr Ward Cooper at Wiley-Blackwell for agreeing that the book should be produced in this more costly way.

Identification to Family is often enough to enable you to spot an insect that might be a potential crop pest or, alternatively, a beneficial insect such as a predator, parasitoid or pollinator. It is surprising how quickly even a limited study of entomology leads to the ability to assign specimens to Families; then you become motivated to track down some you don't recognise, and your knowledge progressively increases!

Insects not only 'matter' to us, they are also amazing and often beautiful creatures who invented many things like camouflage, snorkels, aqualungs and Velcro long before we did. I really hope this book will encourage you to learn more about these fascinating animals. You may even get bitten (not literally!) by the entomology 'bug'?

H. F. van Emden
Reading, January 2012

Acknowledgements

In the Preface, I have already expressed my thanks to Dr Ward Cooper for agreeing to the number and to the scattering of illustrations in the text. Of the many pictures in this book, only a few have been drawn or photographed by me. I have therefore used images from a wide variety of sources. Many images have generously been made available to me by the copyright holder, for which permissions I am most grateful, as I am to the organisations and individuals who have freely put their images on the internet into the public domain.

In all cases where copyright might apply to images in books or on the internet, I have made every effort to establish whether or not such images are free of copyright. If not, I have sought permission, repeatedly if necessary, to reproduce them. I wish to offer sincere apologies to anyone whose copyright has been infringed due to difficulty I have had in obtaining the correct information.

With several books, the original publisher has gone out of business, and the titles are now owned by a different publishing house. However, the original contracts with authors and illustrators are usually no longer to be found, and so current title holders have asked me to make clear that their permission can only extend to the level of copyright that the original publisher held contractually. Again, I apologise where the 'mists of time' have obscured the true copyright position.

I hope I have acknowledged all the images correctly in the legends to the figures. With images from books, the acknowledgement takes the form of author and date of publication in the legend, cross-referencing with the full citation for the publication in the Bibliography at the back of the book.

I am especially grateful for the blanket permission I received to make use of any illustrations from two sources: Blandford Press and the illustrator Bergith Anthon allowed me to use illustrations in the Blandford Press series of books on Life in Field and Meadow, Pond and Stream and Woodland (Lyneborg 1968 and Mandahl-Barth 1973, 1974). Similarly, Bayer Crop Science gave me permission to use any paintings I wished from their Crop Protection Compendium. These illustrations include excellent pictures of damage symptoms.

I should also like to thank colleagues and past students in six continents for providing me with titles of books used locally in teaching agricultural entomology for inclusion in the Bibliography.

Kelvin Matthews and the production team at Wiley-Blackwell deserve my thanks, as well as those of future readers, for the care they have taken in the production of this volume. The extra effort that so many illustrations have involved, including the addition of scales to indicate the size of the insects, must have made their work rather more difficult than usual!

Finally, my love and thanks go to my wife Gillian for the real encouragement she has given me over several years to complete this project and the lively helpful interest she has shown in it.

1 The world of insects

1.1 The diversity of insects

The renowned 20th century geneticist, the late Professor J.B.S. Haldane was on a lecture tour in the USA when he was accosted by a female evangelist with the outburst 'Professor Haldane, in your many years of study of the natural world, you must surely have formed a view of the nature of the Creator'. 'Yes, Madam,' replied Haldane tersely, 'He is extraordinarily fond of beetles'. Figure 1.1 is an unusual visual representation of the living world, in which the size of the organism depicted is in proportion to the number of recorded species. You can see immediately that Professor Haldane had a point! Compare the size of the beetle with the icon for all the world's mammals (the elephant) and that for the entire plant kingdom (the trees). Moreover, beetles are just one group of insects; many of the other insect groups (represented in total by the other insect) similarly put the number of mammal and plant species in the shade.

How many kinds of insects are there? This is an impossible question to answer, since there are many still awaiting discovery and we can only guess at how many. One of many estimates is that there are 1.75 million named organisms in the living world of which 1.5 million are species of insects, with as many of the latter still to be discovered. To understand the concept of so many insect species, think of me reading out their Latin names at the realistic rate of 33 species a minute. My agricultural entomology course at Reading was 25 lectures, each of 52 minutes, and I taught this course for 20 years. Had I done nothing but use all my lecture time over these 20 years just to read out the names of insects (I guess the students would soon have stopped turning up?), I would only have got through rather more than half of just the known species.

The only habitat insects have not conquered is the sea, though they are found right down to the shoreline (e.g. kelp flies). Otherwise you will find them up mountains, in caves, in and on animals and plants both living and dead, in the air to great heights and in rivers and lakes to great depths. Some egg-parasitoid wasps (fairy flies) are tiny, and less than 0.2 mm long and weigh only 0.004 mg; it is said they could fly through the eye of a needle (though surely it would have to be a darning needle?). The largest insect alive today is the Goliath beetle, which is some 12 cm in length and broad with it. It weighs about 50 g. Yet all, however small, are fully functional animals with brains and quite complex behaviours – they are miracles of miniaturisation.

1.2 The impact of insects on us

We have to remember that the evolved diversity of insects when humans first appeared is not greatly different from what it is today. We date the origin of the human race to

Handbook of Agricultural Entomology, First Edition. H. F. van Emden.

Fig. 1.1 The diversity of the living world with icons in proportion to the number of known species. For example, the elephant represents all mammals, the pine trees the whole plant kingdom, the shell all molluscs etc. The beetle and fly (which represents all groups other than the beetles) represent the number of species of insects. (Modified from a cartoon by F.C. Fawcett and Q.D. Wheeler. www.coo.fieldofscience.com, with permission).

about 200,000 years ago, yet by then insects had already gone through well over 400 million years of evolution. Thus they have exploited the new resources created by the entry of the human race into history from that existing diversity far more than by evolving behaviours and properties they did not already possess.

Many people's first reaction to insects is that they eat our crops, and this book on agricultural/horticultural entomology is only likely to reinforce that impression. However, herbivory is actually a rather unusual evolution among insects. We recognise over 30 different evolutionary lines (called 'Orders') of insects, and only nine of these have any herbivorous species. Herbivory presents insects with huge problems. Insects find it hard to sustain their high nitrogen : carbon ratio on such low nitrogen : carbon food as plants provide, and most can only benefit from cellulose if friendly fungi or bacteria do the digestion for them. The waxy surface of plant leaves and the verticality of stems make attachment difficult, especially in strong winds and heavy rain, and life in the open away from the soil surface brings the dangers of desiccation and greater apparency to predators.

However, in spite of the relatively low diversity of herbivorous insects (most are either grasshoppers, bugs, moths, flies or beetles), their impact on people is huge. There is an oft-quoted statistic to the effect that we would need to grow food on only two-thirds of the current acreage if insects did not take so much of what we grow either in the field or in storage – in spite of our efforts to control them. Numbers make up for a lack in diversity. A swarm of locusts may weigh more than a 100 tons and a hectare of sugar beet may host 200 million aphids.

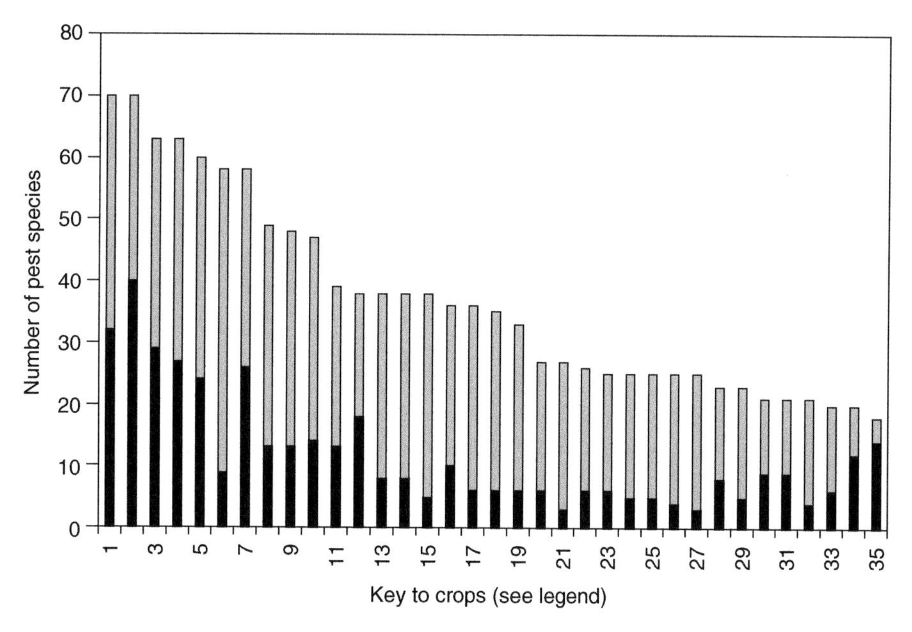

Fig. 1.2 Hierarchy of the 35 world crops with most pest species in relation to the number of major (black) and minor (grey) pests. Key to crops: 1, coffee; 2, rice; 3, citrus; 4, cotton; 5, pulses; 6, wheat; 7, maize; 8, apple; 9, sugar cane; 10, groundnut; 11, brassicas; 12, sorghum; 13, sweet potato; 14, tea; 15, potato; 16, coconut; 17, flowers; 18, cocoa; 19, tobacco; 20, tomato; 21, mango; 22, soybean; 23, banana; 24, castor; 25, guava; 26, pear; 27, fig; 28, peach; 29, strawberry; 30, macadamia; 31, oil palm; 32, sugar beet; 33, currants; 34, capsicum; 35, millet. (Data from Hill 1974).

Chinese cave paintings more than 6000 years old depict pests ravaging crops, and the book of Exodus in the Old Testament describes God visiting a plague of locusts on the errant Egyptians. The description here that the locusts 'darkened the land' (by the swarm blotting out the sun), ate 'every herb of the land' and that 'there remained not any green thing in the trees, or in the herbs of the field' is familiar to farmers suffering from locust swarms today. Insect plagues are certainly not just something from ancient history; famines in the Cameroon have occurred twice in the last 30 years as a result of plagues of armyworms (the caterpillars of a moth). Moreover, it is not only a question of what pests eat themselves; they can cause serious losses in other ways such as vectoring plant diseases, injecting saliva to which the plant may react badly and by fouling the plant with excreta. Figure 1.2 shows the 35 world crops that have the most pests recorded from them ranked in relation to the number of major and minor pests.

There are far more insect Orders with carnivores than with herbivores, though numerically carnivore populations tend to be smaller (often very much so) than those of herbivores. Carnivorous insects can be divided into predators, parasites and parasitoids.

Predators capture, kill and eat several (usually many) individual prey during their lifetime, though not all life stages necessarily share the habit. Thus many ladybird and hover fly species are important predators of aphids but adult hover flies are herbivorous and feed at flowers on pollen and nectar.

Parasites may utilise one or several individual hosts. Although their feeding may debilitate their host or infect it with a disease which may even be lethal, the feeding activity of a parasite is rarely directly lethal. Endoparasites feed on their host internally, and ectoparasites externally. The flea is an example of a parasitic insect.

Parasitoids resemble predators in that they kill their prey before emerging as an adult, but they differ in that they utilise just a single prey individual during their entire development, though one such individual may sustain several parasitoid individuals to maturity. Again parasitoids may be endoparasitoids or ectoparasitoids. The majority of insect endoparasitoids are within one group of the Order Hymenoptera.

Some carnivores are valuable predators and parasitoids of crop pests, but the first insects that humans will have encountered as problems, long before the relatively late development of agriculture in history, will have been parasites – biting and blood-sucking flies and bugs. The book of Exodus in the Bible records the plague of flies visited on the Egyptians by God to punish them.

The early hunter–gatherers will have been well aware of the pain and swellings inflicted by such insects, but it took until the late 19th century to establish that the bite of a mosquito could transmit malaria and incidentally also that insects could transmit diseases of plants. Malaria probably still kills about 2 million people each year, and was the reason for the Victorian explorers referring to West Africa as the 'White man's grave'. Biting flies and the diseases they transmit have had enormous influences on human history, including involvement in the fall of classical empires, and in the ability of people to live, raise livestock and even to wage war in different parts of the world.

The first insects, however, were probably scavengers on dead animal and plant material well before the oldest insect as yet found as a fossil appeared. This is an insect known as *Rhyniognatha hirsti*, which was found in about 400 million-year-old red sandstone in Scotland in 1926. It may even have had wings and perhaps fed on the spore-producing leaves of primitive plants allied to ferns.

Scavenging remains a widely distributed way of life in many Orders of modern insects. They remain important to us in breaking down the vast amounts of fallen leaf material in the tropics and in breaking down cattle dung and burying the corpses of small animals and birds. Insect scavengers are not just useful, they are essential to the cycle of life. However, they can also become pests if we put value on dead plant and animal materials. Thus scavengers such as clothes moths and some beetles attack our clothing and carpets, other beetles bore into the timber of our furniture and buildings; they even feed on the dead insects in museum collections.

Where would we be without bees and other insects to pollinate the many crops that would not produce fruits and seeds without insect activity? Although the pollen of some plants can be distributed by wind and water, many plants rely totally on insect pollination. Some (like snapdragons – *Antirrhinum*) have flowers designed so that only heavy insects like bees can 'trip' the opening of the flower to give access to the nectar and pollen. Cocoa is pollinated almost entirely by small midges. Plants that require insect pollination provide about 15% of the human diet in the USA.

One might suppose that most insects are economically neutral. The trouble is that the only way we are likely to find out is if an insect that seems of no importance to us disappears. For many years, an obscure moth called *Swammerdamia*, which lives on hawthorn, was regarded as economically neutral. However, by pure chance, the discovery was made in 1961 that its caterpillar is an essential overwintering host for an important natural enemy of a potentially serious pest of cabbages in the UK, the diamond-back moth (*Plutella xylostella*). Food webs can be very complex; the stability of an ecosystem depends on its herbivores, carnivores (including wasps!) and scavengers. Consequently, who can be sure that any insect is surplus to requirements?

1.3 The impact we have on insects

1.3.1 World distribution

Humans continue to move insects to new locations through travel and trade activities. Countries regularly pick up new pests in this way and most attempt to prevent, or at least delay, this by inspection and quarantining of potentially infested plants and other materials at points of entry such as docks and airports. The Colorado beetle (*Leptinotarsa decemlineata*) became a pest of potatoes when settlers started growing potatoes near the beetle's weed host in the foothills of the Rocky Mountains in the 19th century. The beetle then spread westwards, causing famine in the USA, and was accidentally introduced into France during World War I (a previous introduction into Germany in 1877 had been eradicated). It spread across Europe, but as yet all appearances in the UK have been eradicated or otherwise failed to establish. In the last 25 years, the Russian wheat aphid (*Diuraphis noxia*) has become a feared invader. It reached the USA in 1986. It occurs in many eastern parts of Europe, and countries such as Germany and the UK are attempting to predict whether it is likely to survive in their country and what its economic impact may be. Further afield, Australia is facing a similar threat.

1.3.2 Climate change

To the degree that humans are contributing to the pace of this phenomenon, they will be having an impact on all aspects of insect biology (especially fecundity and the number of generations per year) that are climate dependent. Also the distributions of insect species will change, either because of effects on their survival or because the distribution of the habitats and food they depend on change. Will malaria become a regular threat in the UK? Pests mentioned above, like the Colorado beetle and the Russian wheat aphid, may well find the UK a congenial environment!

1.3.3 Land management practices

The dramatic changes we make to natural habitats in order to fell trees, graze animals, grow crops and build houses have benefited some insects and disadvantaged others. Those that have benefited are of course insects that can exploit the riches of crop monocultures; most of these are quite rare in more natural habitats. Thus the frit fly (*Oscinella frit*), which can produce pest populations of 30 per m^2 from the stems and 110 per m^2 from the panicles of oat crops, ticks over in wild grasses at only 0.8 per m^2. Workers on cabbage root fly gave up a project to study the fly on wild brassicas because wild host plants were just too few. In another study, 60% of cages with carrot fly (*Psila rosae*) confined over potential umbellifer weed hosts produced no flies at all.

Climax vegetation, even if florally quite diverse, can cover huge areas with uniform light and humidity conditions (e.g. forest, prairie). For example tree felling creates clearings, and the grazing of livestock can create florally diverse grassland. Such 'semi-natural' man-made habitats encourage insect species that would be absent in the climax vegetation for an area. Many of the UK's butterflies rely on such 'artificial' habitats. We provide new habitats for mosquitoes to breed when we flood large areas for rice production and when we discard objects such as empty tins and car tyres, which can fill with water during rain.

Furthermore, many species suffer when we destroy their habitats and when we reduce the floral diversity within a habitat by, for example, the use of weedkillers and the removal of hedges in agricultural landscapes. It is suspected that such destruction has been largely responsible for the loss of four butterfly and 60 moth species and the decline by 70% of other butterfly and moth species during the 20th century.

When I was taught applied entomology in the mid-1950s, virtually none of the really important crop pests in the world today even got a mention. The whole pest spectrum has changed in the last 50 years. I can actually identify that most of the really important pests today have been man-made, either because of insecticides killing their natural enemies, new crop varieties that especially suit them being introduced, fertiliser use increasing or crop rotations and the timing of crops being changed.

1.4 Exploitation of insects

The domestication of honey bees is perhaps the most striking example of the commercial exploitation of insects. Bees not only produce honey, but also other saleable materials such as beeswax, with its many uses for candles, lubrication and polishes, royal jelly and a material (propolis) with antibiotic properties made from gums and resins. In many parts of the world beekeeping is a high-investment and large-scale industry, with pantechnicons transporting hives to rich natural nectar sources. Beehives are also rented out to farmers and growers to aid the pollination of their crops, or even sold for such purposes if the prevailing use of insecticides suggests it is unlikely the colonies can be returned to the beekeeper at the end of the crop season! The yield of some glasshouse crops like tomatoes is greatly increased with improved pollination, and cardboard bumblebee nests are commercially available.

Other insects reared commercially are a range of biological control agents, particularly for release to control pests in glasshouses.

Silkworms have been farmed on mulberry leaves for over 4000 years for their cocoons from each of which 1 kilometre of silken thread can be obtained. Other insects producing useful products are the scale insect *Lactifer lacca*, whose scale provides shellac for French polish and other purposes, a mealybug (*Dactylopius coccus*) whose dried bodies are ground to powder as the food-colouring agent cochineal, and aphids in China that make plant galls from which medicines and pigments for inks are extracted. To this list we can add *Ericerus pela*, the Chinese wax scale insect, the wax of which is used to make candles said to be of even higher quality than those made of beeswax.

Many insects are large and attractive, especially many butterflies and stick insects. 'Butterfly houses' are popular tourist attractions, but there is also a market (much of it now illegal) for such showy insects for collectors, or for using parts of insects (such as iridescent butterfly wings) in pictures and jewellery.

In some parts of the world insects or their products are sold as food. Honey has already been mentioned, but other examples are sweets made from the honeydew excreted by some sucking insects and the eggs of water bugs sold in parts of Central America as an ingredient for cakes. Some years ago a range of tinned insects, including 'fried locusts', 'stewed bumblebees' and 'chocolate-covered ants', was imported into the UK from the Orient; perhaps not surprisingly, the fashion was short lived.

1.5 Other uses humans make of insects

Insects may be collected and eaten as a form of free nutrition. Locusts, large beetle or moth grubs (e.g. the 'witchetty grub' of a swift moth) and honey-pot ants are often important sources of food for poor communities in developing countries.

The ability of maggots to clean wounds in human flesh has been known for centuries and exploited in field hospitals during battles before the advent of antibiotics. Today, the problems of antibiotic-resistant bacteria has caused a resurgence of interest in the use of maggots, particularly in the antibiotic compounds they secrete.

Finally, insects have uses in forensic science. The sequence and age of insects scavenging on human corpses has often enabled the time, and sometimes even the place, of a murder to be determined.

1.6 Insect classification

Like other members of the animal kingdom, the classification of insects follows a sequence of divisions into progressively smaller taxa from phylum to species (and sometimes a further division to subspecies). This is illustrated in Table 1.1 with reference to one particular insect, the peach–potato aphid. At each taxonomic level, a second parallel example has been added in the table.

Names of genera and species are written in italics, and the full citation of a specific name ends (not in italics) with the name of the person who first describes it (the 'taxonomic authority'). Such names have standard abbreviations, but today it is most

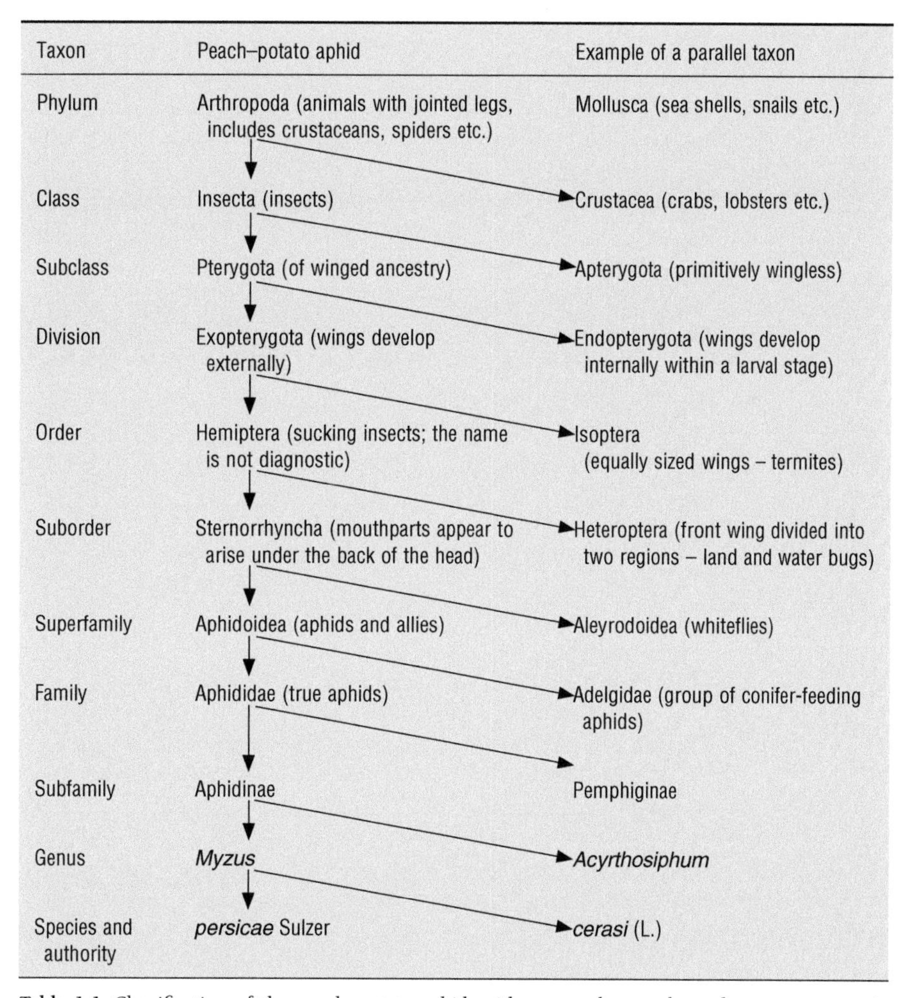

Taxon	Peach–potato aphid	Example of a parallel taxon
Phylum	Arthropoda (animals with jointed legs, includes crustaceans, spiders etc.)	Mollusca (sea shells, snails etc.)
Class	Insecta (insects)	Crustacea (crabs, lobsters etc.)
Subclass	Pterygota (of winged ancestry)	Apterygota (primitively wingless)
Division	Exopterygota (wings develop externally)	Endopterygota (wings develop internally within a larval stage)
Order	Hemiptera (sucking insects; the name is not diagnostic)	Isoptera (equally sized wings – termites)
Suborder	Sternorrhyncha (mouthparts appear to arise under the back of the head)	Heteroptera (front wing divided into two regions – land and water bugs)
Superfamily	Aphidoidea (aphids and allies)	Aleyrodoidea (whiteflies)
Family	Aphididae (true aphids)	Adelgidae (group of conifer-feeding aphids)
Subfamily	Aphidinae	Pemphiginae
Genus	*Myzus*	*Acyrthosiphum*
Species and authority	*persicae* Sulzer	*cerasi* (L.)

Table 1.1 Classification of the peach–potato aphid, with a second example at the same taxonomic level. Note that names of genera and species are written in italics.

common to give the name in full, other than for two especially frequent authorities of the past, Linnaeus and Fabricius. For these we use, respectively, the abbreviations L. and F. The authority is usually added only at the first mention of the species in a publication.

The name of the authority is sometimes found with, and sometimes without, brackets. So you may find both L. and (L.) or Buckton and (Buckton) for different species. Over time, many species have been assigned to a different genus from that originally proposed; the original authority's name is then still used, but enclosed in brackets. Thus *Myzus persicae* Sulzer in Table 1.1 indicates that Sulzer placed the aphid in the genus *Myzus* and that this allocation has not changed. By contrast, *Helicoverpa armigera* (L.) indicates that Linnaeus placed this moth in a different genus (*Heliothis* in fact) when he described it in the 18th century. After first mention in an article or in each chapter of a book, the genus is usually abbreviated to the initial capital (e.g. *M. persicae*) unless this would be ambiguous.

2 External features of insects – structure and function

2.1 Introduction

It is from the external features of an insect that we obtain most of the information needed to describe and identify it.

Much of the early descriptive studies on insects were done in the 17th and 18th centuries, when biology was more of a hobby than a profession. Many of those involved came from two professions which then (totally unlike today!) offered plenty of leisure time. Both doctors and clergymen would have had a classical education, and so derived most of their terminology from Latin and Greek roots. Moreover, they, especially the doctors, were familiar with human anatomy and so sought to find analogies between insect and human structure. For example, the legs of insects have a 'tarsus', 'tibia' and 'femur' but then there are two further sections, and these were given names from the ball joint at the top of the femur. The next section of leg above the femur is the 'trochanter' (from the expanded head of the human femur) and the final one is the 'coxa' (from the coxal cavity, which accommodates the head of the femur in our pelvic girdle).

Sometimes the analogies were completely wrong. For example, the grooves on the head of insects looked like the 'sutures' on the human skull. There they are points of weakness where the bones of the skull meet, but in the insect the skull is actually thickest at these points (see Section 2.4). So here and elsewhere solid plates were regarded as divided. Another problem created by the early entomologists was a failure to identify that structures that looked different were really homologous. The result, for example, is that the dorsal plates on the thorax have totally different terminology from those on the abdomen and that different terminology is sometimes given to what are really identical structures but in different insect Orders.

The terminology has now been so long ingrained that no one is suggesting simplifying the terminology of entomology by creating a new one! So more terminology has to be learnt than would be necessary given our more modern understanding of insect structure. As new terms are introduced they will be highlighted in bold italics to make it easier to use this chapter as a glossary. This convention of highlighting new terms will continue throughout the book.

This book is not intended to cover insect physiology, and readers are referred to the excellent book by Chapman (1998). However, it would be needlessly uninformative to pass over structures without some reference to their function but readers are warned that such information here is very much a 'rough guide' to general principles.

Handbook of Agricultural Entomology, First Edition. H. F. van Emden.

2.2 The exoskeleton

As hinted above, the structure of insects and humans is hardly comparable. Firstly, humans have an internal skeleton of bone supporting the other tissues and organs around it, whereas the skeleton of insects is external with the other tissues and organs contained within it. Secondly, humans have a circulatory system in which blood is pumped through vessels around the body by a heart and which plays an important role in the gas exchange involved in respiration. By contrast, the insect organs are bathed in a body fluid (***haemolymph***), which is moved by body movements and a tubular dorsal ***heart*** – a pulsating tube open at both ends. The respiratory system (see Section 2.6.4) is quite separate.

The insect ***exoskeleton*** is a ***cuticle*** of dead proteinaceous material known as ***chitin***. Every external structure as well as the first part of the gut is lined with cuticle. If we dissolve all the living material by boiling the insect in caustic potash, the exoskeleton that remains shows every detail of every hair and pore on the surface. Beneath this chitinous cuticle is the living skin of the insect, called the ***hypodermis*** because it is 'under' the exoskeleton in contrast to our outer *epi*dermis.

The cuticle has several layers (Fig. 2.1), which together form a barrier to the loss of liquids from within and the entry of liquids from the outside. The latter barrier includes a very thin ***wax layer*** on its surface. A process known as sclerotisation has hardened much of the cuticle, and the hard plates which form the armour of the exoskeleton are known as ***sclerites***. When we identify a part of the exoskeleton with a name, we are almost always identifying a sclerite. Now, a suit of armour plating is of little practical use without joints between the plates of armour, and so there is also cuticle in the form of flexible ***arthrodial membrane*** at the joints, for example between the parts of the leg and different parts of the body.

As the insect grows while a juvenile, it adds weight steadily and gradually. As the exoskeleton is mostly rigid, it has to be shed from time to time and replaced with a larger version. This is akin to replacing the jackets of children as they grow. The child grows continuously, but the jacket appears to grow in jerks!

In insects, the phenomenon of exoskeleton shedding is called ***moulting***, and each actual event is an ***ecdysis.*** The periods between the ecdyses are the stages of the insect known as ***instars***. The number of instars varies, but is usually constant within a species or even within an Order.

Moulting is regulated by ***juvenile hormone***, a compound secreted by glands attached to the brain and acting as the instar timer. The timer is the gradual increase of the

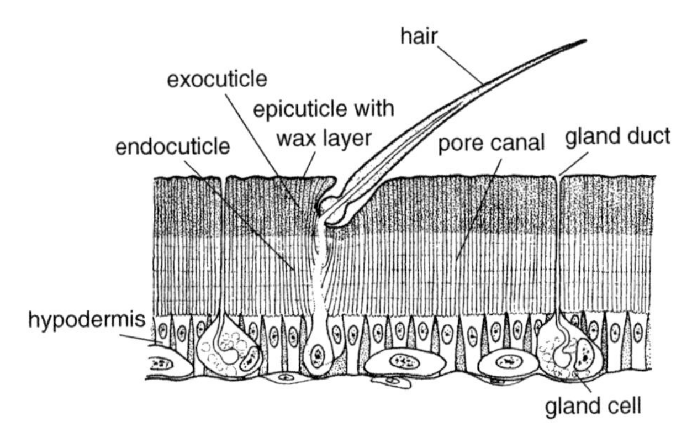

Fig. 2.1 Section through cuticle (from Richards and Davies 1977).

titre of juvenile hormone in the haemolymph and, when it reaches a certain level, the trigger for ecdysis is another hormone (the moulting hormone ***ecdysone***) released from glands in the body just behind the head. Meanwhile, the level of juvenile hormone declines to reset the timer for the next instar. Ecdysone stimulates glands in the hypodermis to secrete an enzyme (***chitinase***) to dissolve the chitin of the old cuticle and weaken it from the inside, while the wax layer on the new cuticle forming on the hypodermis prevents the same fate befalling the new chitin. At ecdysis, and before the new cuticle is sclerotised, the insect breaks out of the old cuticle and fills its gut and any air sacs with air to expand its size while the new cuticle hardens. This makes the new exoskeleton a 'loose fit', allowing further growth of the insect during the next instar. The instar count is complete when the juvenile hormone is no longer destroyed with a release of ecdysone and either the adult or a pupa appears from the old skin as the next life-stage.

2.3 The basic body plan of the insect

The insect body consists of a number of sclerotised 'boxes' called ***segments***. The box has the simple basic construction of four sclerites, again joined by arthrodial membrane (Fig. 2.2). The dorsal plate is the ***tergum*** (plural ***terga*** or ***tergites***), the ventral plate the ***sternum*** (plural ***sterna*** or ***sternites***) and the side plates are each a ***pleuron*** (plural ***pleura*** or ***pleurites***). The words sternum and pleuron (as pleural cavity, pleurisy etc.) probably ring a bell from human anatomy. The boxes (segments) so formed can be thought of as railway carriages coupled by arthrodial membrane, with the gut and nerve cord running through them.

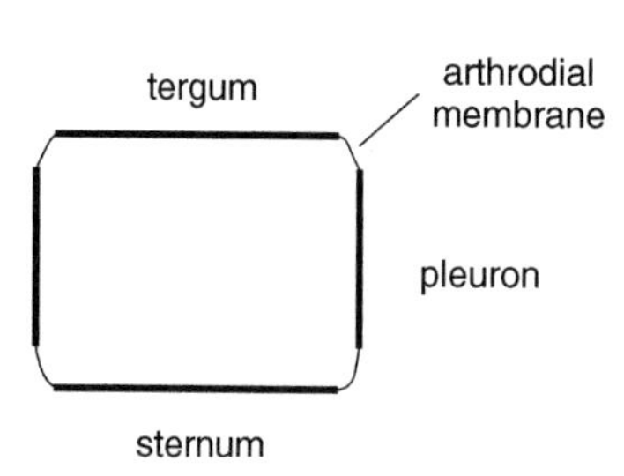

Fig. 2.2 Basic cross-section of the insect exoskeleton.

The segments at the rear of the 'train' form the ***abdomen***. The number of segments forming the abdomen varies with the type of insect (usually with the Order) and is typically six to eleven. It is in the abdomen that the vital functions of digestion, excretion and reproduction occur (see Section 2.6); thus the abdomen is really the functional insect.

This functional abdomen is carried from place to place by the ***thorax***, the engine of the train. This consists of just three segments, which are distinguished by Greek names for 'front', 'middle' and 'back' as the ***prothorax***, the ***mesothorax*** and the ***metathorax***. The thorax is just a large muscle box working the wings and legs – the part of the gut passing through it is not involved in absorbing food, but merely joins the mouth to the abdomen. Jointed legs are found on all three segments; wings only occur on the meso- and metathorax. The terminology for the three ***thoracic*** segments is used to identify structures associated with the thorax, for example the ***prothoracic*** leg.

The ***head*** is the 'train driver' which tells the thorax where it is taking the abdomen. It is therefore well equipped with ***sense organs*** for smell, sound, touch and taste.

Figure 2.3 combines the above information in one simple diagram.

2.4 The head

The head has about six segments fused indistinguishably into a single ***head capsule*** (Fig. 2.4). The capsule is divided into various contiguous regions by analogy with the human skull (in brackets), for example ***gena*** (cheek), ***frons*** (frontal bone), ***vertex*** (temple) and ***clypeus*** (area below the nose). The clypeus joins onto a loose flap, the

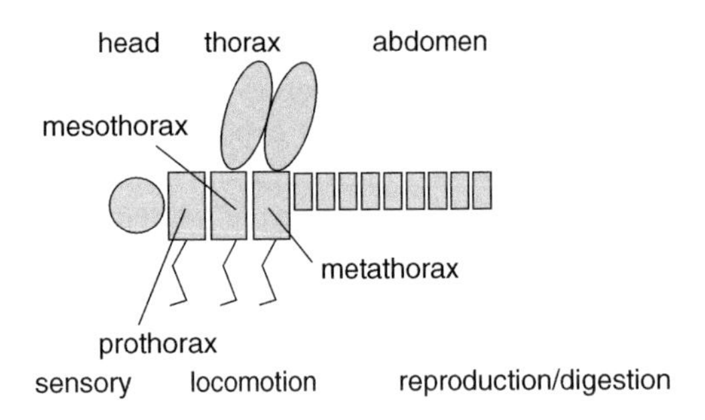

Fig. 2.3 The main divisions of the insect body.

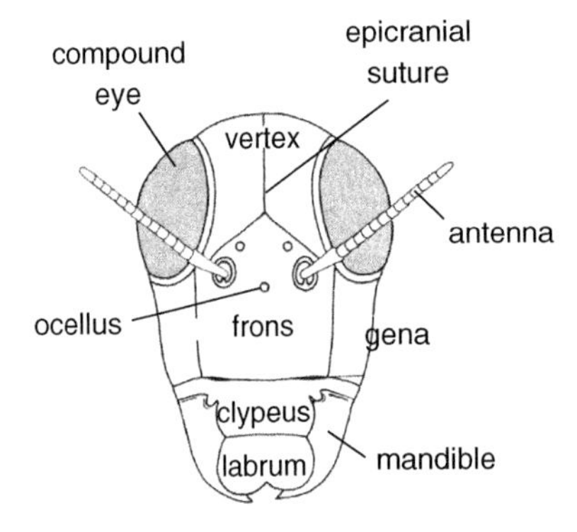

Fig. 2.4 Front view of an insect head, showing the sclerites of the head capsule and other external features (from Chapman 1971, with permission).

labrum (upper lip). These areas enable the position of other features to be established. Thus there may be pale ***frontal*** patches or ***vertical*** bristles (the latter so-called because they are situated on the vertex; they are not necessarily perpendicular).

Grooves on the head are identified as ***sutures*** and by which regions of the head they separate, for example ***frontoclypeal suture***. Common is a Y-shaped suture across the vertex (Fig. 2.4); this is the ***epicranial suture***. In Section 2.1, I pointed out that these sutures were not lines of weakness; they are the outward signs that additional cuticle projects into the head at that point. Indeed, these 'tongues' of cuticle in the head fuse to form internal scaffolding (Fig. 2.5), called the ***tentorium***. The extra cuticle inside the head is not just to provide extra rigidity; the lower struts provide extra area for attachment of the powerful ***adductor muscles*** needed to close the jaws of the insect when biting. Details of the dissected-out tentorium are used in identifying bumble bees – unfortunately the specimen is not much use after identification! The head of course bears the sense organs, linked to a large mass of nervous tissue which can fairly be called the brain. From the brain, a double nerve cord continues rearwards on the ventral side of the body tissues. For this reason (and the dorsal heart) it has often been quipped that an insect is 'a vertebrate lying on its back'.

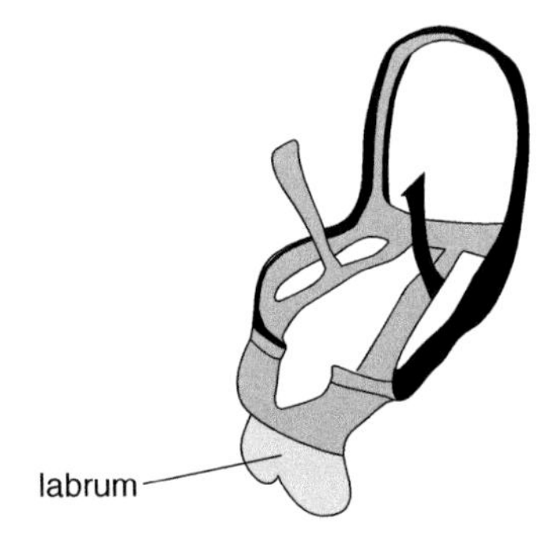

Fig. 2.5 A typical tentorium dissected out from the head capsule.

2.4.1 Sense organs

2.4.1.1 Antennae

The ***antennae*** or 'feelers' have a strong ***olfactory*** (smelling) function, though they may also be used for taste (***gustation***) and in some insects are actually their 'ears'. The structure of a simple antenna is shown in Fig. 2.6. Regardless of any variation in size of the 'segments', the first is identified as the ***scape***, the second as the ***pedicel*** and the remainder in combination as the ***flagellum***.

However, insect antennae are very far from uniform in design (Fig. 2.7) and their variety is a great aid to identification. This introduces a concept that is very important in entomology. Structures can be modified from a basic plan (almost to the point of non-recognition of which parts are which) in adaptation to different life styles. This gives us a whole vocabulary for the appearance of different antennae (Fig. 2.7). This vocabulary can be explained as follows:

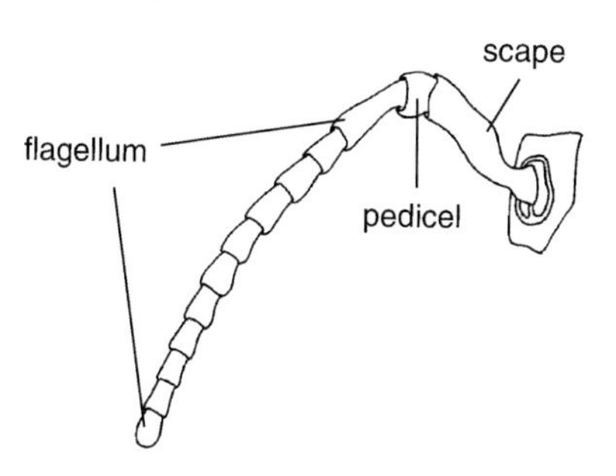

Fig. 2.6 Structure of a simple insect antenna (from Chapman 1971, with permission).

filiform=a string of almost identical segments, each rather tubular in outline;
moniliform=a string of beads, spherical with a clear constriction between the segments;
setaceous=very narrow segments, which together look more like a single bristle;
serrate=an analogy to saw-teeth; the segments are expanded at their tips on one side only;
pectinate=comb shaped, an extreme form of 'serrate'. The expansions of each segment are extremely long;
clavate=the segments gradually increase in thickness towards the tip;
clubbed=similar to 'clavate', but the segments are of even diameter until the last few are suddenly expanded to form an obvious 'club';
geniculate=simple or clubbed, but with a long pedicel from which the flagellum arises at nearly a right angle;
lamellate=often 'geniculate', with the flagellum segments adpressed together like the parts of a fan, and capable of opening and closing in a similar way;
plumose=whorls of hairs project from the joints between the segments;
cyclorrhaphous=named after the taxon of flies with such antennae (see Section 10.4) these antennae have a small scape and pedicel, with the flagellum represented by an obvious lozenge-shaped segment and by the bristle (***arista***) set back on it and projecting from it.

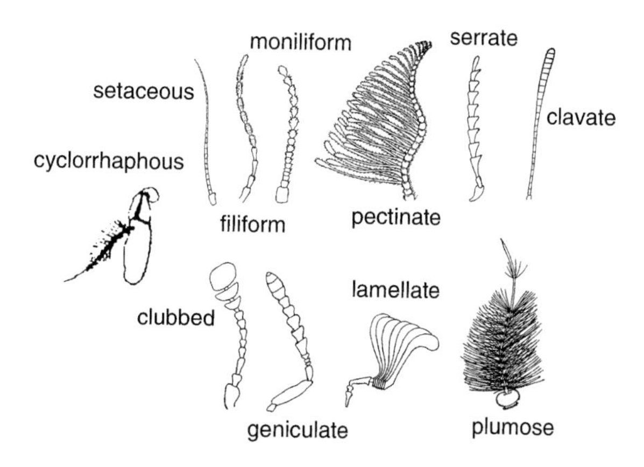

Fig. 2.7 Various types of insect antennae (see text) (from Richards and Davies 1977).

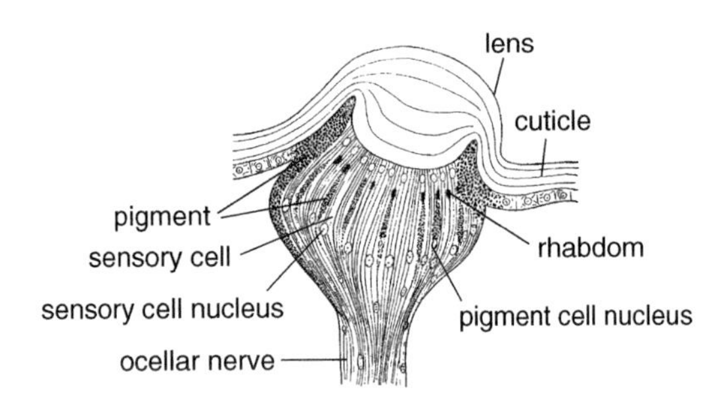

Fig. 2.8 Structure of ocellus (from Richards and Davies 1977).

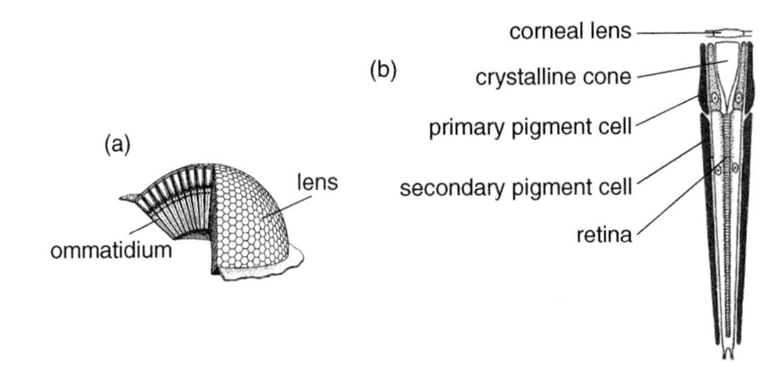

Fig. 2.9 (a) Compound eye section; (b) single ommatidium (from Zanetti 1977).

Examples of the adaptations shown by these antenna types are the cyclorrhaphous antenna, which is streamlined for a rapidly flying insect, and the lamellate antenna, which can provide a large area for picking up odours yet be folded away when in flight or burrowing in cattle dung (seems a good idea!).

2.4.1.2 Ocelli

Only some insects have two or three ***ocelli*** (Fig. 2.4). These very simple light-detection structures may also be found on insect larvae, and are the only 'eyes' caterpillars and grubs ever possess. An ***ocellus*** is a clear dome of cuticle over a pigmented patch of hypodermis supplied with a nerve connection (Fig. 2.8). Ocelli would appear able to act as light meters, but also seem to provide the insect with rather more complex information. The standard technique to determine the functions of a sense organ is to prevent the insect from using it – with ocelli this is easily achieved by painting them over with an opaque wax. If this is done for the ocelli of locusts, the insects cannot take off for flight and bees with their ocelli 'disabled' lose the ability to detect the plane of polarisation of light, a talent which enables them to navigate relative to the sun even under overcast conditions.

2.4.1.3 Compound eyes

Compound eyes are characteristic of most insects, but are never found in caterpillars and grubs. Compound eyes are exactly 'what it says on the tin': they are a dome of individual eyes (each eye is called an ***ommatidium***) with their adjacent lenses making a honeycomb pattern (Fig. 2.9a). The number of individual eyes in a compound eye can be very large; for example, a housefly eye has about 4000 ommatidia.

A tabloid newspaper once featured a girl in a bikini photographed through an area of lenses stripped off from a water beetle's compound eye. The definition of the several hundred adjacent images was pretty good, and the headline ran something like 'Lucky water beetle – we only get *one* image of this beauty'. The water beetle is nowhere as lucky as the paper claimed. It is a mystery why the lenses of insects should be capable of any definition, as none is available in practice. This lack of acuity is for two reasons. First, the focal length of the lenses is so long that the retina would have to be some distance beyond the other side of the head of the insect. Second, ommatidia (Fig. 2.9b) are slender inverted cones with the retina not parallel to the lens, but at right angles to it! The most an ommatidium can provide is information about the light level falling on the retina through the lens. Accordingly, the brain receives a pattern of 'pixels' of different darkness (Fig. 2.10), but the trade-off for limited definition is that the many lenses in a domed structure are ideal for detecting the direction of a change in light intensity, for example the shadow of an approaching predator.

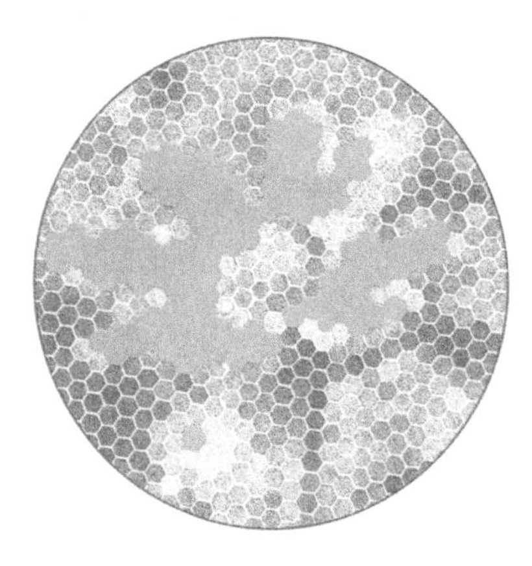

Fig. 2.10 Impression of how an insect eye sees a flower (from Zanetti 1977).

In day-active insects, each ommatidium is screened from stray light through adjacent lenses by a shroud of pigment. Such eyes are called ***apposition eyes*** in contrast to the ***superposition eyes*** characteristic of nocturnal insects. Here the pigment is missing, giving brighter 'night vision', but of course at the expense of clarity. Some insects can expand and withdraw the pigment to switch their eyes between apposition and superposition, depending on ambient light levels.

2.4.2 Basic structure of the mouthparts

Like the antennae, the mouthparts of different insects have been adapted for different feeding methods, such as sucking and licking, to an extent where it is often hard to identify which parts are homologous with the various parts of the simple ancestral biting structures. Examples will be found at the start of Section 7.5 and in Section 10.3.1; in this chapter I shall only describe the basic biting mouthparts (Fig. 2.11) as found in insects such as cockroaches and grasshoppers.

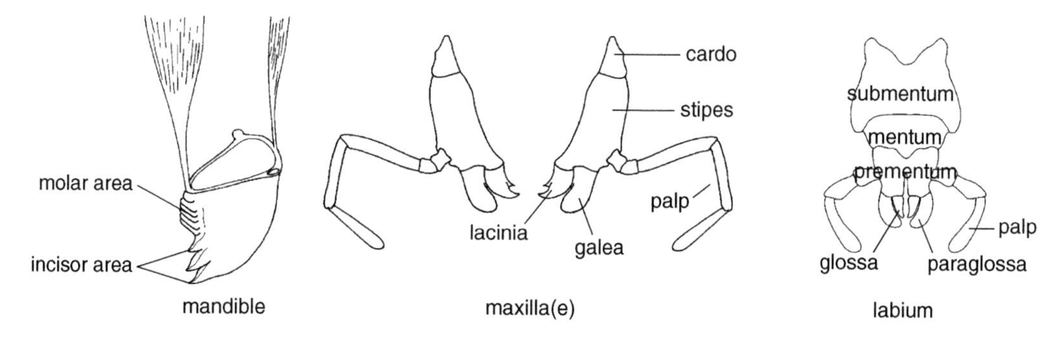

Fig. 2.11 Basic structure of simple biting mouthparts (from Chapman 1971, with permission.)

Three pairs of legs from the segments fused to form the insect head have been modified to provide the mouthparts. These lie under the upper lip or labrum (Fig. 2.4) and consist of the ***mandibles*** (=human upper jaw), ***maxillae*** (=lower jaw) and ***labium*** (=lower lip).

2.4.2.1 Mandibles

These are identifiably two heavily-sclerotised opposing jaws with an apical toothed ***incisor area*** for cutting and often a basal grooved ***molar area*** for grinding. The jaws have two ***condyles*** (ball-shaped protrusions), which articulate in sockets at the base of the head capsule. At right-angles to the hinge so formed, large powerful ***adductor muscles*** are attached near the cutting/grinding face of each mandible to close the jaws; behind the condyles at the back of the mandibles are attached the much smaller ***abductor muscles*** that open the jaws.

2.4.2.2 Maxillae

These are located between the mandibles and the labium (see Section 2.4.2.3). Each ***maxilla*** has a basal part, consisting of a usually smaller ***cardo*** articulating with the head capsule, and a larger ***stipes*** which forms the attachment for the main functional parts of the maxilla.

The innermost such attachment is a hard, sclerotised and toothed ***lacinia***, outside of which is a softer lobe, the ***galea***, equipped with a battery of sense organs. The lacinia holds the food when the mandibles let go as they open to take the next bite. As the food is fed into the mouth by the galeae, it is continuously monitored and its flavour assessed by the sense organs on the latter. Similar sensory organs are found at the tip of a segmented and mobile ***maxillary palp*** mounted near the apex of the stipes.

2.4.2.3 Labium

This forms the lower lip of the mouth and is another structure equipped with sense organs to monitor the food before the maxillae move it forwards towards the mandibles. Unlike the mandibles and maxillae, the labium looks like a single structure. However, the symmetry of its left and right halves are the give away that the labium has been formed by two 'legs' of very similar design to maxillae, but fused together at the mid-line.

The ***submentum*** is a plate which, like the cardo of the maxilla, articulates with the head capsule. Forward of the submentum are the ***mentum*** and ***prementum***. The prementum, like the stipes of the maxilla, carries two lobes and a ***labial palp*** on either side of the mid-line. The innermost lobe is the ***glossa*** (which is also the name of the human tongue). The outer and larger lobe is the ***paraglossa***. Labial palps, glossae and paraglossae are all equipped with sense organs for monitoring the food before it is eaten.

Look at Fig. 2.11. It is easy to envisage a mid-line along which two 'maxilla-like' mouthparts have fused to make a single lower lip.

The structure of the mouthparts as illustrated in Fig. 2.11 does vary in detail in different insects with this basic biting equipment. For example, there is variation in how the base of the labium (shown as submentum, mentum and prementum in Fig. 2.11) is divided, and in locusts you will find the glossae asymmetrically very unequal in size.

2.5 The thorax

2.5.1 Thoracic sclerites

The sclerites of the thorax (Fig. 2.12) provide a particularly good example of how the early entomologists imposed a whole new terminology on what are effectively the same four sclerites as found in the basic insect cross-section (Fig. 2.2). Only the sternum retains the same name as the ventral sclerite on the abdomen, and so – with the

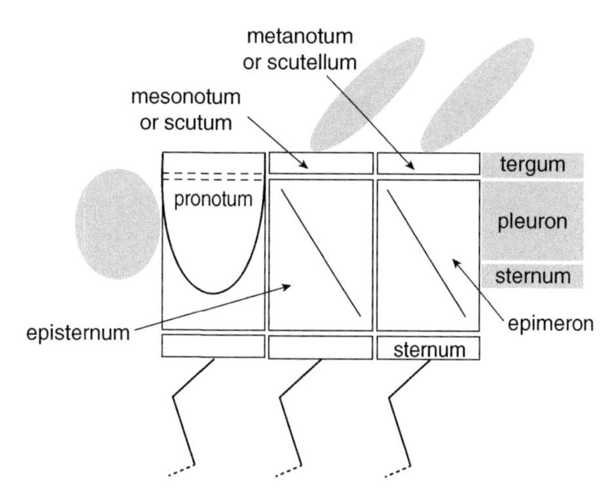

Fig. 2.12 The sclerites of the insect thorax.

appropriate prefix for the three segments of the thorax – we can identify them as ***prosternum***, ***mesosternum*** and ***metasternum***.

The terga are not obviously the same simple dorsal plates as on the abdomen, and so were given a totally new nomenclature around the word 'notum'. The tergum of the prothorax is particularly large in most insects, and is often expanded laterally and downwards to obscure most of the side of the prothorax. As a result, the ***pronotum*** is often a more obvious and distinct structure than the ***mesonotum*** and ***metanotum***, particularly as the wing articulations on the meso- and metathorax leave less room for the sclerites, which as a consequence are greatly reduced in size and often partly hidden by the wings at rest. In some Orders of insects the metanotum is visible as a triangle between the wing bases, and often has yet another name, the ***scutellum*** ('little shield'). Similarly, the mesonotum may be referred to as the ***scutum*** in the Hymenoptera. However, a 'little shield' also called the scutellum is part of the mesonotum in other Orders (Hemiptera and Coleoptera).

The pleuron may indeed be referred to as such on the thorax. However, because of the need for extra areas of sclerotised cuticle for muscle attachments for the legs in the thorax, we find grooves on the outside of the meso- and metapluron, which apparently split these pleura into two parts. These have been given different names. Figure 2.12 shows how the visible groove leads to the anterior part encompassing the junction of the pleuron with the sternum, leading to its label of ***episternum***. The posterior part is called the ***epimeron***. So in identifying an insect, you may be asked to look for a bunch of hairs on the ***metaepisternum***, or for patterns in the cuticle of the ***mesoepimeron***.

The thorax is effectively a group of three muscle boxes, providing large surface areas of heavily sclerotised chitin for attachment of the muscles to work the legs and wings. Even the section of gut that passes through it has a chitinised lining and no digestive function until the abdomen is reached (see Section 2.6). However, the locomotion of the insect requires considerable nervous co-ordination, and the ventral part of the thorax has large ***ganglia*** of nerve material in the nerve cord from the brain. Just inside the thorax is another important ganglion, the ***prothoracic ganglion***, concerned with the movement of the mouthparts.

2.5.2 Legs

The basic leg structure (Fig. 2.13) includes a joint that looks like the human knee. The early entomologists therefore started at the 'knee' and used human anatomy terms in both directions. Upwards there was the ***femur*** (plural ***femora***), but thereafter no more human leg sections exist. As pointed out earlier, the remaining two insect leg segments used names from the articulation of the human femur and pelvic girdle – first the ***trochanter*** and then the ***coxa*** for the top segment that articulates with the thorax. Downwards from the knee there are usually many more than two segments. The first obviously became the ***tibia*** (=human shin bone) and everything left over was called the ***tarsus*** (=foot). The number of ***tarsal segments*** varies between insects, and can be a useful aid to identification. With some beetle families, the number of tarsal segments varies between the three pairs of legs, giving a ***tarsal formula***, for example 3–4–5 (see Section 12.1.1).

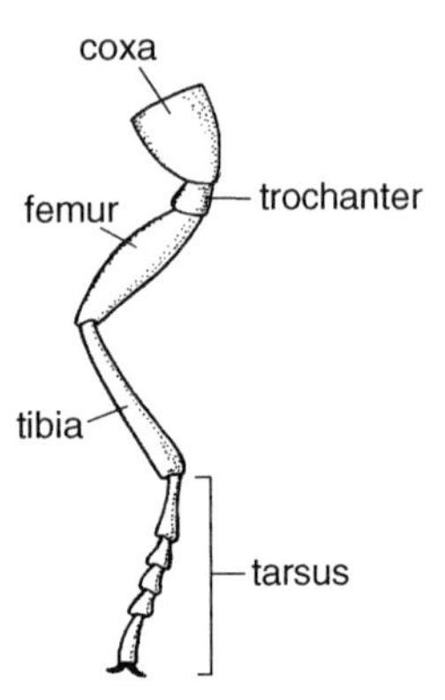

Fig. 2.13 Parts of the insect leg (from Zahradník and Chvála 1989).

The tarsus usually ends with structures for gripping surfaces (Fig. 2.14). Together these form the ***pretarsus***. ***Claws*** are common, verging on the universal, but between the claws there may be soft pads. An ***empodium*** consists of two pads with Velcro-like outer surfaces, and a ***pulvillus*** is a single bladder-like pad that seems to give adhesion in a manner like that of a vacuum cup.

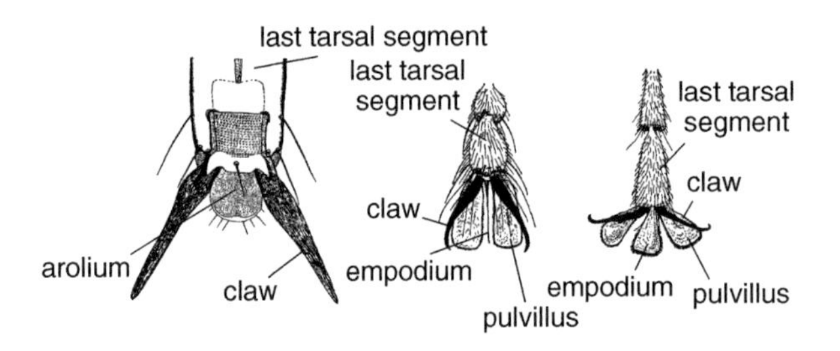

Fig. 2.14 The pretarsus of three different insects to show different pretarsal structures: claws, pulvillus and empodium (from Richards and Davies 1977).

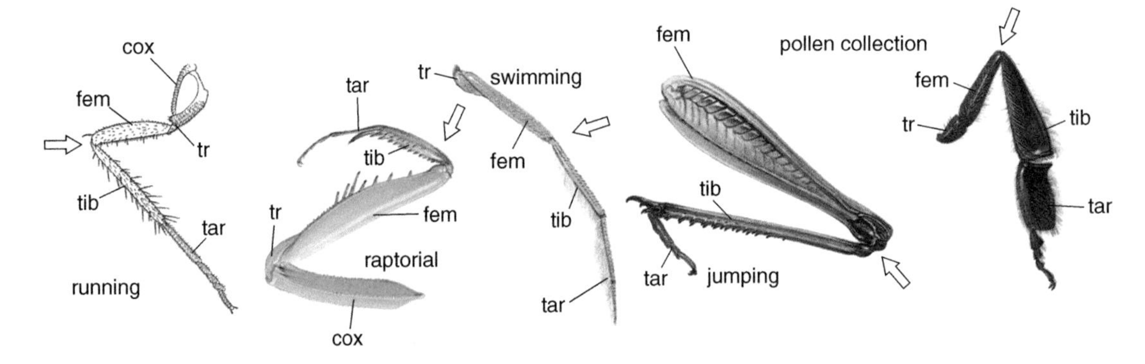

Fig. 2.15 Insect legs adapted for the different purposes indicated; in each example the knee joint is arrowed: cox, coxa; tr, trochanter; fem, femur; tib, tibia; tar, tarsus (running leg from Richards and Davies 1977; others from Zanetti 1977).

Like the antennae and mouthparts, the legs of insects are highly modified in appearance for different purposes (Fig. 2.15). Often the front or hind legs will differ markedly from the others. Thus in Fig. 2.15 it is the ***forelegs*** of the praying mantis that are modified for catching prey, and in the locust it is only the ***hindlegs*** that have enlarged femora to accommodate the muscles needed for jumping. In working out the segments of any type of leg, the secret is to find the knee joint (arrowed in Fig. 2.15) and then from there to identify the segments in both directions.

2.5.3 Wings

Where present, the wings arise as outgrowths of cuticle from the meso- and metathorax. Wing ***veins***, which may fork along their path, fan out from the base of the wing. These veins are pumped full of body fluids when the newly moulted adult has to expand its wings from small buds to their full size; thereafter, the veins are empty rigid tubes which give the wing rigidity as the spokes do to an umbrella. The pattern of veins (a pattern called the ***venation***; Fig. 2.16a) differs between taxa and is much used in identification. In places, the radiating veins are linked by cross veins like the vertical mortar joints in a brick wall. The areas of the wing bounded on all sides by veins (like 'bricks') are called ***cells***; the presence and shape of particular cells is again used in identification.

It is the veins on the forewing of insects that are mainly used in classification, and these veins are divided into a number of sectors, though how completely they are

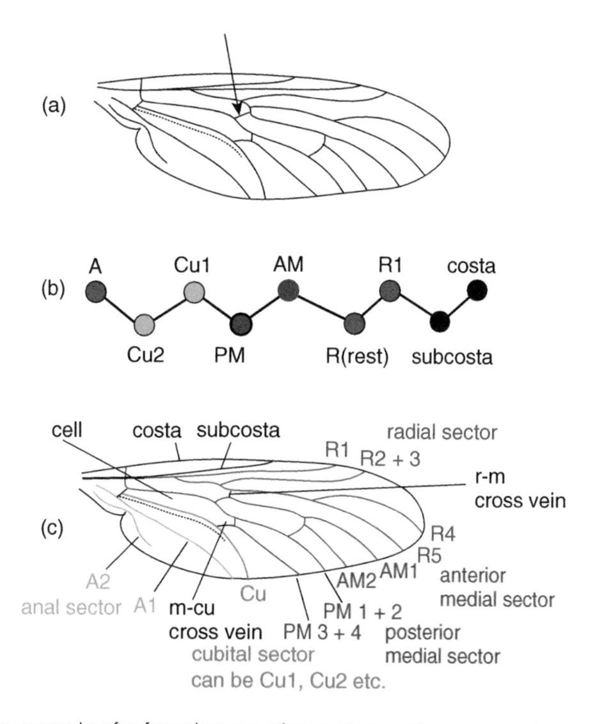

Fig. 2.16 (a) An example of a forewing venation pattern, with a cross vein arrowed; (b) identification of + and – veins (sectors: R, radial; AM, anterior medial; PM, posterior medial; Cu, cubital; A, anal); (c) interpretation of the venation shown in (a).

represented varies greatly with taxon. The sectors starting from the leading edge of the wing (and their abbreviations) are:

costa (C) the vein at the leading edge and ***subcosta*** (Sc);

radial sector (R) which may subdivide into the first radial vein, second radial vein etc. (R1, R2 etc.);

medial sector (M) which may subdivide into the ***anterior medial sector*** (AM or AM1, AM2 etc.) and the ***posterior medial sector*** (PM or PM1, PM2 etc.);

cubital sector (Cu) which may fork to Cu1, Cu2 etc.;

anal sector (A) which may have several branches (A1, A2 etc.).

Cross veins link sectors and are named by the sectors they link, for example the ***r-cu*** and ***m-cu cross veins***.

Theoretically, such cross veins should help us sort out the sectors of veins on a wing, but do they? Look at the arrow in Fig. 2.16a. Does it point to a cross vein or merely where a vein forks? It is, in fact, safer to work out the sectors in a different way, and only then identify the cross veins.

If you look at an insect wing under sufficient magnification, it becomes apparent that the wing is not like a flat sheet of paper, but more like a concertina-folded paper with the veins at the apex and troughs of folds. Those at the apices are called +veins and those in the troughs −veins. This is illustrated in Fig. 2.16b; you will note that each time we move from a trough to an apex we also move to the next sector. This is a convention for all insects and with it we can identify the sectors for the pattern of venation of Fig. 2.16a to arrive at Fig. 2.16c.

In the higher insect Orders such as the Lepidoptera and Hymenoptera, the hind wing is usually noticeably smaller than the forewing. In flight the power comes from the forewing and the hind wing is linked to it in order to increase the wing surface area. However, when the insect is at rest, the wings are de-coupled and the forewing slides over the hind wing on the back of the insect. The forewing, as it swings into the flight position, automatically 'picks up' the leading edge of the hindwing by a ***coupling mechanism***. Figure 2.17 shows three such mechanisms viewed from underneath. A ***jugum*** is a simple projecting flap which lifts the hind wing with the forewing, whereas a ***frenulum*** (found in some moths) consists of a long bristle sliding in a bridge of cuticle on the underside of the forewing. ***Hamuli*** are little hooks in a row on part of the leading edge of the hind wing which engage in a groove on the trailing edge of the forewing as the latter opens. This is the same mechanism as the human 'sailor's grip', if you know what that is.

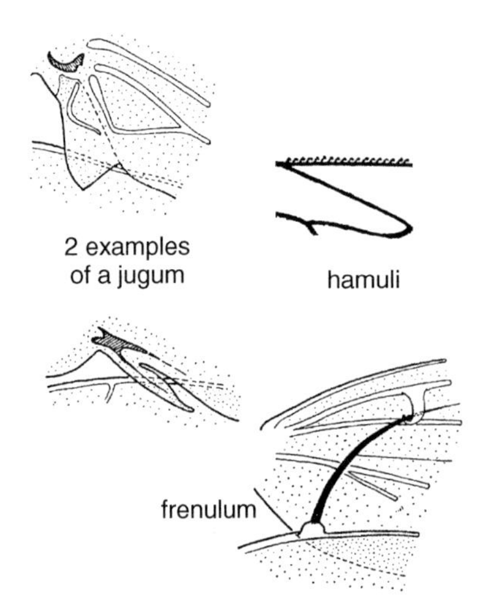

Fig. 2.17 Examples of wing coupling mechanisms (hamuli from Comstock 1940; others from Richards and Davies 1977).

If you dissect the thorax at the wing articulation, you will find a muscle attached to the wing. That is not surprising, but there is the surprise that the muscle is so small. The reason is that this ***direct wing muscle*** does not provide the power for the wing stroke; its job is only to change the angle (pitch) of the wing so that this can be 'feathered' on the upstroke to avoid cancelling out some of the lift given by the down stroke. If the insect were merely to beat its wings at a uniform pitch with uniform power, the upstroke would force the insect down after the downstroke had forced it up! So as well as feathering the wing on the upstroke, the power is reserved for the downstroke.

Where does this power come from if not from muscles attached to the wings? The power actually comes from the sclerites of the notum as well as from some very elastic cuticle called ***resilin*** around the wing articulation. This resilin is 'wound up' by (i.e. it stores the energy of) the contraction of large ***indirect flight muscles*** which also distort the notum sclerites as springs, and all this stored energy is released suddenly when the muscles relax. The indirect flight muscles do not connect an anchor of solid cuticle to the wing, but connect two areas of cuticle to each other. These areas of cuticle are the sternum and the lid of the segmental box, the ***notum***. The principle of insect flight can thus be crudely illustrated by Fig. 2.18.

The indirect flight muscles are of a particular type known as ***involuntary muscle***. This type of muscle receives nerve impulses that trigger a programmed activity. No impulses for 'wings up' and then 'wings down' could arrive fast enough to create the speed of the wing-beat of most insects; for example the wing beat frequency of the honey bee is close to 200 per second, while that of a mosquito is about 600 per second. So the flight muscles automatically alternate contraction and relaxation when triggered to 'start' until triggered to 'stop'.

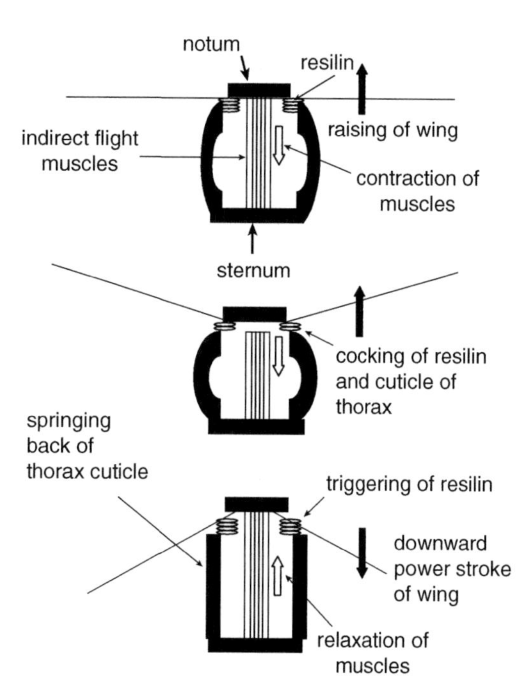

Fig. 2.18 Cartoons of the principle of insect flight, with a little spring as the icon for resilin. The top two cartoons represent the 'cocking' of the resilin and deformation of the thorax cuticle by the contraction of the flight muscles, with the bottom cartoon showing the triggering of the release of the stored energy to force the wing downwards when the flight muscles relax.

2.5.4 Spiracles

These are openings in the cuticle which allow the entry of air into the body as part of the respiratory system of the insect. They are found on segments of the thorax, most usually only on the mesothorax. All this will be described in the next section (the abdomen).

2.6 The abdomen

At the start of the abdomen (Fig. 2.19), the cuticular lining of the gut (which also has to be shed at ecdysis) comes to an end, and from that point onwards the gut wall secretes a continuous tube of permeable cuticle, the ***peritrophic*** (=around the food) ***membrane***. This forms a sort of 'sausage skin' around the food, which prevents it abrading the gut wall as it passes through the abdomen while allowing enzymes secreted by the gut wall to digest the food and nutrients to pass into the gut lumen and be absorbed by the gut wall. The faeces of some insects (e.g. locusts and grasshoppers) in fact take the form of 'sausages' of excreta still enclosed by peritrophic membrane.

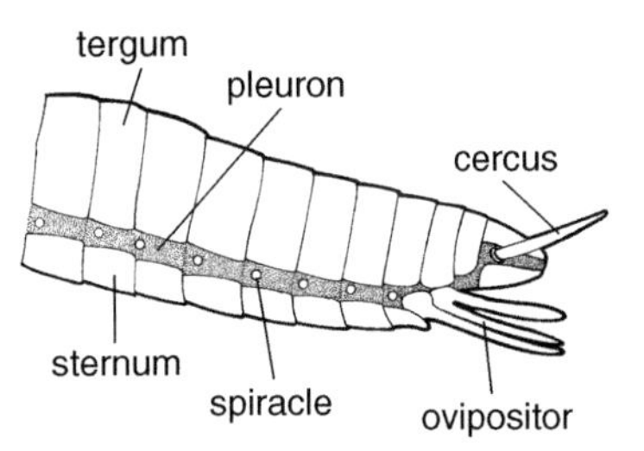

Fig. 2.19 External features of the abdomen (from Zanetti 1977).

Joining the gut in the abdomen where the peritrophic membrane starts are the openings of the numerous ***Malphigian tubules***, which ramify into the haemocoel and remove from it waste products for excretion *via* the gut. The abdomen also houses the gonads, the ***testes*** or ***ovaries*** with the ducts carrying their products to the exterior, usually at the end of the abdomen. The double ventral nerve cord of the thorax continues into the abdomen. Commonly, there is a concentration of neural material, a ***ganglion***, in each segment; from this ganglion nerves issue laterally to process impulses to and from the tissues of the segment. The number of segments in the abdomen varies with taxon, but is usually more than five with a maximum of 11.

2.6.1 Abdominal sclerites

It is in the abdomen that the basic sclerite structure of insect segments is simplest and most obvious. There is a dorsal tergum, a ventral sternum and lateral pleura, usually less sclerotised than the terga and sterna.

2.6.2 Appendages

True segmented walking legs as found on the thorax are not found on the abdomen. In most adult insects the only abdominal appendages are found on the last few segments. Frequently, a pair of ***cerci*** (singular ***cercus***) are found projecting from segment 9. Cerci may be inconspicuous or quite long, unsegmented or of several segments. They are often referred to as 'tail-feelers', inferring that they may provide the insect with sensory information when it is reversing. In some insects, vibration of the cerci provides the insect with auditory information.

2.6.3 Genitalia

In males, the sclerotised ***claspers*** used in copulation are normally only protruded from the abdomen during mating. For some insects, such as certain closely related species of Lepidoptera, they provide the only consistent distinguishing characters. This makes it necessary to dissect out these structures before identification is possible, and so one often finds that museum specimens have a small piece of card with the genitalia glued on it and mounted on the same pin as the insect from which they came.

Many female insects that insert their eggs into a substrate rather than simply laying eggs on it, insert a tube (the ***ovipositor***) made up of four ***valves***. Pairs of valves arise from the eighth and ninth abdominal segments and fit snugly together to form a tube. They are usually not fused, and the tube formed is narrower than the diameter of the eggs that pass down it. Accordingly, if you watch the oviposition process, you will see the valves temporarily separated at the point where the egg is being squeezed from the tube. Similarly as for the antennae, mouthparts and legs, the basic ovipositor plan is subject to many variations. According to the hardness of the substrate and the depth to which the eggs are to be placed, the ends of the valves may be equipped with teeth for sawing or drilling and the length of the tube may be quite short or extremely long, projecting far beyond the end of the ***abdomen*** (Fig. 2.20). The sting of wasps and bees is a modified ovipositor.

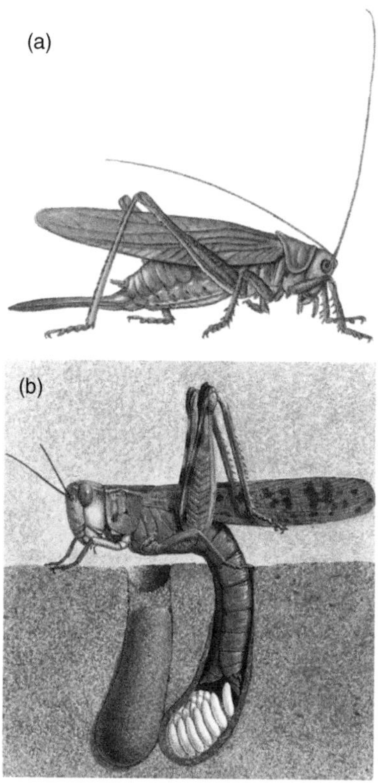

(a)

(b)

Fig. 2.20 Contrasting ovipositors in grasshoppers. (a) the long ovipositor of a long-horned grasshopper (from Mandahl-Barth 1974, with permission); (b) the short ovipositor with extended abdomen of a locust ovipositing (from Zanetti 1977).

2.6.4 Spiracles

Commonly, insects, particularly when adult, have ***spiracles*** on most abdominal segments. These are situated in the pleuron on either side of the body, or in the tergum if there is no distinct pleuron. The spiracles are segmental openings of the respiratory system to the outside environment, allowing the entry of oxygen and the expulsion of carbon dioxide. These openings are more complex than simple holes (Fig. 2.21). They may have 'lips' like a mouth or be covered with a perforated plate with the struts making various patterns. ***Cribriform plates***, where the openings are elongated ovals and the whole plate looks like loose-woven basketwork, are another type. If we could miniaturise ourselves and enter the insect through one of the spiracles, we would find ourselves in an air-filled tunnel of delicate cuticle, strengthened to prevent collapse by ribs of thicker cuticle, making the tube rather like a flexible vacuum cleaner hose. We would be standing in the largest diameter tunnel of a system of trunking (Fig. 2.22), branching to ever smaller-diameter tubes supplying all tissues in the segment with oxygen and carrying away the carbon dioxide product of respiration of the tissues. The larger tubes are the ***tracheae*** (singular ***trachea***) and the finer branches are the ***tracheoles***. Where a fine tracheole finally reaches its tissue destination (e.g. part of a muscle), the tracheole forms a net of fine tubes spreading over a substantial area of the tissue, and capillarity draws some body fluid into the end of the tracheole.

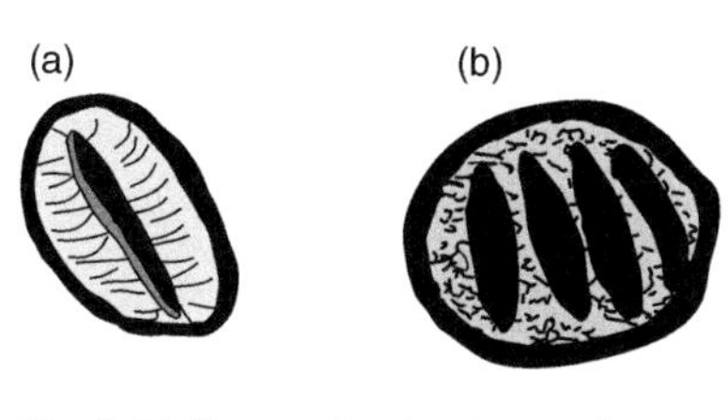

Fig. 2.21 Types of spiracle openings: (a) simple spiracle; (b) cribriform plate.

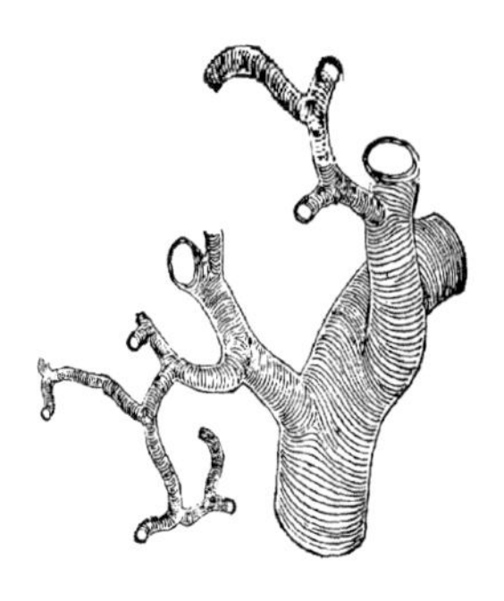

Fig. 2.22 Tracheal trunking.

Since the tracheolar system is made of cuticle, it is moulted together with the rest of the cuticle, including the lining of the gut from the mouth to the start of the abdomen.

The respiratory system of insects is therefore totally different from that of most other animals, including ourselves, where gas exchange is into and out of a circulation of liquid serving all the tissues and the final exchange with the outside air is centralised in an organ such as our lung or the entire skin in some other animals (e.g. worms).

The unusual tracheal respiratory apparatus is found in mites; otherwise it is unique to insects. This is actually rather important to us humans, who can swat noxious insects with ease. But imagine unswattable mosquitoes the size of golden eagles or a caterpillar the size of bus! Indeed there was once a horror film where the traditional 'mad scientist' had mutated a house fly to the size of St Paul's cathedral and the fly then escaped to terrorise London and just crushed the fire-engines sent to deal with it beneath its tarsi.

Fear not, you are pretty safe! The tracheal system sets a strict limit to body size. The whole system is ventilated by movement of the body, and diffusion of the gases is slow. Thus any tissue too far from a spiracle will be starved of oxygen. It is therefore inevitable that all the tissues of an insect have to be near a spiracle, and this sets a limit on insect size. If an insect is large, such as the 'giant stick insects' or large dragonflies, then their bodies have to be long and thin (see Figs 5.6 and 6.15). Giant dragonflies with 32 cm wing spans flew in the Carboniferous era 325 million years ago; again they had slender bodies. The bulkiest insect is probably the Goliath beetle (*Goliathus goliatus*), which I mentioned in Chapter 1 as about 12 cm long and weighing up to 50 g. The Goliath beetle represents the limit of what can be sustained by a tracheal system; it moves very slowly and takes only a few steps before apparently finding it necessary to pause and get its breath back!

How small can insects be? I have already mentioned the less than 0.2 mm long fairy fly in Chapter 1. The lower limit of insect size is set by the difficulty a small insect has of escaping from wetting. It has to succeed in breaking the surface tension of a drop of water that has surrounded it, for example in heavy rain. Some minute insects avoid the problem by living inside a protective material. Thus some beetles live and feed inside the caps of mushrooms; they are effectively surrounded by blotting paper. The most vulnerable are small insects with wings, which wetting may glue down onto the abdomen. You will find that tiny winged insects, such as fairy flies, thrips and some minute beetles, all have one thing in common. Instead of a large wing area, they have the wing reduced to a central strap from which long hairs, to which water will not adhere, radiate (Fig. 2.23).

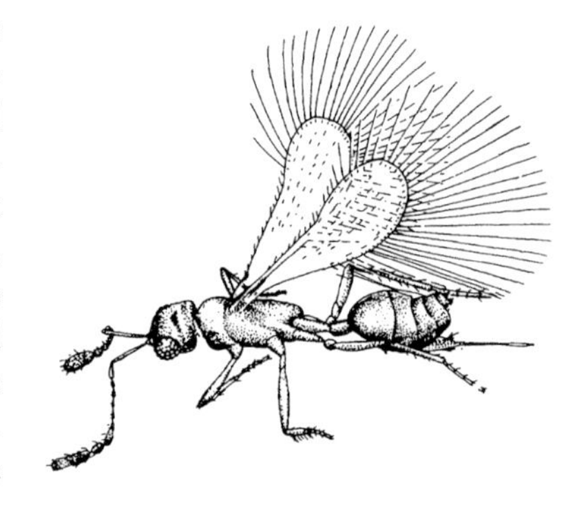

Fig. 2.23 Fairy fly showing fringed wings (from Ross *et al.* 1982, with permission.)

3 The major divisions of the Insecta

3.1 Introduction

As already mentioned in Chapter 1, the Insecta are a Class of the Phylum Arthropoda (=animals with jointed legs). The other Classes of the Arthropoda are the Crustacea (crabs, lobsters etc.), Arachnida (scorpions, spiders and mites), Chilopoda (centipedes) and Diplopoda (millipedes).

The traditional, first major division of the Insecta into two is that those animals that do not have wings and are not related to insects with wings are split off as the Apterygota (a=absent, pteron=wing) from all the others, the Pterygota. This division is given Subclass status in this book on entomology in order to justify its inclusion, although there are credible arguments that Class status is more appropriate.

3.2 Class Insecta, Subclass Apterygota or Phylum Arthropoda, Class Entognatha

Today, many zoologists believe that the traditional apterygote Orders (Collembola, Protura, Diplura and Thysanura) should be taken out of the Class Insecta and a new arthropod Class of Entognatha (mouthparts submerged in a pocket) has been erected for them.

The apterygotes are regarded as primitive Orders, which split off from the main evolutionary line leading to the pterygote insects before wings developed in evolution. It is important to recognise these as 'primarily wingless', in contrast to the many secondarily wingless insects. The latter either are wingless forms of winged insects (examples would be wingless ants and many life history stages of aphids) or have clearly lost wings their ancestors possessed. Thus bed bugs are clearly a group of the otherwise winged Order Hemiptera. As recently as 1955, the public insect collection in the Natural History Museum in London included fleas as Apterygota, though by then the larvae had already been found half a century earlier, clearly placing fleas in the higher group of winged insects called the Endopterygota (see Sections 3.3.2 and 8.3).

The apterygotes are distinguished by having a life history where, after hatching from the egg, there is really no change in appearance as the insect grows and periodically moults. Adult apterygotes look just like larger juveniles (Fig. 3.1a), and many continue to moult as adults (which never happens in the Pterygota). Thus the only way one knows an apterygote has become adult is either to count the instars or observe the onset of mating behaviour or egg laying. The noticeable change of appearance at the end of the juvenile period of winged insects (Subclass Pterygota) is referred to as metamorphosis. We therefore say that the Apterygota show 'no metamorphosis'. The

Handbook of Agricultural Entomology, First Edition. H. F. van Emden.

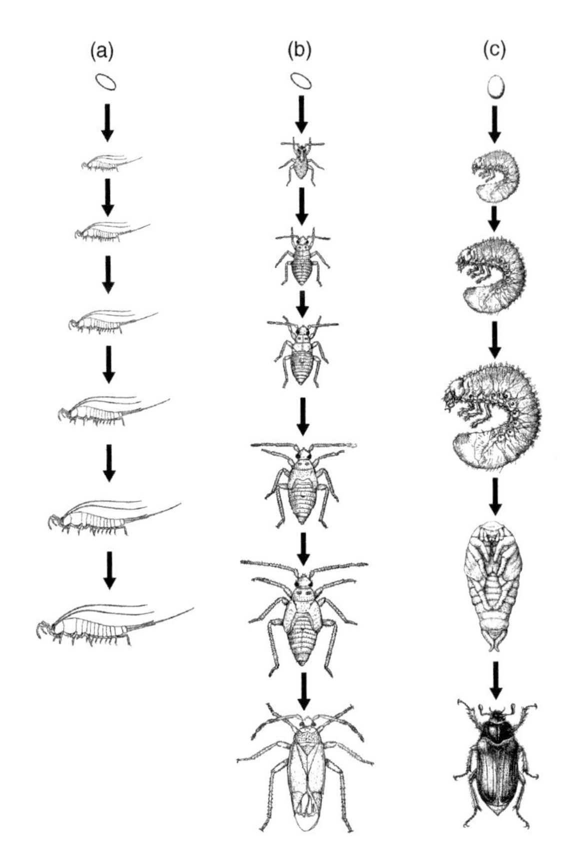

Fig. 3.1 Insect life cycles: (a) Apterygota (from Ross *et al.* 1982, with permission); (b) Exopterygota (from Richards and Davies 1977); (c) Endopterygota (from Zahradník and Chvála 1989).

technical term for showing no metamorphosis is 'ametabolous', so the apterygotes are often called the Ametabola (metabol=transformation).

3.3 Subclass Pterygota

These are the primarily winged insects though, as pointed out above, species or life-cycle stages may have lost their wings in adaptation to their lifestyle. Pterygota also show a marked change in appearance during their development (a metamorphosis) and the wings develop throughout the juvenile stages.

The Pterygota are divided into two Divisions based on whether the wings develop externally or internally during development.

3.3.1 Division Exopterygota (exo=outside)

Here, after the egg stage, the wings develop externally as wing pads (Fig. 3.1b), which grow with every moult. However, the final increase in wing area to the adult state happens dramatically at the moult to adult, where often there are other changes in appearance and colour. A young plant bug has a body form very like an adult one, but has small wing pads extending from the meso- and metathorax over the dorsal surface of the abdomen. These wing pads grow at each successive instar, though a final large increase in the size of the wings occurs at the moult to adult. The colours of juveniles may be less pronounced than those of the adults. Adults and juveniles more often than not are found together, sharing the same food resource.

There is therefore an identifiable metamorphosis, but one which is far less total than the change from grub to adult that occurs in butterflies, for example. The exopterygote metamorphosis is therefore identified as 'incomplete metamorphosis'. Such insects are described as 'hemimetabolous' (hemi=half) or as the Hemimetabola.

To be scientifically accurate, we should call the immature stages of the exopterygotes 'larvae', but I will follow common practice and use the word 'nymph' since these look nothing like caterpillars or grubs, for which the term 'larva' is standard English usage.

For convenience, the exopterygotes are divided into three groups (Box 3.1).

Box 3.1 Classification of insects used in this book.

Class Insecta, Subclass Apterygota or Phylum Insecta, Class Entognatha
- Order Thysanura (silverfish)
- Order Diplura
- Order Protura
- Order Collembola (springtails)

Subclass Pterygota

Series Exopterygota

Palaeopteran Orders
- Order Ephemeroptera (mayflies)
- Order Odonata (dragonflies)

Orthopteroid Orders
- Order Plecoptera (stoneflies)
- Order Grylloblattodea
- Order Mantophasmatodea (gladiators)
- Order Zoraptera (angel insects)
- Order Orthoptera (grasshoppers and crickets)
- Order Phasmida (stick and leaf insects)
- Order Dermaptera (earwigs)
- Order Embioptera (web spinners)
- Order Dictyoptera (cockroaches and mantids)
- Order Isoptera (termites)

Hemipteroid Orders
- Order Psocoptera (book lice)
- Order Mallophaga (biting lice)
- Order Anoplura (sucking lice)
- Order Hemiptera (bugs)
- Order Thysanoptera (thrips)

Series Endopterygota
- Order Mecoptera (scorpion flies)
- Order Siphonaptera (fleas)
- Order Neuroptera (lacewings and allies)
- Order Trichoptera (caddis flies)
- Order Lepidoptera (butterflies and moths)
- Order Diptera (true flies)
- Order Hymenoptera (ants, bees and wasps)
- Order Coleoptera (beetles)
- Order Strepsiptera (stylops)

3.3.1.1 Palaeopteran Orders (example: dragonflies)

These are reminiscent of the fossil insects of millions of years ago, with a very complex wing venation with many cross veins, and an inability to fold the wings over the abdomen at rest; instead they meet together vertically above the body.

3.3.1.2 Orthopteroid Orders (example: grasshoppers)

These can fold their wings over or alongside the abdomen; they have biting mouthparts and single- or many-segmented cerci.

3.3.1.3 Hemipteroid Orders (example: plant bugs)

These can also fold their wings over or alongside the abdomen, but have sucking mouthparts and no cerci.

3.3.2 *Division Endopterygota* (endo=inside)

Here wing development is not visible externally in the juveniles, but takes place internally. The life history stages are the familiar egg, larva, pupa and adult of, for example, butterflies (Fig. 3.1c), with the larval stage having several instars. The larva and adult are very different in appearance, showing a complete metamorphosis. Thus they are 'holometabolous' (holo=fully) and sometimes known as the Holometabola.

Not only do larvae and adults differ in appearance, but they also usually have distinct behaviours. Larvae are (like the nymphs of the exopterygotes) flightless, although you may be surprised to learn they do have wings. Actually you should not be surprised; I did describe the Endopterygota has showing internal wing development. You will find wing buds projecting internally from the walls of the thorax if you dissect a larva (Fig. 3.2). Moreover, the adult (often called the imago) and the larva of a species more often than not feed on different things. Thus many caterpillars of the Lepidoptera feed on leaves and have mouthparts designed for biting. By contrast the adult butterfly sips nectar from flowers through a long coiled tube. The larval mosquito is aquatic and filters small organic particles from the water; yet the adult is terrestrial and has piercing and sucking mouthparts for imbibing warm vertebrate blood.

Such a total transformation between larva and adult is made possible by the pupal instar that occurs between larva and adult. The pupa is a largely immobile instar in which the larva is restructured into an adult.

Sometimes the larva spins a silken case around itself in which it eventually pupates, often after passing several months as an immobile larva (a prepupa). This silk case is called a *cocoon*.

Two types of pupa are distinguished. The majority of endopterygotes have ***exarate*** pupae. In external appearance, these are not too dissimilar from adults with very shortened wings. These 'wings', as well as the mouthparts, antennae and legs, mostly hang free from the main body of the pupa. By contrast there is the ***obtect*** type of pupa, with all the free parts of the exarate type soldered down to make a smooth outline; however, all the structures are demarcated with lines on the cuticle. This type is exemplified by the ***chrysalis*** of the Lepidoptera.

What happens inside the pupa is quite remarkable. Much of the larval tissue is broken down into simple compounds such as amino acids, leaving a fluid-filled pupa except for clumps of cells called the 'imaginal buds'. The internal wing buds referred to above are among these, as are groups of cells forming the foundation for the brain, the gonads (sex cells) etc. These migrate to their relative adult positions and then

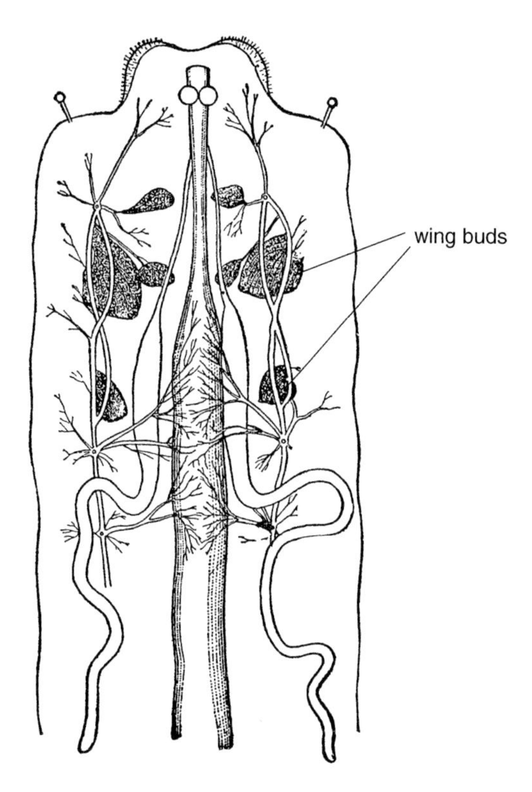

Fig. 3.2 Caterpillar dissected to show wing buds developing internally (from Folsom and Wardle 1934).

new adult tissues are built up progressively from them, using the materials in solution. The cuticle of the pupa is more adult than larval, with a template of adult antennae, wings, legs etc. It seems analogous to crystal growth from a concentrated solution around a nucleus! Another analogy might be a set of Meccano or Lego, where you can dismantle say an aeroplane you have built and use the parts to make an ambulance. While the pupa contains liquid material it cannot respond to stimuli; once the adult muscles have formed, the segmented abdomen of the pupa will twitch when touched. I still find the whole thing a mystery, especially how a pupal cuticle with its adult features is laid down inside a larval one before the last larval ecdysis. Sometimes the new pupa has to hook itself into a silk pad (laid down on a plant stem by the larva) using hooks (the ***cremaster***) at the end of its abdomen. There is therefore a split second of total non-attachment of the pupa as the end of its abdomen divests itself of its larval skin and before the cremaster hooks into the silk pad. In the early stages of pupation, liquefaction of the larval tissues has still to happen, so the hooking of the cremaster into the silk pad is an action of larval muscle inside a pupal shell escaping from a larval skin. Thus before the last larval ecdysis, the pupal cuticle would appear to be an 'adult' cuticular filling in a larval sandwich.

A complete metamorphosis enables a single species to occupy two ecological niches. The two feeding life stages of larva and adult are not in competition. Different forces of natural selection will act upon them, making it unlikely that both stages will suffer together from any temporary reduction in the quality of the habitat. By contrast, the typically similar structure (especially of the mouthparts) of nymphs and adult exopterygotes means that they compete and are both affected by any changes to the habitat. The downside for endopterygotes, however, is that an immobile and therefore vulnerable pupal stage is necessary.

4 Subclass Apterygota

4.1 Introduction

There are four small Orders of insects that are not only wingless, but are considered to have separated from the evolution of insects before the other Orders evolved wings. As mentioned earlier (Section 3.2) they are now often regarded, not as insects, but as a separate Class of the Arthropoda. The Apterygota are therefore 'primarily' wingless in contrast to 'secondarily' wingless insects which, in adaptation to specific lifestyles, have lost the wings possessed by their ancestors (see Section 3.2; examples are fleas and bed bugs). Other features of the Apterygota that contrast with the Orders in the Subclass Pterygota are that some have abdominal appendages other than cerci and genitalia and that they show no metamorphosis (see Fig. 3.1). They may continue moulting as adults and the immature and adult stages are almost identical except for size. Another characteristic of several of the apterygote Orders is that the transfer of sperm to the female takes place externally.

4.2 Order Diplura (two-pronged bristle-tails) – *c.* 675 species

Diplura are small creatures, usually just a few millimetres long, though exceptionally larger species occur with a length approaching 2 cm. Compound eyes and ocelli are absent, and tarsi are one-segmented. With their long segmented cerci and antennae of very similar appearance and size, the Diplura have very symmetrical front and back ends (Fig. 4.1). However, the Superfamily Japygoidea differs in that the cerci are short, and in some species are strongly sclerotised, looking not unlike the forceps of earwigs (Fig. 4.1, inset). The Campodeidae (e.g. *Campodea;* Fig. 4.1) is a large and cosmopolitan Family; other Diplura are mainly subtropical and tropical. Diplura live under stones, among fallen leaves, in dead wood or in the soil. Moulting may occur 30 times in a lifespan of up to 3 years, with sexual maturity being reached after eight to ten moults. They feed on a variety of food, from live prey (such as mites and springtails) to fungi and dead organic matter. Most species show a bizarre form of external sperm transfer. The male lays spermatophores that are elevated from the ground on short stalks. The female then collects the spermatophores into her genital opening behind the eighth abdominal segment.

4.3 Order Protura – *c.* 170 species

These (Fig. 4.2) are minute (<2 mm) animals, which can occur in large numbers (densities of more than 90,000 per square metre have been recorded) in moist soil, leaf litter and turf. They are also found under stones or bark, and in animal burrows. Like the Diplura, they have no eyes or ocelli, though they have an ocellus-like structure on each side of the head. This structure is called the ***pseudoculus***, and may have

Handbook of Agricultural Entomology, First Edition. H. F. van Emden.

Fig. 4.1 Order Diplura (from Zanetti 1977). Inset: end of abdomen in the Japygoidea (from Richards and Davies 1977).

Fig. 4.2 Order Protura (from Zanetti 1977).

some simple light receptor function, though some experts consider it more likely that it is a chemoreceptor. There are no antennae, but their sensory function is served by the forelegs, which are longer than the other legs and are held forward. The Protura probably feed by sucking the juices from fungal hyphae, and indeed the mandibles are rather needle like. They may also feed on decaying vegetable matter. The tarsi are again one-segmented but, unlike the Diplura, the Protura have no cerci. The first three abdominal segments bear very small appendages (***styli***).One peculiarity of the Order is that, although the animals hatch from the egg with an eight-segmented abdomen, more segments are added at the rear in subsequent moults until the full adult complement of up to 12 (depending on species) is reached. There is usually one generation a year.

4.4 Order Thysanura (silverfish) – *c.* 330 species

These have a long central segmented tail in addition to the long segmented cerci (Fig. 4.3), and so are sometimes called 'three-pronged bristle-tails'). They are probably the largest of the Apterygota at 1–2 cm long. The antennae are also long and many segmented. Compound eyes may or may not be present, and tarsi are two to five seg-

Fig. 4.3 Order Thysanura (from Zahradník and Chvála 1989).

mented. The abdomen is 11 segmented, and carries a variable number of styli (cf. Protura above). Many are covered in scales with a metallic sheen (hence the name 'silverfish'). *Lepisma saccharina* is common in the UK. Silverfish are considered pests in dwellings, but their presence is indicative of a more serious problem, since they feed on the moulds associated with dampness and therefore with water penetration of walls or absence/failure of a damp course. They can damage wallpaper and the backs of pictures, as they are able to digest cellulose. They also feed on cereals and dried meat. In warmer climates they are not restricted to dwellings as they are in the UK, and live outdoors under stones or bark; many also occur in the nests of ants and termites. Although Thysanura have external genitalia they do not copulate. Instead, sperm transfer is external. Most commonly, and not that different from in the Diplura, the male spins a thread between a surface and an upstanding object and deposits droplets of sperm on this. This sperm is then picked up by the female when the male coaxes her to encounter the thread. In other species, the procedure is less bizarre, and the female picks up the spermatophores the male has simply deposited on the ground. However, in either case, any sperm picked up is then lost at the next moult. Asexual reproduction by parthenogenesis has also been deduced from the rarity of males. Development through 14 larval instars can take 3 years, and the adults may live for a further 2 years, during which they continue to grow with very many (e.g. 52 in one species) further moults.

4.5 Order Collembola (springtails) – *c.* 2000 species

Collembola are normally less than 6 mm long, with six or fewer abdominal segments. This is the apterygote Order of main interest in agriculture and horticulture, and is also the largest. The name relates to the jumping mechanism these insects possess on their abdomen (Fig. 4.4), and which enables them to leap high and far in relation to their size. The mechanism consists of a forked appendage called the ***furcula*** which is

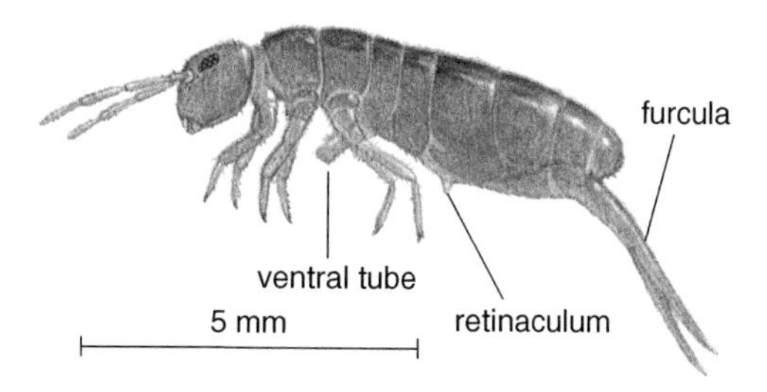

Fig. 4.4 Order Collembola, Suborder Arthropleona (from Zanetti 1977).

attached to the fifth abdominal segment. At rest, the distal arms of the furcula are clipped around a central 'press stud' (the ***retinaculum***) on segment 3.

In order to jump, the insect pumps blood into the furcula, building up tension in the structure until the capacity of the retinaculum to hold it in place is exceeded. The furcula then flies downwards to impact with the substrate, forcing the insect off the ground in a mighty leap. The furcula 'follows through' and projects rearwards from the abdomen while the insect is in the air.

Another appendage on the abdomen (first segment) is a central projection known as the ***ventral tube***. This soaks up any drops of water encountered by the insects, which are soft-bodied and easily become desiccated. Collembola are therefore usually concealed in humid environments such as soil and leaf litter, where they can be very abundant. The top 20 cm of soil of a typical hectare of meadow may contain 400 million Collembola.

The ventral tube is a good example of how much and in what detail the study of insects has progressed. Here I have dealt with the ventral tube in three sentences. In 1988, I was at one of the 4-yearly International Congresses of Entomology, on this occasion held in Hamburg, Germany. In one of the lecture rooms, a one and a half day symposium was held with 14 speakers all discussing just one thing – the function of the ventral tube in the Collembola!

Compound eyes are absent. There is great variety of colouration from blue-black to yellows and greens. Copulation occurs in the Suborder Symphypleona, but otherwise spermatophores are deposited externally by males in response to female pheromone. Parthenogenesis is common in species living buried in the soil. Sexual maturity may be obtained after only 5 moults, well before maximum size has been reached.

4.5.1 Suborder Arthropleona

These are the straight-bodied springtails, as exemplified in Fig. 4.4. They are the springtails we encounter most commonly in leaf litter, under stones and under the bark of trees, etc. They are often used in agricultural studies as indicators of the population effects on soil organisms of events in the soil such as tillage and the application of pesticides. The Arthropleona feed on fungal spores and fallen pollen grains.

Some Arthropleona have a very short furcula, projecting from the back of the abdomen and non-functional as a jumping organ.

4.5.2 Suborder Symphypleona

These have a more humped and spherical appearance (Fig. 4.5). They are phytophagous (i.e. plant feeders) and the lucerne flea (*Sminthurus viridis*) is a green member of the Suborder which can be a pest of clover and alfalfa in Europe and many other

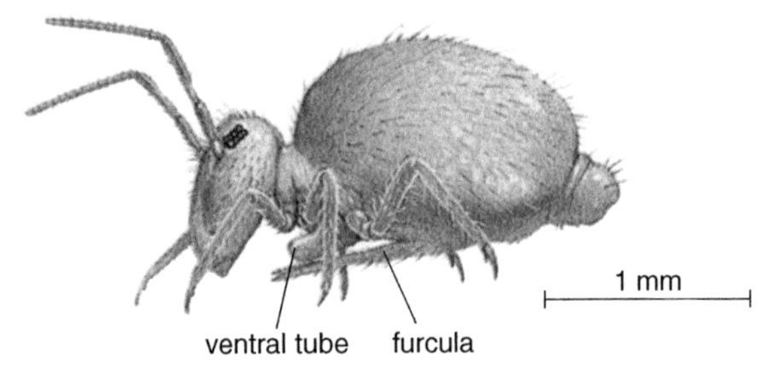

Fig. 4.5 Order Collembola, Suborder Symphypleona – the lucerne flea (*Sminthurus viridis*) (from Zanetti 1977).

parts of the world. You can often find little 'shot-holes' in clover leaves – that is damage caused by the lucerne flea. *Sminthurus betae* attacks the seedlings of beets such as sugar beet and spinach. Precision-drilled seedlings of sugar beet of monogerm seed (to avoid the labour cost of thinning as necessary for the normal polygerm seed) can suffer serious losses. *Sminthurus* spp. breed more or less continuously if low temperature and humidity are not limiting. There are eight instars, with one moult after maturity has been attained.

5 Subclass Pterygota, Division Exopterygota, Palaeopteran Orders

5.1 Introduction

Two Orders are grouped as Palaeoptera (= ancient wings) since they are regarded as the most primitive of the Exopterygota, resembling early fossil insects in having very complex wing venation with numerous cross veins and in being able to move their wings only up and down vertically. They cannot sweep their wings round to lay them flat or roof-like over the abdomen, and so hold them vertically above the body or out sideways when at rest.

5.2 Order Ephemeroptera (mayflies) – *c.* 2100 species

The adults (Fig. 5.1) are easily recognised by the very complex venation of the wings, which are held vertically above the body at rest, and with the hind wing very much smaller than the forewing. Some species have no hind wing at all. The muscle provision for the forewing results in the mesothorax being very large in comparison with the pronotum. The abdomen ends in two or three very long and many-segmented cerci. The antennae are short and bristle like, and the mouthparts are biting in design but vestigial.

The 'Ephemer' bit of the name of the Order refers to the short adult life, particularly of the male, which may die only a few hours after emergence. This 'short life but a gay one' (gay=merry rather than the more modern usage of the word!) has been an inspiration for romantic poets, but they forget that the adult may already have lived as a larva for more than a year. This places Ephemeroptera among the insects of greater than average longevity.

The larvae are aquatic with external gills (usually seven pairs) projecting from the sides of the abdomen (Fig. 5.2), and have well-developed compound eyes and ocelli. They are herbivorous (feeding on plant detritus and algae), though a few species take small animal prey. Ephemeroptera larvae can be distinguished from the rather similar larvae of the Plecoptera (Section 6.2) by having three long cerci rather than two, even in those mayfly species where in the adult the number is reduced to two. Larvae of different species show considerable morphological variation in terms of the length and position of the gills, mouthparts and robustness of legs and claws, all adaptations to the substrates and speed of the water flow in their habitat, and whether or not the species burrows into mud.

Like many other aquatic insects, the larva has to swim to the water surface (where many are taken by fish) before the adult can emerge. However, Ephemeroptera are unique among insects in having two adult instars. The larva first moults to a winged ***preimago*** (what fishermen call the 'dun') at the water surface. This then flies to land

Handbook of Agricultural Entomology, First Edition. H. F. van Emden.

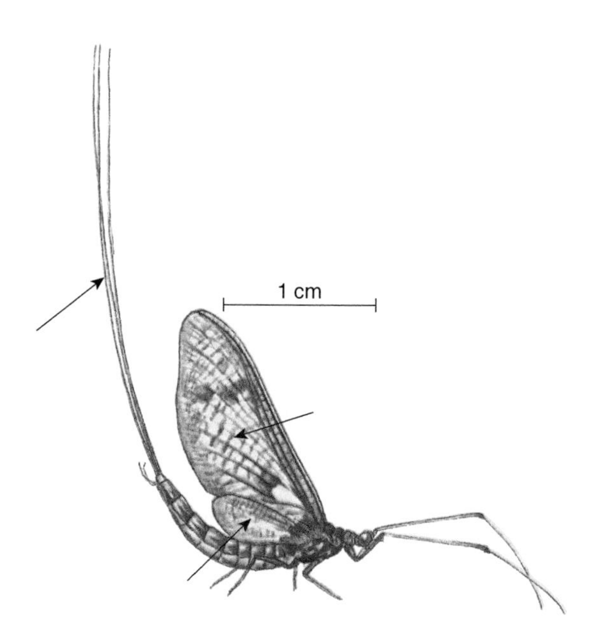

Fig. 5.1 Ephemeroptera (mayfly) adult. Note: in the species shown, the forelegs are very long and held out rather like antennae; the real antennae are very short and just visible between the bases of the forelegs (from Mandahl-Barth 1973, with permission).

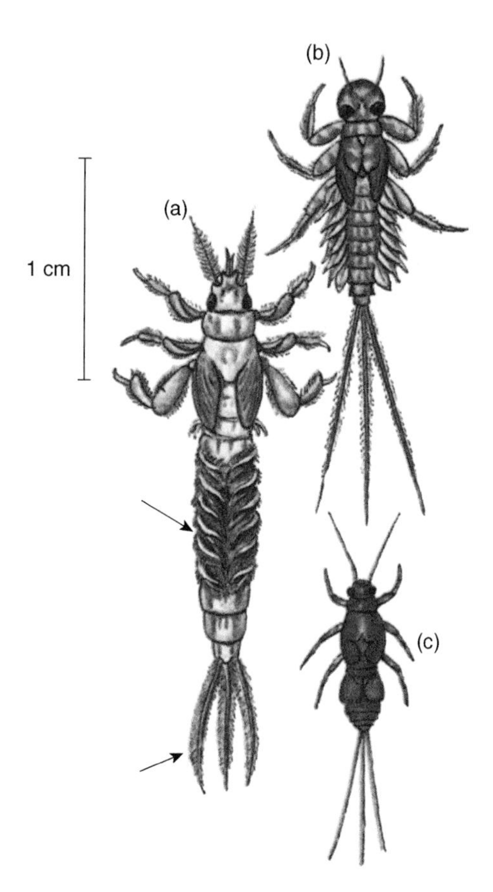

Fig. 5.2 Ephemeroptera nymphs adapted to: (a) tunnelling in mud; (b) gripping stones in fast-flowing water; (c) living on sandy bottom (all from Mandahl-Barth 1973, with permission).

and may take 24 hours before it moults to the final and short-lived imago (adult) instar. In one Family (the Palingeniidae) the subimago is already sexually mature; there is no final imago instar.

Mayfly larvae require water of reasonably high quality; thus they are useful pollution indicators if they die when lowered in a gauze-ended tube into a polluted river.

5.3 Order Odonata (dragonflies) – *c.* 5000 species

The adults are predators, with large compound eyes and two ocelli. The front and hind wings are of roughly equal size and with very complex, net-like venation. There is often a dark pigment spot (***stigma***) near the tip of the forewing. Antennae are filiform but very short, and the abdomen is long and thin. The larger Odonata (the true dragonflies) hawk on the wing, and the sclerites of the thorax point diagonally forwards (Fig. 5.3) so that the legs can catch small insects in the air. Indeed, the forelegs seem to be attached just behind the mouth. The adults are sun-lovers, and are often found basking on foliage with the wings stretched out sideways. The insects are elegant and often brightly coloured, with frequent sexual dimorphism in colouration, especially in damselflies. Here, overall colour and whether the wings are clear or dark can be totally different in males and females.

The sexes mate with the female receptive organs and the male penis arranged in the characteristic posture shown in Fig. 5.4, and will fly from perch to perch in this position while continuing to mate. The gravid female then hovers over the water to lay her eggs in rafts on the water surface. The hatching larvae descend to the bottom of the lake or stream and complete their development in 1 year (damselflies) or over a period of up to 5 years or even longer (dragonflies), involving 10–15 moults. The larvae are carnivorous and have a labium in the form of a folded ***mask***, which can be suddenly straightened to shoot out forwards (Fig. 5.5) and catch prey in the jaws at the end. The compound eyes are so arranged that there is a patch of visual acuity only at the point to which the jaws of the mask project. Accordingly, any prey coming 'into focus' will be exactly where the projected mask can seize it.

When development is complete, the last instar larva seeks a plant stem growing at the water's edge and crawls up this till it arrives well above the water surface. The

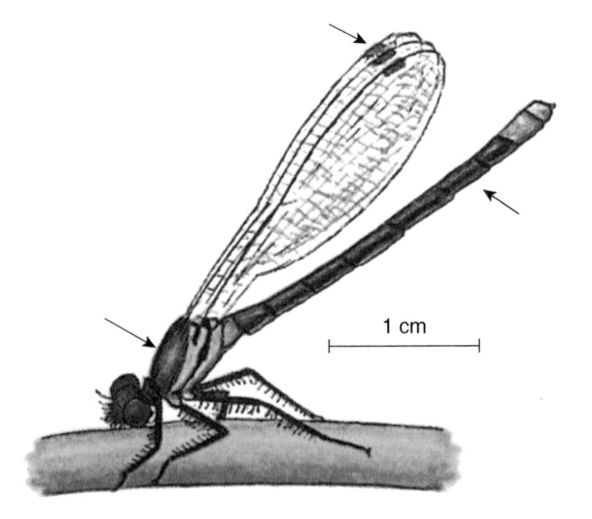

Fig. 5.3 Odonata adult (a damselfly) (from Mandahl-Barth 1973, with permission).

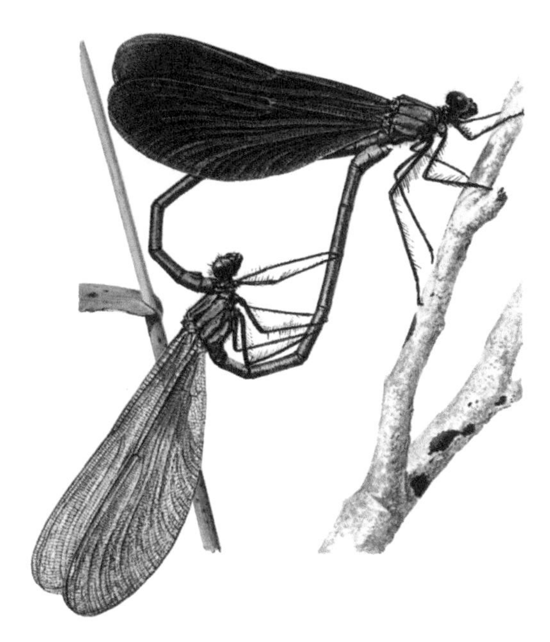

Fig. 5.4 Damselflies mating (from Zanetti 1977).

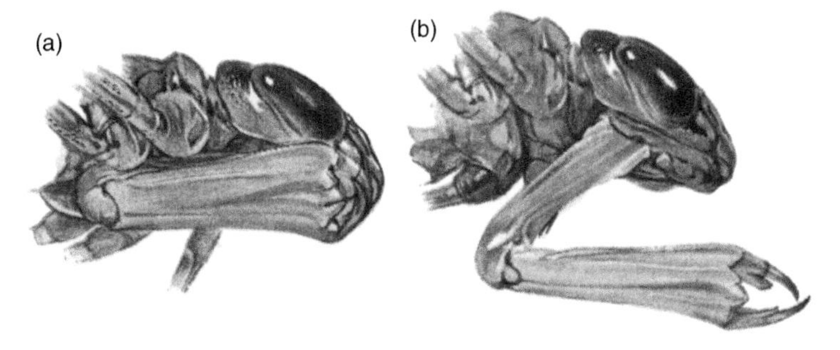

Fig. 5.5 Labial 'mask' of dragonfly nymph in (a) retracted and (b) extended position (from Zanetti 1977).

insect then anchors itself, the cuticle splits open and the adult emerges, expanding its wings and remaining on the vegetation until the wings have hardened.

5.3.1 Suborder Zygoptera (damselflies)

These are the smaller and slender-bodied odonates, which are often brightly coloured with clear or coloured wings held vertically above the body at rest. The eyes are well separated on the head (both as adults and larvae) and the larvae have three external gills projecting at the rear of the abdomen (Fig. 5.6). The adults are carnivorous and mainly hunt on submerged vegetation rather than on the wing in the way that dragonflies do.

Although normally much smaller and more slender than the Anisoptera, the damselflies do include the largest of all the Odonata. This is *Megaloprepus caerulatus* of the forests of Central and South America, and which can have a wing span of 19 cm.

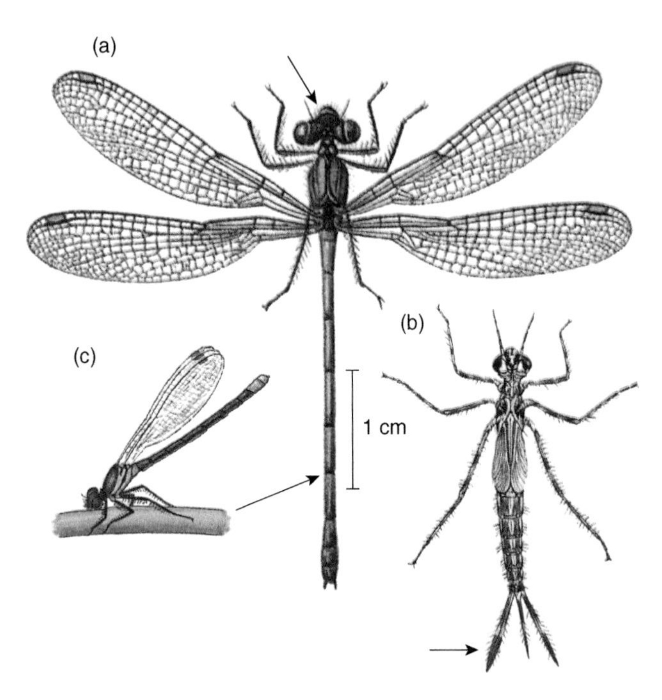

Fig. 5.6 (a) Zygoptera (damselfly) adult (from Zahradník and Chvála 1989) and (b) nymph on same scale (from Zanetti 1977). (c) Normal resting position (from Mandahl-Barth 1973, with permission).

This is still much smaller than the giant Odonata found as fossils from the Upper Carboniferous (350 million years ago), which had wing spans of 75 cm.

5.3.2 Suborder Anisoptera (dragonflies)

These are usually larger than damselflies with wider, more robust abdomens. The eyes are much larger and (in both adults and larvae) almost touch each other at the front of the head (Fig. 5.7). The wings are clear or lightly tinged, and have an obvious and complex venation; they are held out sideways at rest. The larvae are, like the adults, sturdier than in the Zygoptera and the gills are not visible externally. They are internal, retracted into a cavity at the end of the abdomen. Zygopteran larvae move slowly on the mud at the bottom of a lake or river. They are carnivorous and take prey up to the size of fish fry.

Most Anisoptera hawk for prey by patrolling up and down a regular 'beat', but the members of one Family (the Libellulidae) are known as 'darters' from their very different habit of ambushing passing prey such as flies by 'darting' out from a perch on vegetation.

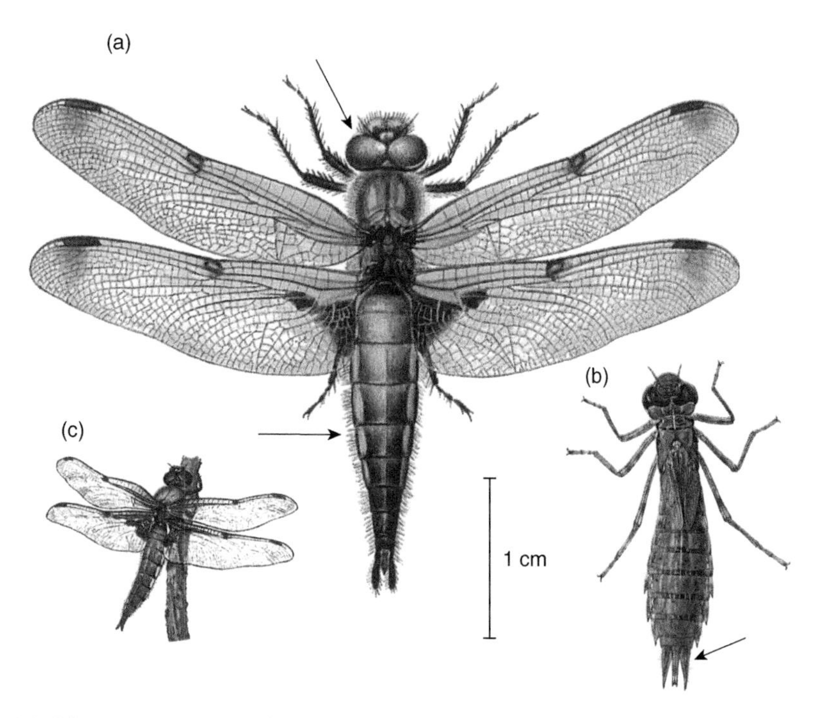

Fig. 5.7 (a) Anisoptera (dragonfly) adult (from Zahradník and Chvála 1989) and (b) nymph on same scale (from Zanetti 1977). (c) Normal resting position (from Mandahl-Barth 1973, with permission).

6 Subclass Pterygota, Division Exopterygota, Orthopteroid Orders

6.1 Introduction

The majority of the Exopterygote Orders fall into this group. They have a much reduced wing venation compared with the mayflies and dragonflies, and can fold them laterally as well as vertically, either flat or like a pitched roof over the abdomen. The Orthopteroid Orders are distinguished from the Hemipteroid Orders by their biting mouthparts unmodified from the basic biting pattern (see Fig. 2.11) together with their possession of cerci. They also have a large ***anal lobe*** on the hind wing.

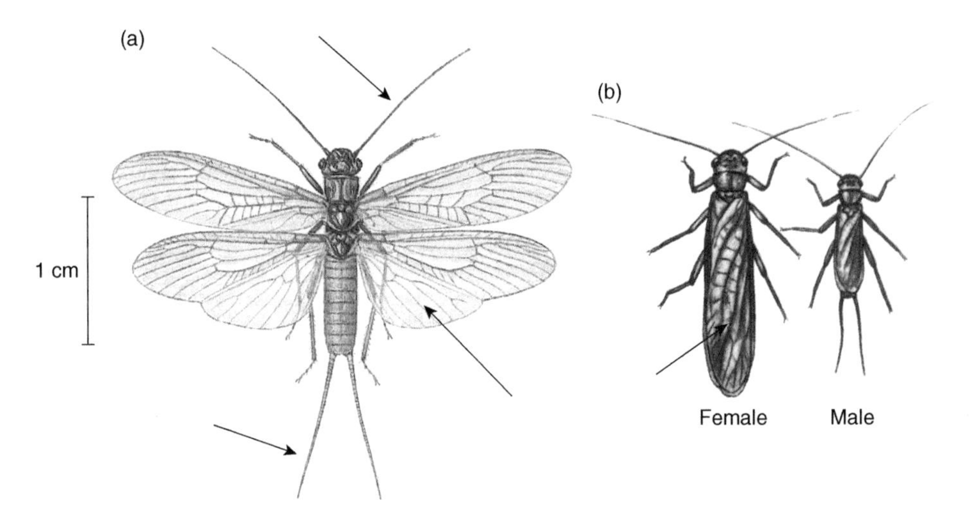

Fig. 6.1 Plecoptera (stonefly) adults (a) showing the features of the Order (from Zahradník and Chvála 1989); (b) female and male of a different species in the normal resting position (from Mandahl-Barth 1973, with permission).

Handbook of Agricultural Entomology, First Edition. H. F. van Emden.

6.2 Order Plecoptera (stoneflies) – *c.* 1700 described species

These are one of the most primitive groups of the Orthopteroid Orders, and close relatives have been found fossilised as early as the Carboniferous. Adult stoneflies are weak fliers with the hind wing rather larger than the forewing (Fig. 6.1); the wings are folded flat on the back of the abdomen when at rest. Males are noticeably smaller than females. There are some brachypterous and a very few wingless species. Like the mayflies they roost on vegetation near water, but they differ in that the antennae are long and filiform and always there are just two cerci (long or short – even just one segment) with no central tail filament. Females lay hundreds of eggs in a slime-ball, which they carry on the abdomen till they deposit the ball in water when the slime dissolves to free the eggs. Adults live a few weeks; some never feed and the others are herbivores, often feeding on lichens and algae. The nymphs (Fig. 6.2) live under stones in clear water, especially if it is fast flowing; like those of mayflies, they require unpolluted water, and so are also indicators of high water quality. Most species, like mayfly nymphs, have external gills on either side of the abdomen; but they can also breathe through the skin and some have no gills at all. Stonefly nymphs are aquatic carnivores or herbivores and may take up to 4 years to develop to adult, moulting between 10 and over 30 times in the process. Then, again like mayflies, they moult to adult at the water surface.

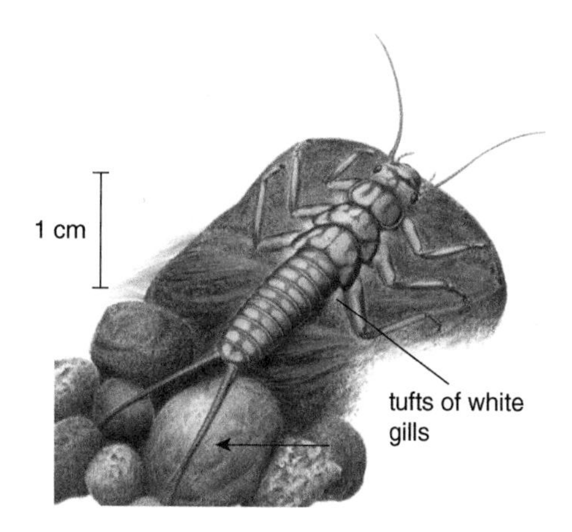

Fig. 6.2 Plecoptera (stonefly) nymph (from Dipper and Powell 1984).

6.3 Order Grylloblattodea – 25 described species

This Order comprises just one Family (Grylloblattidae) with five genera. These insects are clearly related to the other orthopteroids but are secondarily wingless. The adults are 2–3.5 cm long with long feelers, which can have over 40 segments, and long cerci with eight segments (Fig. 6.3); eyes are absent or reduced, and there are no ocelli. Tarsi are five-segmented, and females have an obvious ovipositor. Grylloblattodea are nocturnal predators and detritus feeders, hiding under stones by day on cold mountain tops. The low temperatures they prefer lead to a very slow development. Eggs incubate for about a year and the nymphs then take another 5–6 years to pass through eight instars.

The first specimens were found in the Canadian Rockies in 1914, and were described as *Grylloblatta campodeiformis*, a right chimera of a name signifying that "this cricket or perhaps cockroach looks like a genus (*Campodea*) in the apterygote Diplura"!

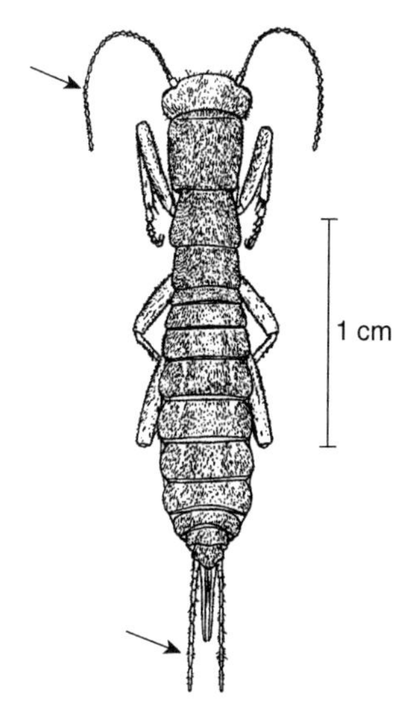

Fig. 6.3 Grylloblattodea adult (from Richards and Davies 1977).

6.4 Order Mantophasmatodea (gladiators or heelwalkers) – 14 described species

This Order of very rare insects was discovered in 2002, and is the first new Order to be described since the Grylloblattodea (to which they are clearly related) in 1914. Originally described from Namibia as two species in the genus *Mantophasma*, they have also been found in 45 million-year-old Baltic amber (*Raptophasma kerneggeri*). The gladiators are nocturnal predators on other insects and spiders, and by day live in rock crevices. The Order is secondarily wingless (Fig. 6.4), and the insects have slender filiform feelers, five-segmented tarsi and short one-segmented cerci. Three Families have been identified so far.

Fig. 6.4 Mantophasmatodea adult, as used in the logo for the International Congress of Entomology held in Durban, South Africa in 2008 (courtesy of the Congress and the Entomological Society of Southern Africa).

6.5 Order Zoraptera (angel insects) – 30 described species

This is the third smallest insect Order and also one of the most abstruse. The first specimen was only described in 1913, and most entomologists only know the Order from the same oft-repeated illustration of *Zorotypus guineensis* from Africa (Fig. 6.5). In fact, Zoraptera are found in the tropics and subtropics of the USA, Africa and south-east Asia. They are delicate insects about 3 mm in length with triangular heads and moniliform (= like a string of beads) nine-segmented antennae. They live in colonies of 15–120 in organic substrates such as rotting wood and leaf litter; they are sometimes found in termite nests, and are superficially quite similar to termites in appearance. Zoraptera feed on fungal mycelium and spores, but are also carnivorous on microarthropods such as small mites. Also like termites (see Section 6.11), the wings (when present) can easily be shed along fracture lines near the wing bases. Most species of Zoraptera show alary polymorphism where both males and females can be macropterous or apterous. The latter form is much more common, pale and without eyes or ocelli, in contrast with the winged form. In some species no macropterous individuals have as yet been seen. Wing venation is reduced with often a dark spot near the tip of the forewing; it has been compared with the venation of the Psocoptera (see Section 7.2). An ovipositor is absent. The first instar has a cuticular egg-burster on the head which breaks the egg shell, and there are four nymphal instars.

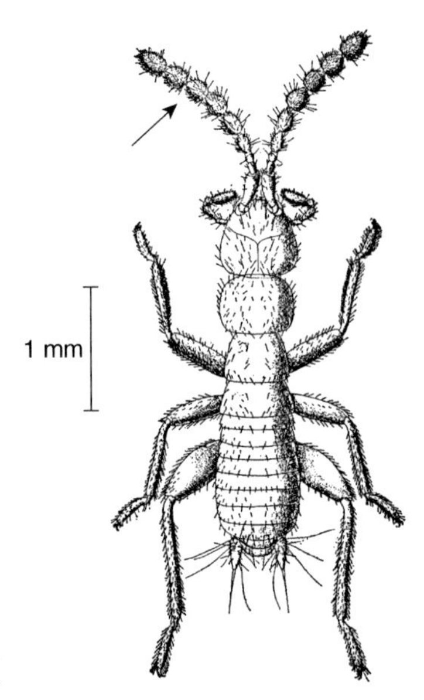

Fig. 6.5 Zoraptera adult (from Richards and Davies 1977).

6.6 Order Orthoptera (grasshoppers and crickets) – 20,000 plus described species

These are, by contrast with the preceding three Orders, reasonably familiar insects. The pronotum is usually large and saddle-like over the prothorax. Tarsi are usually four- or five-segmented, though rarely with three, two or one segments. The hind legs are often adapted for jumping (see Fig. 2.15) and forewings are very often narrow and leathery compared with the larger and membranous hind wings. Such leathery forewings protecting the more delicate and flight-functional hind wings are called ***tegmina***

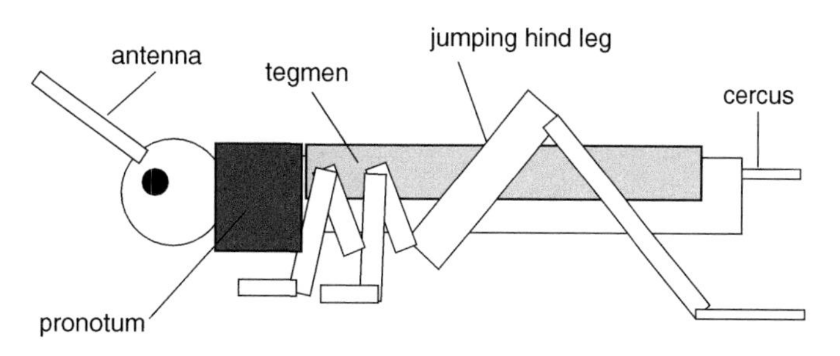

Fig. 6.6 Cartoon of orthopteran external features.

(singular ***tegmen***) (Fig. 6.6). During development, the wing buds undergo an unusual reversal whereby the hind wings are outside the forewings until the moult to adult. Stridulation ('singing') is a common behaviour, especially but not exclusively of males, coupled with structures specially adapted for hearing. The ovipositor is usually well developed and obvious. The cerci may be long or short, but are never segmented. The Order Orthoptera contains some economically very important insects.

I shall describe this Order in terms of its main Superfamilies (ending '-oidea').

6.6.1 *Suborder Ensifera* (= sword bearer)

These Orthoptera are so called because of the generally obvious, curved ovipositor, a bit reminiscent of a scimitar. They have long segmented antennae (more than 30 segments) and usually a long ovipositor. When present, the ***tympanal organ*** (auditory) is on the front legs, and stridulation (a behaviour of males only) is by rubbing the wings together (see Sections 6.6.1.1 and 6.6.1.2).

6.6.1.1 Superfamily Tettigonioidea (long-horned grasshoppers) – the main Family is the Tettigoniidae

The common name of 'long-horned grasshopper' is highly descriptive. Tettigoniids are often quite large and sometimes beautifully coloured (especially bright green) and delicate insects with long slender antennae reflexed over the body and often exceeding it in length (Fig. 6.7). Tarsi are four-segmented. The often curved ovipositor is very conspicuous and robust compared with that of other grasshoppers. Eggs are inserted into plant tissues or into the soil. The Superfamily is mainly carnivorous, but includes some herbivores. The more delicate carnivores in the Superfamily are arboreal. 'Bush-cricket' (Fig. 6.8) is a common name often applied to the tettigoniids, which live in herbage and low shrubs in the warmer parts of the world. This is a misnomer, as true crickets are a different Superfamily (see Section 6.6.1.2). At dusk, bush-crickets stridulate actively, and the noise can verge on the deafening where there are large populations. In the USA, they may be referred to as 'katydids'; apparently to American ears their stridulation sounds like a repetitive staccato 'katy-did, katy-did . . .'. When the males stridulate, the sound is produced by a row of teeth (the 'file') on part of vein Cu (the 'stridulatory vein') on the left tegmen rubbing against the hind margin (the 'scraper') of the right tegmen. The sound is amplified by the resonance of a circular area, known as the 'mirror', defined by branches of the cubital veins and most developed on that right tegmen. Commonly, there are four or five nymphal instars. Some species have no hind wings and the tegmen reduced to just the area needed for stridulation.

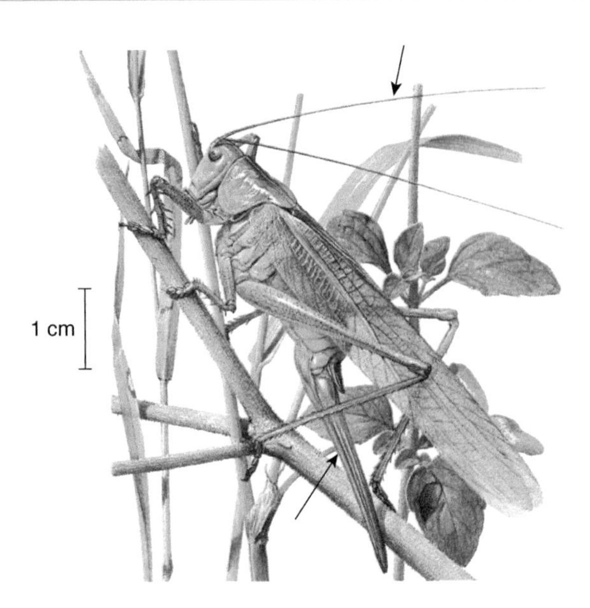

Fig. 6.7 Tettigoniidae (long-horned grasshopper) (from Zanetti 1977).

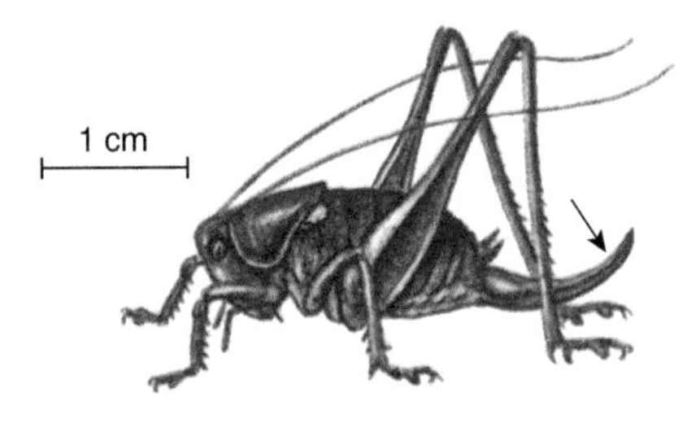

Fig. 6.8 A bush-cricket (from Mandahl-Barth 1974, with permission).

Among the herbivorous species is a wingless and quite polyphagous pest in the USA called *Anabrus simplex* – with the common name of the 'Mormon cricket' though it is a grasshopper. Its mode of attack is similar to that of a locust; when it descends on crops as a swarm great destruction results. The name 'Mormon cricket' derives from a remarkable story of a swarm marching in to destroy the wheat crop providing the main food supply to the Mormons settled at Salt Lake City in Utah in the mid-19th century. The Mormons, at a time before insecticides, gathered in church and prayed for help, whereupon a huge flock of seagulls came out of the sky. They landed on the crop and consumed the entire pest population. This is still the only recorded example of biological control by divine intervention, all the more remarkable because Salt Lake City is more than 1000 km from the coast.

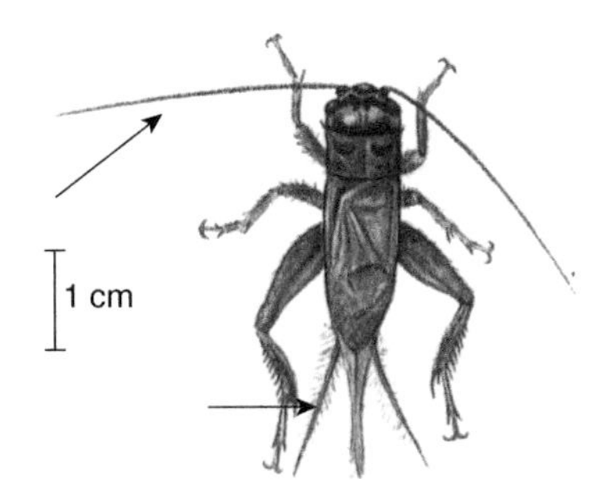

Fig. 6.9 House-cricket (*Acheta domestica*) (from Lyneborg 1968, with permission).

6.6.1.2 Superfamily Grylloidea (crickets) – the main Family is the Gryllidae

These are mostly ground-living scavengers, or sometimes herbivores. Many species are wingless. The typical cricket (the 'cricket on the hearth'; Fig. 6.9) has long antennae, often reflexed over the body, and long unsegmented cerci, between which females have a needle-like ovipositor. The wings are folded flat over the abdomen. The stridulatory apparatus is similar to that in the long-horned grasshoppers, except that the 'mirror' is nearer the tip of the tegmina, giving space for a 'harp' area with several cross veins in the cubital sector. When stridulating, males raise the wings to 45° and rub the file against the scraper. Although the tegmina and wings do not differ between left and right sides, it has been said that the 'left-handedness' of the tettigoniids is reversed in crickets, in that it is always the file on the right that is rubbed by the scraper on the left. It is claimed that the stridulation of the cricket *Brachytrupes megacephalus* can be heard over a mile away. However, many species have no forewings.

Species of the genus *Acheta* are known as the field crickets. They live in the soil, are nocturnal and feed on most parts of many different herbaceous crops. Seedlings may be completely cut at the base and then eaten, and woody plants may even have their bark gnawed. However, damage is rarely enough to warrant control. The elongated curved yellow eggs are laid in batches in the soil, and each female can lay 2000 eggs. There can be several generations a year in warm climates, though one generation is the norm in temperate regions. Field crickets are omnivorous, but appear to require some animal food (other insects) for normal development.

Crickets are also cannibalistic. This behaviour led to a sport akin to cock-fighting, developing in China and Japan, with male crickets being pitted against each other. In the 19th century, a legendary cricket with the unlikely name of 'Ghengis Khan of Canton' won fights with stakes as high as £30,000 (equivalent to over £2 million today).

Mole-crickets (Family Gryllotalpidae) are very different from other crickets in appearance. They are large, dramatic and distinctive crickets (Fig. 6.10) adapted to burrowing in the soil, where they breed in an underground 'nest'. The antennae are shorter than the body. Mole-crickets usually have very reduced wings behind a large prothorax covered in short brown felt-like hairs (looking remarkably like the fur of a mole) and distinctive large front legs adapted for digging. The eyes are small and the ovipositor is tiny. In the south-west USA, mole crickets can cause a lot of damage to crops, particularly wheat. They graze the roots of young seedlings, but the damage is magnified by the seedlings they 'uproot' as they tunnel along the softer earth of the seed drill. They also feed on other insects and worms. Some mole-crickets are fully winged and can fly surprisingly well. In Africa, Asia and Australasia, the African mole-cricket (*Gryllotalpa africana*) attacks many field crops, especially as seedlings, and species in the genus *Scapteriscus* do such damage in the New World.

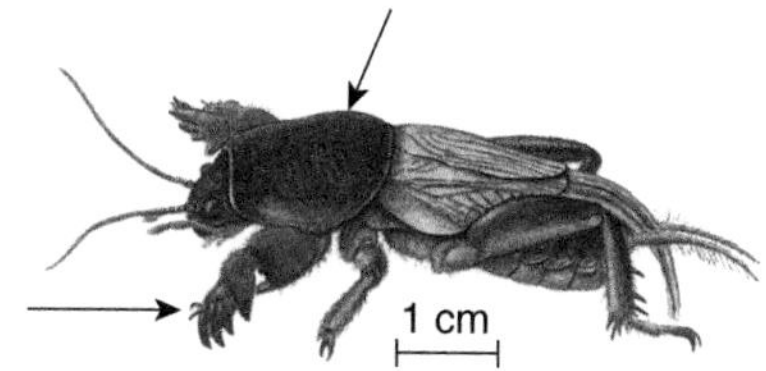

Fig. 6.10 Mole-cricket (from Lyneborg 1968, with permission).

The eggs are laid in a 'nest' well below (10–15 cm) the soil surface, and the nymphal duration (probably about 10 instars) is about 2 years. The first instars remain in the nest and are fed by the female. Later instars stay in the nest by day, but leave it at night to forage for food.

Gryllotalpa gryllotalpa is a mole cricket that occurs in the UK – quite the opposite of a pest insect, it is an endangered and protected species; I have only seen a living specimen once, in North Wales, and in 2005 eight individuals were reported to have been found in a compost heap in Oxfordshire.

6.6.2 Suborder Caelifera – meaning 'relief bearing', probably referring to the sculptured cuticle

These insects are typical grasshoppers with antennae of less than 30 segments and which are shorter than the body. Cerci are short and unsegmented. The wings are held along the sides of the abdomen at rest. When present, the auditory organ is a tympanum (see Section 6.6.2.1) at the base of the abdomen.

6.6.2.1 Superfamily Acridoidea (short-horned grasshoppers) – the main Family is the Acrididae

These are the herbivorous grasshoppers and locusts (Fig. 6.11). Most have greatly enlarged hind femora for jumping and an obvious eardrum (***tympanum***) on either side of the first abdominal segment near its junction with the thorax. This is associated with the common practice of sound production (e.g. for mate finding) by stridulation, which grasshoppers accomplish with the violin principle. There are three different systems in the detail, but the commonest system is that a row of pegs (equivalent to the violin bow) on the inside of the wide hind femur is scraped on the prominent and hardened radial vein (equivalent to the strings) on the tegmen (forewing). Tarsi mostly have three segments. Females in some species also produce sound, but with a much reduced stridulatory system. The ovipositor is not conspicuous but is able to excavate a hole in soft (often sandy) ground as the end of the abdomen is extended into the deepening hole until the eggs are laid in a chitinous pod (Fig. 2.20b).

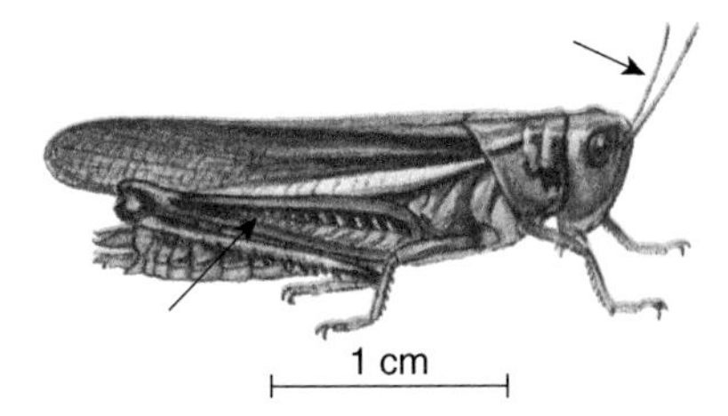

Fig. 6.11 A grasshopper (from Lyneborg 1968, with permission).

Some large species of grasshoppers have the locust behaviour of forming large swarms, which then destroy crops over a wide area. These are so well known that they have attracted the common name of 'locusts' even though this title identifies a behaviour shared by some insects in other quite different Orders (e.g. the armyworms in the Lepidoptera).

There is a huge literature on locusts, but here follows what one might call a 'rough guide'. Locusts spend most of the time at low density in poor-quality and often parched natural grassland. These sites are called the 'outbreak areas' because it is here that outbreaks, which are often infrequent, originate from these low-density grasshopper populations. The insects in the outbreak areas are in the physiological phase known as 'solitary' – they are dispersed, loath to fly and of a rather drab colouring well camouflaged to the background (Fig. 6.12a). They can stay like this for several years, ticking along and breeding slowly. The outbreak areas may then flood from nearby rivers or swamps, or extreme drought may also reduce the amount of green vegetation. Such events force the solitaries to crowd together on the small amount of grass or shrub that remains.

The close contact between individuals and released pheromones then causes hormonal changes and a switch from the solitary to the 'gregarious' phase. These locusts are more fecund, more flight ready and brightly coloured with black, orange and yellow (Fig. 6.12b). It is these gregarious-phase insects that swarm and fly to devastate crops, often far from the outbreak areas. Where they land they breed, and swarms of locust nymphs (known as 'hoppers') then march across the landscape, eating anything green in their path. Eventually, sometimes only after a few years, the resultant shortage

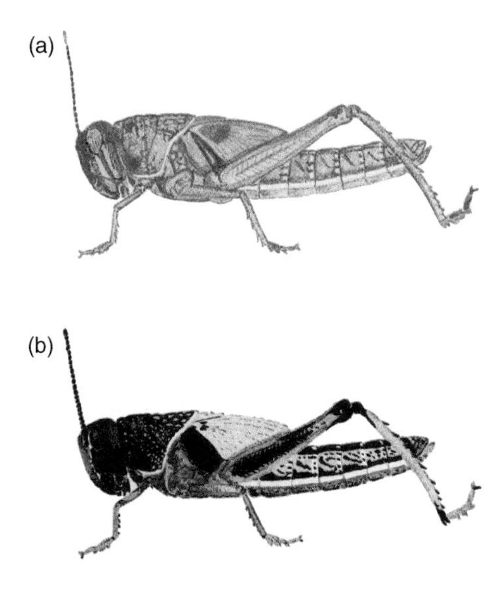

Fig. 6.12 Locust (*Schistocerca gregaria*) hoppers in (a) solitary and (b) gregarious phase (from Uvarov 1966, courtesy of Natural Resources Institute).

of food brings the plague to an end and all that remains of the locust species are then the solitaries in the old outbreak areas.

The most important locust grasshoppers (Fig. 6.13) are:

Locusta migratoria (the migratory locust; Fig. 6.13a). Although this species is polyphagous, it shows a marked preference for grasses and cereals. The outbreak areas of this species are mostly to be found in dry grassy areas adjacent to swamps in the flood plains of the river Niger. This species tends to develop outbreaks at a time of year when the intertropical convergence zone of winds develops. The swarms are then carried into southern Europe, to many parts of Africa and eastwards along the convergence zone across Asia and even as far as Australasia. Locust swarms can extend over areas from less than one to several hundred square kilometres, and contain 40 to 80 million locusts per square kilometre. There is one generation per year.

Nomadacris septemfasciata (the red locust; Fig. 6.13b). As the name suggests, this is a red-coloured locust, which does not travel far from its outbreak areas and is a pest primarily in southern Africa, particularly in Zambia and Tanzania. Like the migratory locust, the red locust is polyphagous with a preference for grasses and related crops. Again there is one generation per year.

Schistocerca gregaria (the desert locust; Fig. 6.13c). This species is very polyphagous in comparison with the two species described above though, like them, it is particularly fond of cereals. The desert locust breeds in sand hills and flies huge distances all over Egypt, north Africa, the Mediterranean and north India. There may be up to 100 eggs in each of the four to five pods laid by a female. There are several generations a year.

Without doubt, locusts are extremely important pests, but their outbreaks are sporadic and often relatively infrequent. By contrast, non-swarming grasshoppers can be a serious occurrence every year, and many tropical farmers regard such grasshoppers

Fig. 6.13 Three important species of locusts: (a) migratory locust (*Locusta migratoria*) (photo by Jonathan Hornung); (b) red locust (*Nomadacris septemfasciata*); (c) desert locust (*Schistocerca gregaria*) (photo by Arpingstone).

as a greater problem than locusts. However, locusts are 'box office' and international agencies and governments will often provide insecticides for their control, which the farmers then use against their local grasshoppers instead! Species of the genus *Zonocerus* are common throughout Africa, and feed on most broad-leaved crops and finger millet, though other cereals and grasses are not attacked. They are frequently recorded as damaging cotton and legumes, but also seedling coffee, cocoa and cassava. There is one generation a year. *Zonocerus*, which is often called the 'elegant grasshopper', is indeed a beautiful insect (Fig. 6.14), basically green but with large black, yellow and orange patches. Brachypterous specimens are common.

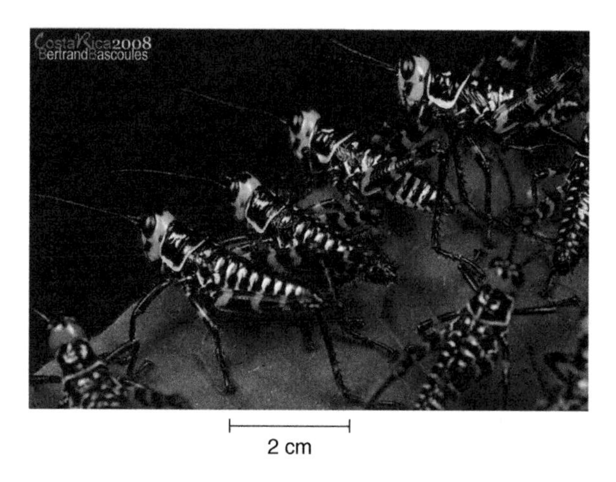

Fig. 6.14 Nymphs of the grasshopper *Zonocerus* (courtesy of Bertrand Bascoules).

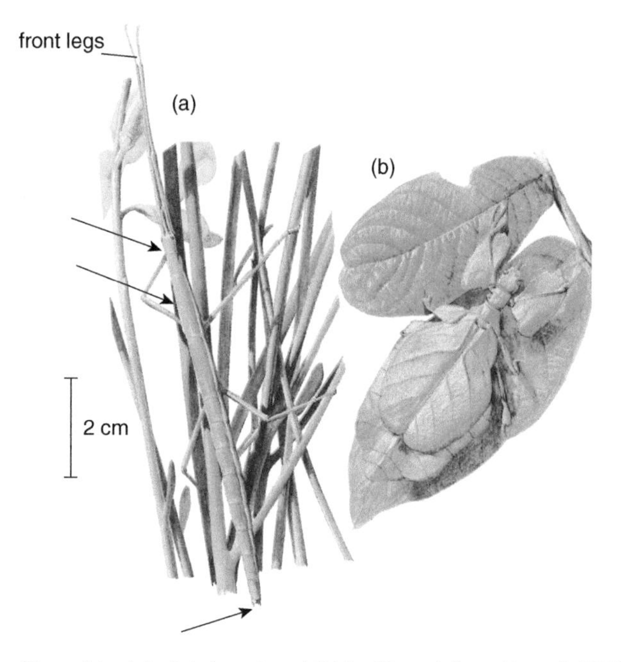

Fig. 6.15 Order Phasmida: (a) stick insect and (b) leaf insect (from Zanetti 1977).

6.7 Order Phasmida (stick and leaf insects) – *c.* 2700 described species

Stick insects (Fig. 6.15a) are familiar to most people, with their long thin bodies and spindly legs. The prothorax is short, but the meso- and metathorax are usually elongated. The long mesothorax makes for a large gap between the front and middle legs. Phasmids are herbivorous, mainly on shrubs and trees. Stick insects are often wingless, and when winged they usually have very short horny front wings, though some tropical phasmids have longer horny forewings (tegmina) and large hind wings not dissimilar from the wings of locusts. The cerci are short and unsegmented. Many stick insects are brown or green, but can change their colouration to match their background. Stick insects can be up to 26 cm long, possible because their slender bodies make tracheal breathing (see Section 2.6.4) possible. Their name derives from their similarity to twigs; this and their ability to change colour make them very cryptic; they can be hard to spot as they sit in vegetation.

Reproduction is often parthenogenetic, that is asexual, the eggs developing although unfertilised, and males are rare in many species. Eggs are laid singly, very often just dropped by the walking female without stopping. The ovipositor is small and concealed.

Leaf insects (Fig. 6.15b) are phasmids with broad flattened bodies and flat expansions on the segments of the leg. They mimic the shape of leaves, even to the veins and, more remarkably still, they may even have patches that mimic holes as made by insects feeding on leaves. They are usually a green to match the foliage they are on, and many leaf insects living on deciduous trees change colour in the autumn to match senescing leaves. It has even been claimed that the individuals increasingly develop reddish colour in proportion to the green and senescing leaves as the season progresses.

Phasmids are found in tropical and subtropical climates, though two species have established as far north as the Scilly Isles and a few places in southern England. Although herbivorous, they are not usually included as plant pests, though numbers of some species on eucalyptus in Australia can be so large that serious damage occurs. These same species show phase differences not dissimilar to locusts (see Section 6.6.2.1) with colour and biometric differences between solitary and crowded individuals.

6.8 Order Dermaptera (earwigs) – *c.* 1200 described species

This is another familiar Order of insects, recognisable by the terminal unsegmented forceps (the cerci), which are used mainly in defence and courtship, and the very short tegmina leaving much of the dorsal surface of the abdomen exposed. The cerci of the sexes differ, with the female ones less curved and thus more parallel (Fig. 6.16). Under the horny tegmina are large, many times folded hind wings. For many years it was

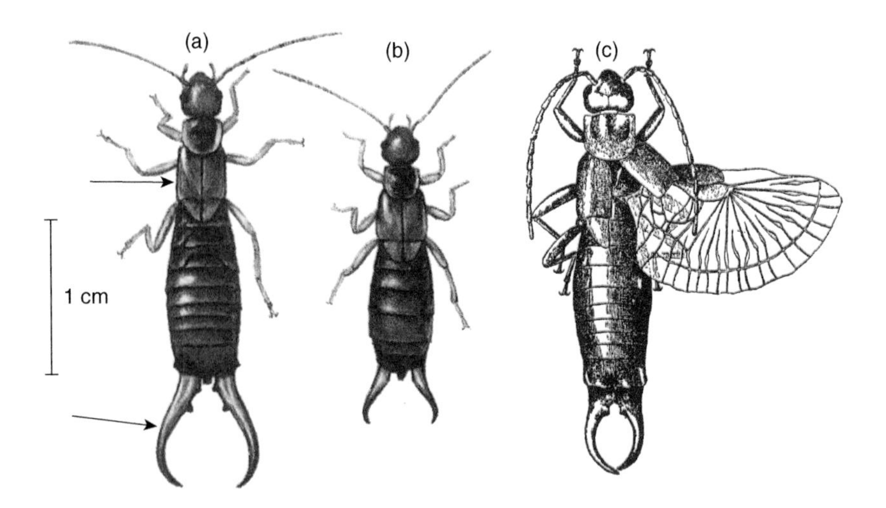

Fig. 6.16 Earwigs: (a) male and (b) female (from Lyneborg 1968, with permission); (c) male with right wing extended (from Richards and Davies 1977).

thought earwigs never flew, but they are in fact excellent flyers. What is amazing is how, on landing, the hind wings (which mainly represent a large anal area) are folded away under the tegmina in a fraction of a second. It is a real skill of origami, using the direction of folds and exploiting the tension in the veins of the open wing. Indeed, if you open out the hind wing of a dead specimen and try to repack it you will find it more difficult than repacking a parachute! However, apterous species are common in some Families. The antennae are long and slender, with perhaps as many as 50 segments. The head bears large compound eyes, but no ocelli. Tarsi are three-segmented. The ovipositor is small or absent.

The common earwig (*Forficula auricularia*) is typical of the Suborder Forficulina. It is found throughout Europe as well as in most other temperate regions of the world. There are many legends concerning earwigs, including that they will crawl into human ears and penetrate the brain. Crawling into ears is, I suppose, a possibility. Being nocturnal, earwigs shun the light and thus seek out dark, moist places by day. Additionally, they are strongly thigmotactic, meaning they seek to feel contact with something on all sides. Thus, especially in the autumn, they crawl into hollow plant stems (a tunnel like the ear canal) to overwinter, and move upwards till they become wedged.

They are primarily harmless and pretty omnivorous scavengers, but will also take bites out of fruit such as apples and make holes in the petals of flowers, including prized garden blooms. They also take animal food and are regarded as valuable predators of aphids and small caterpillars in orchards and hop gardens. Rolled corrugated cardboard is sometimes hung in the trees as 'earwig traps', not to trap the insects in order to kill them, but to retain them in the orchard overwinter so that they are on site as predators early the next spring.

This willingness to take animal food is reflected in their cannibalistic behaviour, cannibalism which in the male even extends to eating his own offspring. Earwigs lay up to 100 eggs in cells in the soil. The eggs are guarded by the female even after they have hatched particularly because they would otherwise be attacked by her carnivorous husband. There are usually two generations a year.

There are two other Suborders of the Dermaptera. The Arixeniina are cave-dwelling earwigs scavenging on bat guano whereas the Hemimerina are ectoparasitic on rats.

6.9 Order Embioptera (web spinners) – *c.* 360 described species

This Order is mainly tropical, though some species get as far as southern Europe. In particular they are found in South America. The dark brown insects live gregariously in silk tunnels beneath stones or loose tree bark. A 'colony' may contain hundreds of individuals. They are very small, being only 4–8 mm long. They can move backwards and forwards in these tunnels with equal speed. The walls may have the added protection of plant fragments, soil particles and cast skins. The fore tarsi are strongly enlarged (Fig. 6.17) to accommodate spinnerets from which the insects produce the silk with which they build their tunnels. Cerci are two-segmented and of unequal size (i.e. asymmetrical) in males, and females are much rarer than males. There is in fact marked sexual dimorphism, with the males usually winged and the females usually apterous. In winged web-spinners, the front and hind wings are like those of termites (see Section 6.11) of pretty much identical size. The insects emerge at night to forage for any plant material, fresh or dry, leaves, flowers, roots or twigs. Adult males probably do not feed. Their feeding habits make them pests very occasionally, for example when they feed on the roots of commercially grown orchids. There are some signs of parental care in that the little urn-shaped eggs are licked and moved around in the tunnels; in

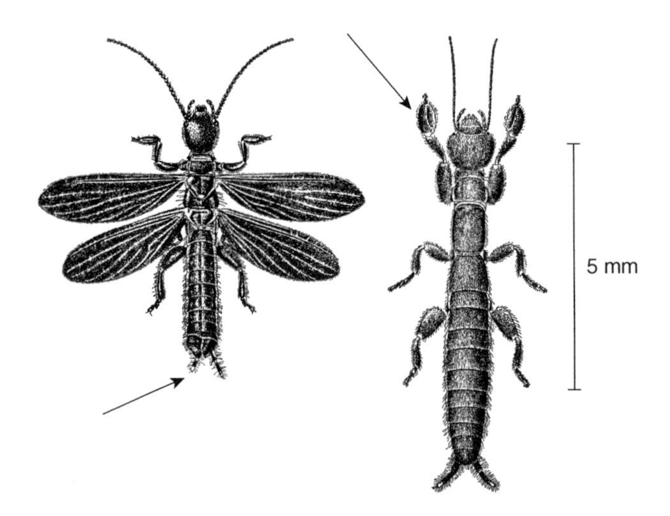

Fig. 6.17 Embioptera. Left, male and right, female (from Richards and Davies 1977).

some species the adults even provide fragments of food for the new hatchlings. Parthenogenetic reproduction is not uncommon.

6.10 Order Dictyoptera (cockroaches and mantids) – *c.* 5700 described species

Antennae are many segmented and filiform. The coxae are rather large and close together, tarsi are nearly always five-segmented. The cerci are unusual for an orthopteroid Order in being many segmented. The forewings are somewhat hardened (tegmina); the ovipositor is not obvious. The eggs are contained in a hard ***ootheca*** (see Fig. 6.19), formed inside a chamber at the end of the oviduct by secretions from the accessory glands.

6.10.1 Suborder Blattaria (cockroaches)

These are well known as fast running flattened omnivores with long antennae and elongated striated cerci. The head is almost completely covered in dorsal view by the large shield-shaped pronotum (Fig. 6.18a). The front wings are horny tegmina; wings are reduced in the females of some species. Cockroaches are active at night in buildings, particularly where food is stored (e.g. kitchens, bakeries and university halls of residence). The legs are long for fast running and the tibiae have longish spines. Cockroaches are frequent domestic pests. They eat almost anything (except cucumber and putty, apparently). They have benefited greatly from the appearance of drink vending machines, where running along a water pipe leads into a cornucopia of food such as dried milk, cocoa and coffee powder.

Cockroaches are general scavengers. The number of eggs in an ootheca varies from 12 to 50 depending on the species. The ootheca is often carried around for a while, hanging from the end of the abdomen until it drops off as the insect runs around. When the nymphs are ready to emerge, the ootheca splits along its dorsal edge.

The main Family of economic importance is the Blattidae. The mid and hind femora have downwardly pointing spines; the large hind wings are folded like a fan under the harder tegmina.

Two species in this Family have been transported around the world and are now cosmopolitan (as well as metropolitan!) pests:

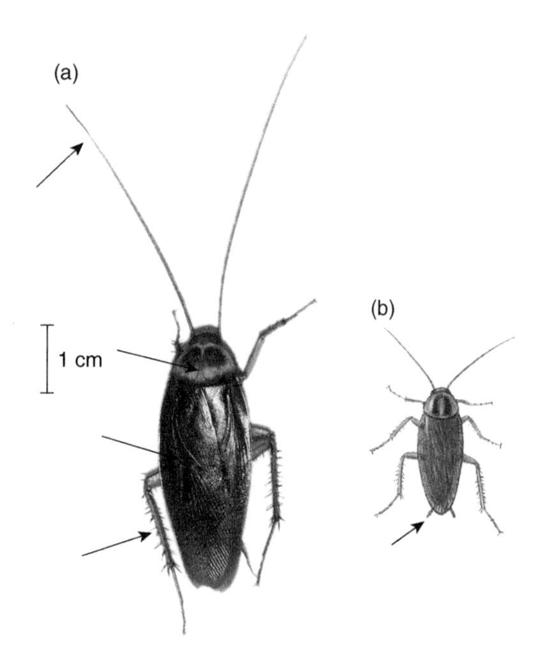

Fig. 6.18 (a) American cockroach (*Periplaneta americana*) (photo by Gary Alport); (b) German cockroach (*Blatella germanica*) (from Lyneborg 1968, with permission).

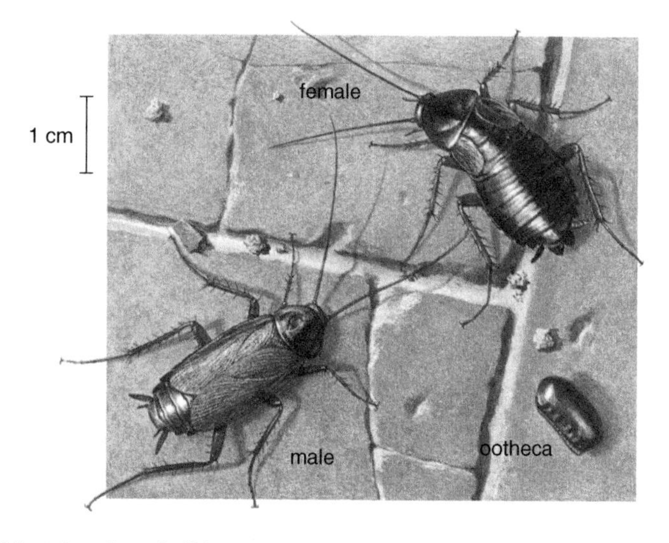

Fig. 6.19 Oriental cockroach (*Blatta orientalis*) (from Zanetti 1977).

Blatta orientalis (the oriental cockroach; Fig. 6.19) is a shiny brownish-black insect about 2.5 cm in length. The females have short, almost rudimentary tegmina, while those of the male cover about three-quarters of the abdomen and hide the fan-folded membranous hind wing. The male makes flight-assisted jumps rather than real flights. These cockroaches prefer damp shady areas and the life cycle takes almost a full year.

Periplaneta americana (the American cockroach; Fig. 6.18a) is a much larger cockroach at 4 cm long. It is clearly brown rather than black, and the pronotum has a much lighter rim, especially transversely at the back edge. Both males and females are fully winged; in the male the tegmina project well beyond the end of the abdomen. Like *B. orientalis*, outdoor populations prefer damp shady places and the species also has a 1-year life cycle.

In a different Family (the Blatellidae), with all femora spiny and again a fan-folded hind wing, is a similarly transported pest species, *Blatella germanica* (the German cockroach; Fig. 6.18b). Unlike the two other cosmopolitan cockroaches mentioned above, *B. germanica* is found almost entirely associated with human shelter and activity, as the species cannot survive at low temperature. It is probably the species most responsible for the pest status of cockroaches. The adult is 10–15 mm long, and is dark brown in colour with two longitudinal black stripes on the pronotum with a pale area between them. Both sexes are fully winged, but in *B. germanica* it is the female whose tegmina project beyond the rear of the abdomen. The life cycle is only somewhat over 3 months, and breeding is continuous.

With the ubiquity and high profile of these cosmopolitan pests, it is easy to forget that many species of smaller cockroaches (such as members of the genus *Ectobius* – Family Blattelidae) occur naturally under bark and leaf litter and make no economic impact on man. Additionally, a few semiaquatic species are known. At the other end of the humidity extreme, there are species that live in deserts; such cockroaches have especially long spines on their legs that enable them rapidly to dig themselves into sand.

6.10.2 Suborder Mantodea (mantids)

These striking carnivores (Fig. 6.20) are mainly tropical in distribution and are known by the way they hold their predatory forelegs (see Fig. 2.15 'raptorial') while in ambush mode – hence the common name 'praying mantis'.

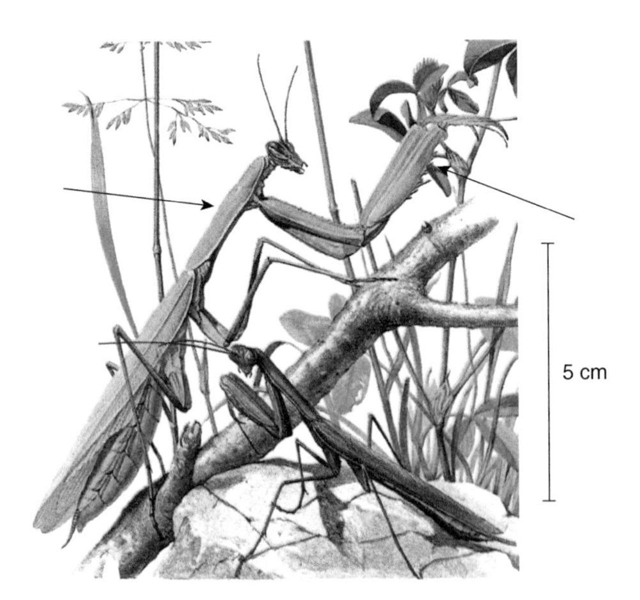

Fig. 6.20 Praying mantis (from Zanetti 1977).

They are often a bright green and the adults have large hind wings under the horny forewing. They can fly well, though in some Families there are brachypterous species or ones where only the females have this characteristic. The head projects clearly in front of the pronotum, which carries the striking raptorial forelegs referred to above. The pronotum is very long, so that the predatory forelegs are some distance in front of the mesothoracic legs. Males are very much smaller than their females and the sexes have a bizarre courtship behaviour. The male does not so much seek a female as avoid being ambushed by one. When she has caught her man, the first action of the female is to bite off his head. This removes the inhibitory centre of the brain, and the thorax and abdomen of the male then meekly walk round the female and copulate. When this is finished, the female eats the rest of her mate, using his protein for the yolk of her eggs. It almost typifies the biblical quote 'Greater love hath no man than this . . .'.

There are five Families, of which the Family Mantidae is the largest and most diverse. There is one European species, *Mantis religiosa*.

6.11 Order Isoptera (termites) – *c.* 1200 species

Isoptera means 'equal winged', and winged termites are unusual among the Exopterygota in having very similar front and hind wings, both in size and venation (Fig. 6.21). By contrast, most other Exopterygota have a small forewing (often as a tegmen) and a much larger hind wing, and the mayflies (see Section 5.2) have a large forewing and a small hind wing. A characteristic structure in the head is the ***frontal gland***, the secretions of which are known to be defensive. The gland is therefore most developed in the soldier caste (see below in this Section) where it may open in a pale-coloured cuticular depression (the ***fontanelle***), at the end of a prominent ***frontal tubercle***, or have this tubercle enlarged as a long 'snout' in nasute soldiers (see below in this Section). In most Families, and all those mentioned later, the tarsi are four-segmented.

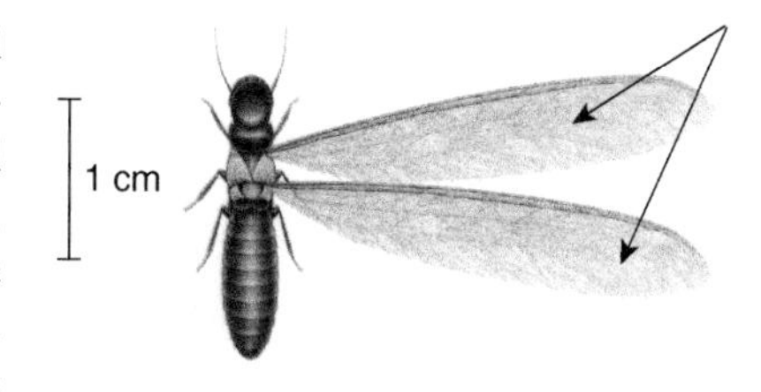

Fig. 6.21 Adult winged termite (wings shown only on right side) (www.cowleys.com).

Termites have evolved from wood-boring cockroaches and many species are serious pests because they are so destructive to wood and therefore to many buildings. They often weaken timbers before it is realised that damage is occurring. When stiletto heels were first available in Africa, many of the proud new owners of this latest fashion came home and promptly fell through their living room floors! Of course termites are also bibliophiles, and the books in libraries in the tropics have to be protected with residual insecticide. Once in Nigeria I was staying in the home of a scientist while he was away on leave. I was impressed by his fine library of leather-bound English classic novels. However, when I pulled one out, I only had half a book in my hand. It had rested on only half a shelf, and there was this large gap where the parts of shelves and books against the wall had been eaten away. Perhaps a beneficial aspect of wood-boring termites, depending on your musical tastes, is that the aborigines in Australia sought out logs hollowed out by such termites for their iconic musical instrument, the didgeridoo.

Termites are highly specialised, building elaborate nests and having a social structure (see below in this Section) based on polymorphism. This means that the adults

Fig. 6.22 A termitarium (termite mound) (photo by Discott).

of a colony (i.e. individuals of one genotype) exist in more than one phenotype; each is known as a caste.

Some termites build relatively simple nests in the soil or in fallen trees, but others build free-standing structures (termitaria) of a material as hard as concrete (Fig. 6.22). The insects mix earth, fragments of wood, their excrement and saliva and place one little ball of this material onto another until the termitarium (which can be up to 9 m high) is built. The interior of the termitarium is a maze of chambers and corridors and the population in one *Nasutitermes* nest only 0.5 m high and 1.3 m wide has been estimated at 1.8 million individuals. Termites also use this material for civil engineering, for example building bridges to cross the liquid in an old oil drum as a moat around a wooden support of a house and specifically designed to prevent invasion by termites.

Termitaria built in crop fields are a real nuisance; they occupy land and impede cultivation machinery. They are thus one component of the pest status of termites. They are tough structures and can even resist onslaught by mechanical diggers; blowing the mounds up with explosive may be the only practical option.

A great deal of crop damage is caused by the termites removing leaf portions and taking them back to their nest. Here the leaf material forms the substrate for culturing fungi in 'fungus gardens'. Cellulose is much more easily digested by fungi than animals, and in this way the termites recover the nutrition of the plant material they have brought back to the nest indirectly by grazing the fungus. Enzymes that can digest cellulose, and produced by symbiotic bacteria in the digestive system, have been

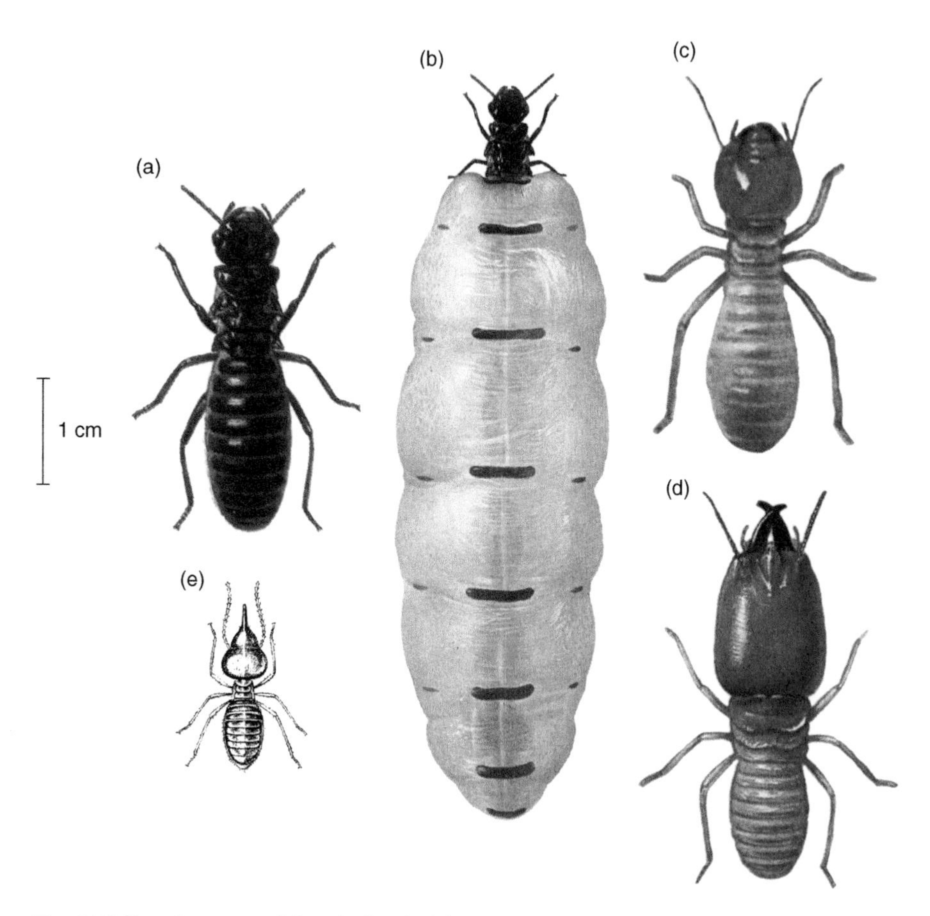

Fig. 6.23 Termite castes: (a) male (king); (b) gravid female (queen); (c) worker; (d) soldier; (e) nasute soldier ((a–d) from Zanetti 1977; (e) from Richards and Davies 1977).

identified in termites, but the insects seem to place greater reliance on their 'fungus gardens' for cellulose digestion than on such enzymes.

Termites are essential service providers in tropical and subtropical ecosystems, especially in areas of low rainfall, by cycling organic matter in a similar way as earthworms are important in temperate climates. Without termites, as can occur on isolated tropical islands, fallen leaves just accumulate to a considerable depth without rotting away.

As mentioned above, termites have quite an elaborate caste system (Fig. 6.23). Males and females (the reproductive castes) that are released from termite nests mate in the air, and then bite off their wings along a line of weakness soon after landing. Release in an area is synchronised by weather conditions, and numbers are such that the ground may appear carpeted with termite wings. Pairs then excavate a simple nest and the female soon begins to lay eggs. They have to feed and rear the first generation of offspring, which become the first 'workers' (Fig. 6.23c), a caste of wingless sterile males and females whose reproductive potential is suppressed by the food they are given and by pheromones released by the female. The male and female become the royal pair ('king' and 'queen') of the colony, and they quickly devolve their foraging, rearing and nest-building activities to their workers. The queen may live for 60 years, and in the Family Termitidae may lay up to a staggering 2000 eggs a day, which are removed and tended by the workers. The latter also feed the royal pair; one of their most essential other duties is to clean the queen's cuticle of fungal spores. The queen's reproductive powers soon cause her to expand her abdomen to a point where she can no longer move or reach much of her body with her legs for cleaning; she would soon die of fungus infection if not cleaned by the workers. A termite queen's abdomen (Fig. 6.23b) then looks like a pale, rather translucent sac with widely spaced narrow brown bars – these are the original tergites before the arthrodial membrane between them became greatly stretched to accommodate the ovaries and eggs. The king (Fig. 6.23a) lives nowhere near as long as the queen; but she fertilises her eggs with sperm received into her spermatheca, and so can continue ovipositing even in the absence of her mate.

Another caste of wingless sterile male and female termites are the 'soldiers' (Fig. 6.23d). These are also fed by the workers. They defend the nest and also escort the foraging workers. Different species have different types of soldiers. Many simply have large opposable jaws, but some specialised ones have scissor-like jaws for literally cutting ants in half. Ants can be aggressive attackers of termite nests, and another specialised 'anti-ant' soldier is the 'nasute' (Fig. 6.23e). These operate underground or in wood and are blind. Their head is prolonged into a snout from which they can squirt a quick-setting cement to immobilise their enemies (see above in this Section).

Finally, some brood are kept aside and reared as 'supplementary reproductives'. These are the only caste that develops wing buds, and they are reared at particular times for release from the nest for mating flights. A few such supplementary reproductives are always maintained to replace the king or queen should either die. Before they have completed their development, supernumerary supplementaries can revert to workers, and are then known as pseudoergates.

Many years ago Karl Escherich, an early 20th century German entomologist, depicted most of the termite castes in a famous pictorial impression of the main chamber of a termite nest. It is worth reproducing (Fig. 6.24).

That termites have the incomplete metamorphosis of the exopterygotes makes it possible for caste determination to be delayed well into development compared with the social insects (ants, bees and wasps) in the Endopterygota.

The economically most important Families are as follows.

6.11.1 Family Kalotermitidae (dry-wood termites)

Ocelli are present, but the fontanelle is absent. The pronotum is usually wider than the head.

The Family feeds and forms colonies in dry, dead wood, mainly in the tropics. *Cryptotermes brevis* (West Indian dry wood termite) is the most widespread of the dry wood termites. The species infests what were originally sound hardwoods and softwoods, including items of furniture, picture frames etc. External symptoms are the 1 to 2-mm holes from which the fine faeces (hence the alternative name 'powder-post termites') is expelled.

6.11.2 Family Hodotermitidae

These live in nests in wood or soil. The nests in soil show above ground as conical earth mounds often surrounded by an area of bare soil from which the termites have harvested the vegetation. Ocelli and the fontanelle are absent, and the pronotum is narrower than the head.

Fig. 6.24 Karl Escherich's idealised representation of the 'royal chamber' in a termite mound. The huge queen lies in the centre with the king beside her. Workers reach up all around them to clean the queen's cuticle, while another throng of workers circle around her in an anticlockwise direction, some presumably bringing food and others carrying away eggs from the end of her abdomen. Meanwhile a ring of soldiers stands guard, their jaws pointing outwards. (Reproduced from a poor photograph of an illustration from the 1870s.)

Hodotermes mossambicus (the harvester termite) harvests all grass species, and is an important pest of grassland in south and east Africa, especially where the grass has already been depleted by overgrazing. The species is also known to be an occasional problem in cotton crops, where the stems are bitten off just above ground level.

6.11.3 Family Rhinotermitidae (wet-wood termites)

These nest in damp dead wood such as that of tree stumps below ground. They do not build mounds. The Rhinotermitidae are the termites most damaging to dwellings and other human wooden structures, though some species attack growing plants. Both fontanelle and ocelli are present.

Reticulotermes is a genus of subterranean termites, which will eat anything made of cellulose. The genus is very important in the essential ecological process of the breakdown of dead wood, but of course is also particularly destructive to timber dwellings. Different species are found in various parts of the Holarctic region – North and Central America, Europe (as far north as Paris) and the Far East.

Species of *Coptotermes* can be found throughout the tropics and the Far East, and damage trees as well as both monocotyledonous and dicotyledonous food crops. With trees, the bark and roots are eaten, and young trees may be killed by ring-barking. *Coptotermes formosanus* can be a problem in south-east Asia.

6.11.4 Family Termitidae (mound-building termites)

The largest Family of termites, the Termitidae, have diverse food habits, and most build the large mound termitaria described earlier. Wing venation is reduced compared with the other Families, and workers and soldiers have a narrow pronotum.

The genus *Macrotermes* is highly polyphagous, and attacks many crops including rice, sugar cane, groundnuts, orchard and forest trees as well as plantation crops such as cocoa and coffee. *Macrotermes* is found between the two tropics in Africa and Asia. The workers forage in tunnels built of plant fragments, soil and saliva. Under this protection they feed on the bark and may even kill the tree if it becomes ring-barked.

Root damage may also be serious. Leaf material is taken to the nest for fungus gardens (see Section 6.11), as these termites cannot digest cellulose themselves.

Colonies of the genus *Odontotermes* are smaller than *Macrotermes*, and their mounds are also smaller; yet they are probably the most numerous termites in tropical soils. They attack established plants and seedlings of many crops, including sugar cane, tea and coconut. They feed similarly to *Macrotermes,* but under protective sheets (only sometimes seen with *Macrotermes*) rather than under tunnels. Many enter woody plants at pruning cuts. Like *Macrotermes*, fungus gardens are grown in underground chambers of the nests.

7 Subclass Pterygota, Division Exopterygota, Hemipteroid Orders

7.1 Introduction

The Hemipteroid Orders are distinguished from the Orthopteroid Orders by the mouthparts being modified from the basic pattern of Fig. 2.11, either less obviously by the fusion of structures or more obviously by being modified for sucking. The hind wings have no anal lobe, and the insects do not possess cerci (but do not misinterpret the cornicles of aphids (see Figs 7.44 and 7.50) as cerci).

7.2 Order Psocoptera (booklice) – *c.* 1650 described species

The booklice are small rather drab insects about the size of aphids or smaller, and indeed they bear a passing resemblance to aphids. They fit into the hemipteroid group of Orders because of the absence of cerci and the highly modified maxillae. The maxillae have a fused cardo and stipes, with a lacinia modified to a strongly sclerotised rod, which it has been suggested may support the head while the insect is feeding. The labial palps are reduced to small lobes. More obvious for identification is the diagnostic inflated clypeus (Fig. 7. 1), and some winged adults have a large dark spot (the ***pterostigma***) on the front wing; some species have marbled wings. However, many species have wingless (= apterous) adults or ones with short (brachypterous) non-functional wings. Asexual reproduction by parthenogenesis (eggs complete development without fertilisation) is common.

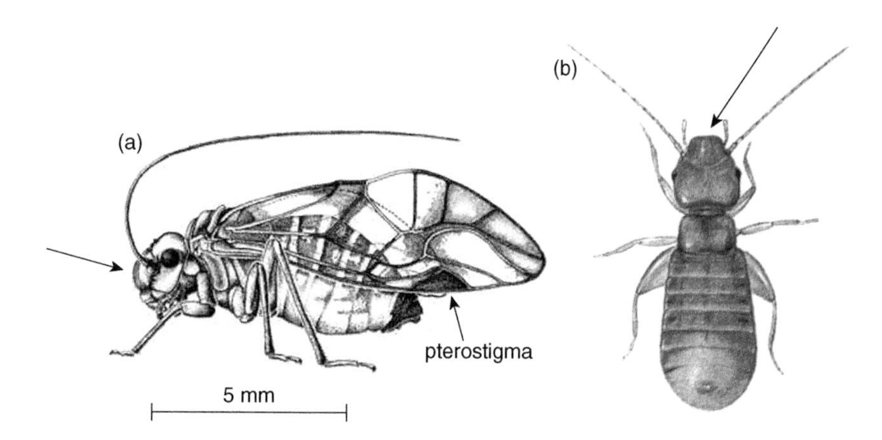

Fig. 7.1 Order Psocoptera: (a) winged adult (courtesy of the University of Queensland); (b) wingless adult (from Zanetti 1977).

Handbook of Agricultural Entomology, First Edition. H. F. van Emden.

The Psocoptera are known as the 'booklice' because they feed on fungal spores and moulds, including those that develop on the glue used in bookbinding. Thus they are commonly found in books or corrugated cardboard, and are regarded as a minor pest in these situations. Indeed, I was once asked to provide expert mediation, between a firm producing tea packets and a warehouse storing the empty outer corrugated packaging after printing, as to the source of a psocid problem.

However, the vast majority of species are free living out-of-doors, particularly on tree foliage and in and under bark. The Family Liposcelidae are flattened apterous psocids and some of these are minor pests feeding on stored products.

7.3 Order Mallophaga (biting lice) – *c.* 2700 described species

The Mallophaga are ectoparasites, mainly on birds (the bird lice) but sometimes also on mammals. They are dorsoventrally flattened and have no cerci (Fig. 7.2). However, although they are clearly related to the other hemipteroid Orders, they differ in having highly modified mandibulate mouthparts rather than ones adapted for sucking. The maxillae are single lobed and not differentiated into sclerites; the labial palps are just small lobes or are absent. Minute rods have been reported, which are probably homologous with the laciniae of the Psocoptera; indeed the Mallophaga probably evolved from the Psocoptera. Mallophaga complete their entire life cycle on one host and, in adaptation to their ectoparasitic habit, the insects are secondarily wingless. They scrape skin and chew feathers and may feed on the blood from the wounds they have inflicted.

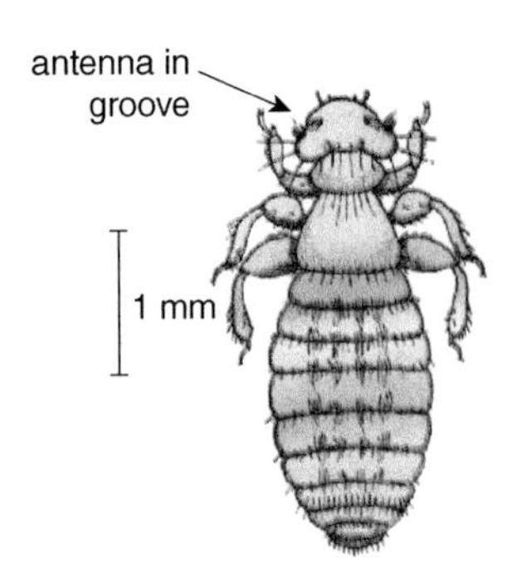

Fig. 7.2 Common chicken louse (*Menopon gallinae*) (reproduced with permission of CSIRO, Australia).

The eggs are cemented onto the feathers or hairs of the host, and there are three nymphal instars. The lice are rather host specific and their transfer between hosts is mainly by contact. However, some Mallophaga species are less than 1 mm in length, and so they also disperse by phoresy (which is 'hitching a lift' on another insect likely to travel to a new host). Thus they have been found travelling on flies, even ones as small as mosquitoes.

The Suborder Amblycera (Fig. 7.2) have segmented maxillary palps and the antennae are concealed in grooves on the side of the head. They include some ectoparasites of rodents and some species over 10 mm long live on raptorial birds.

The Suborder Ischnocera have no maxillary palps and free antennae. Different Families specialise on hyraxes, lemurs and another on cattle, horses, camels and deer. Mallophaga in a large Family, the Menoponidae, are all ectoparasites of birds, and include one of the most damaging bird lice, the common chicken-louse (*Menopon gallinae*).

The third Suborder of Mallophaga, the Rhynchophthirina, contains just two species that specialise on elephants.

7.4 Order Anoplura (= Siphunculata) (sucking lice) – *c.* 500 described species

These are again dorsoventrally flattened ectoparasites with no cerci, but this Suborder of lice has sucking mouthparts consisting of a serrated proboscis containing three stylets, with which the insects suck the blood of their hosts. The mouthparts can be withdrawn into the insect's head. The Suborder is easily recognised by the tarsi which are rather like karabiners, with the claw opposed to the tip of the tarsus to complete a ring and grip a hair of their host (Fig. 7.3). Anoplura are blood-sucking ectoparasites of mammals, and two-thirds of the species are ectoparasitic on rodents.

As with the Mallophaga, eggs are usually cemented to the hairs of the host. When found on the head of a human, the eggs are called 'nits'. They will have been laid by *Pediculus humanus capitis* (the head louse), another and bigger louse (*P. humanus*) sucks blood in other areas of the body of humans. These lice are vectors of diseases of mankind, especially the disease typhus. *Haematopinus suis* is the pig louse, which can be a problem for pig farmers.

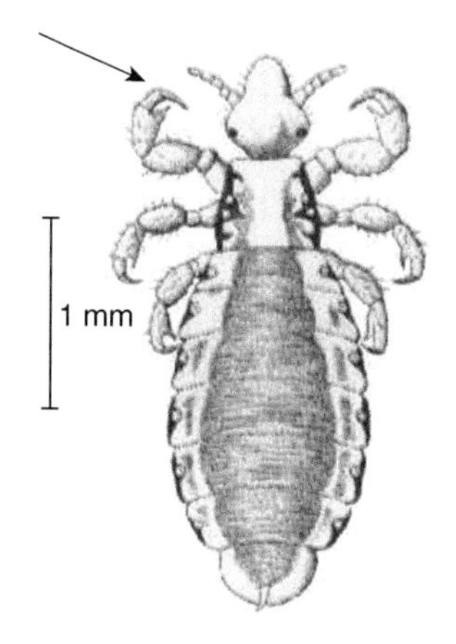

Fig. 7.3 Head louse (*Pediculus humanus capitis*) (reproduced with permission of CSIRO, Australia).

7.5 Order Hemiptera (true bugs) – *c.* 57,000 described species

This is one of the five most important Orders of insects in economic terms (the others being the Lepidoptera, Hymenoptera, Diptera and Coleoptera). The Order contains many important crop pests, many of which vector plant diseases, predators valuable in biological control and there are also medical and veterinary pests, which again may transmit diseases, of animals including humans.

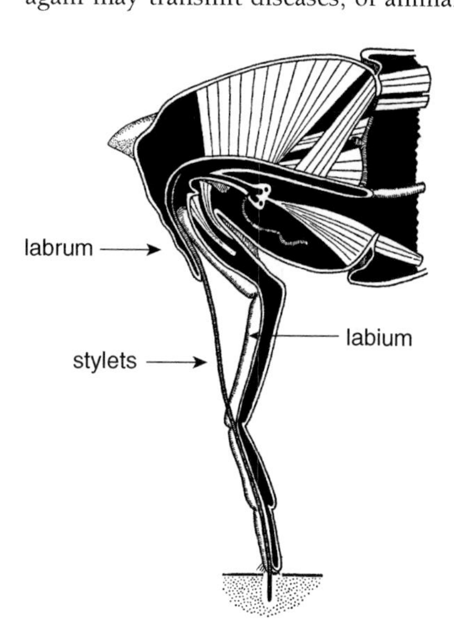

Fig. 7.4 Vertical section of mouthparts of Heteroptera (from Richards and Davies 1977).

The Hemiptera are very distinct from any of the Orthopteroid Orders in having elongated mouthparts (the ***rostrum***) adapted for sucking, and also in having no cerci. The mandibles and maxillae are modified into ***stylets*** lying in a grooved, segmented labium (Fig. 7.4). The stylets are only slightly longer than the labium, and insertion into the substrate is achieved either by bending back a section of the labium (Fig. 7.4) to shorten the distance from the base of the head to the substrate (Suborder Heteroptera) or by telescoping the labial segments (Suborders Auchenorryncha and Sternorryncha). A cross-section of the rostrum (Fig. 7.5) shows how the two maxillae interlock to create two canals running the length of the stylets – a larger ***food canal***, up which food travels to a cavity (the ***pharynx***) in the head, and a smaller ***salivary canal***, down which the insect secretes saliva from the salivary glands. Fluid can be drawn up the food canal by a powerful ***pharyngeal pump*** expanding and contracting the volume of the pharyngeal chamber.

The tips of the mandibles and maxillae can easily be distinguished (Fig. 7.6). The former have barbed tips for anchorage in the substrate in order to allow the maxillae to slide between them. The latter have blunter tips with openings through which liquid and particles can pass into the food canal, or through which saliva can be secreted from the salivary canal. The mandibles are themselves pierced by another canal, which carries a simple nerve system capable of detecting touch but not capable of detecting taste. Tasting is achieved by drawing liquid up the food canal to sensors in the walls

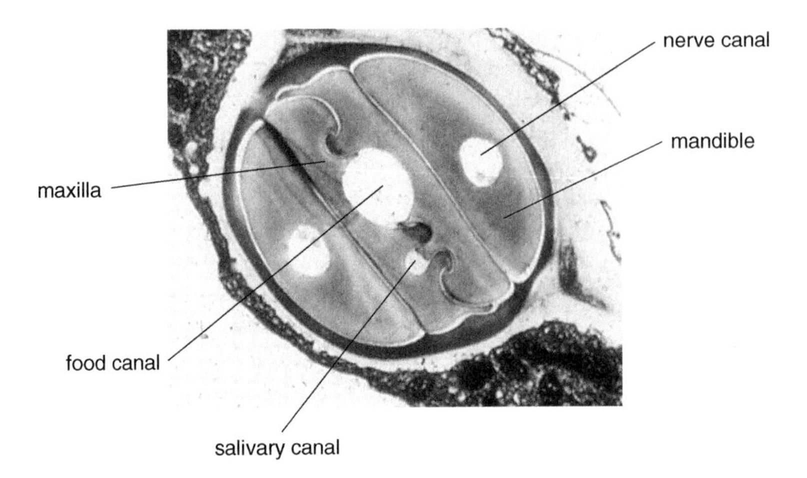

Fig. 7.5 Section across stylet bundle of an aphid (courtesy of F.W. Tjallingi).

of the pharynx and, if necessary, expelling it again without imbibing any. The nerve connection in the mandible is related to the way the insect navigates the tips of its stylets in the substrate (Fig. 7.7). Physical information from the tip of the mandibles and taste information from the pharynx determines which way the maxillae should go when contacting a barrier such as a plant cell wall. By keeping the tips of the maxillae together or by advancing one ahead of the other the stylets can be made to pierce straight ahead or to slip round to the left or to the right. This movement of the two maxillae is accomplished from the head through the way they are held together by a tongue and groove arrangement (Fig. 7.8); this enables them to slide vertically on each other without affecting the integrity of the food and salivary canals. I have often thought of using this principle to invent a device (such as a cath-

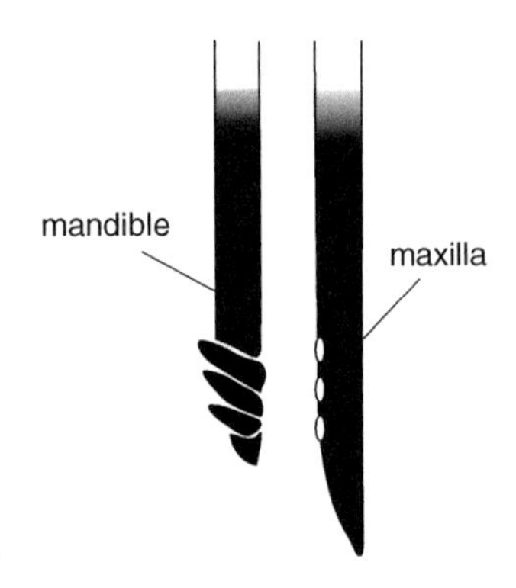

Fig. 7.6 Barbed tip of mandible and blunt tip of maxilla (with holes for passage of fluids into the food canal) of the cotton stainer, *Dysdercus fasciatus* (drawn from a microscope preparation).

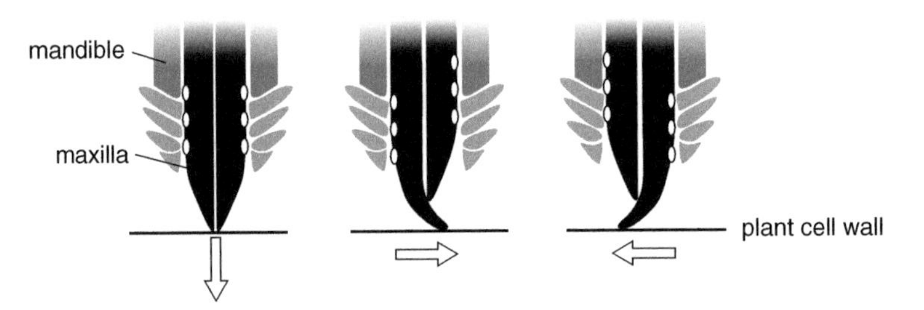

Fig. 7.7 Cartoon of how a bug navigates its stylet tips between or through plant cells by anchoring the stylet bundle with the barbed mandibles and then 'remotely controlling' from the head the direction of penetration of the stylets, by advancing one maxilla on the other. Arrows show the respective direction of penetration.

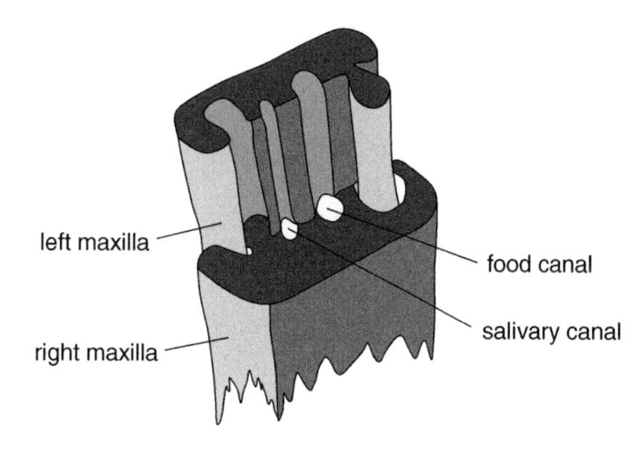

Fig. 7.8 Diagrammatic representation of the tongue-and-groove arrangement which enables the two maxillae to slide vertically along each other.

eter, endoscope or even a drain-clearing rod) that can be directed as to which way to go at a junction in a tube.

The Hemiptera show a complete gradation from, at one end, active insects that occasionally stop to feed to, at the other, the highly adapted plant lice that almost become part of the plant's 'plumbing'.

At the former end are the Heteroptera. These are either parenchyma feeders emptying one cell at a time, or they are carnivores stabbing prey and killing them or ectoparasitically sucking their blood. Some Auchenorryncha are also parenchyma feeders but others, and all the Sternorryncha, navigate the stylets to the conducting tissues (predominantly the phloem but also at times the xylem). There they feed on the fluids in the vessels and in the surrounding tissues. Penetration to the conducting tissues may be intracellular (directly through cells) or intercellular (navigating between the cells without puncturing them).

7.5.1 Suborder Heteroptera (land and water bugs)

The term Heteroptera (= different wing) refers to the division of the forewing into two parts. The proximal part nearest the attachment to the thorax is relatively stiff and horny, while the apical part is a more normal wing membrane. This wing is described as a half-wingcase, a ***hemielytron***. By contrast the other two Suborders, previously grouped together as the Suborder Homoptera (= same wing) have a forewing of uniform texture, either fully horny or fully membranous. The hemielytron is made up of several regions (Fig. 7.9) demarcated by lines along which the hemielytron can flex. The main part of the hardened basal section is the ***corium***, which may have a further division (the ***embolium***) at its leading edge. This edge also forms one of the edges of the apical ***cuneus***, which has another boundary with the membranous part of the wing. A little indentation (the ***cuneal notch***) on the leading edge of the wing often identifies the boundary between the cuneus and the corium/embolium. An extension of the bottom edge of the corium is the ***clavus***.

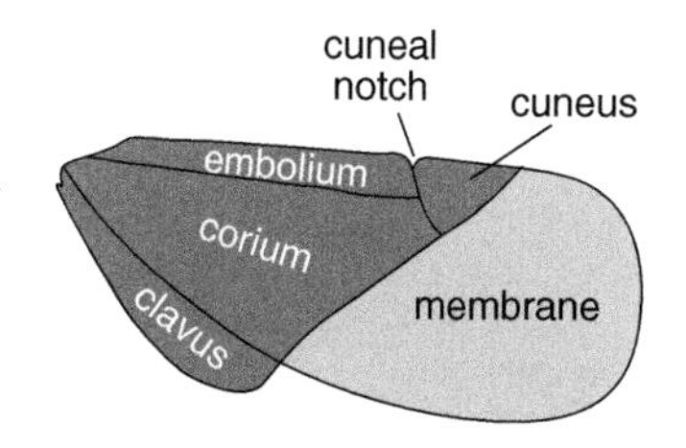

Fig. 7.9 Divisions of the hemielytron of the Heteroptera.

How is it possible to distinguish the Heteroptera from the other Hemiptera in the nymphal stage, when the wings are insufficiently developed to identify that there is a hemielytron? Well, at least the land bugs (Gymnocerata, see Section 7.5.1.2) mostly have the distinctive facies of, when viewed from above, an almost pointed triangular head, with partly protruding spherical eyes and obvious antennae composed of a few long cylindrical segments of about equal length.

Additionally, the forewings of the Heteroptera lie flat on the abdomen with the membranes overlapping (as in Fig. 7.13), except in the Family Scutelleridae, where often a large area (***scutellum***) of the dorsal surface of the mesothorax protrudes between the clavi (Fig. 7.23). The pronotum is usually large, transverse and obvious. By contrast, the wings in the other Suborders are held against the sides of the body rather more like a pitched roof (as in Figs 7.38 and 7.40), and the pronotum tends to be smaller and more rounded.

7.5.1.1 Series Cryptocerata (water bugs)

Cryptocerata means 'hidden antennae', and this characteristic clearly separates them from the land bugs (Gymnocerata='free antennae') with their very obvious antennae. The Cryptocerata comprises several Families of carnivorous water bugs, the 'water boatmen' and 'water scorpions' and allies.

Water boatmen (Families Notonectidae and Corixidae)

These carnivorous water bugs (Fig. 7.10) swim with a rowing action of their long oar-like middle legs. The Notonectidae (greater water boatmen) swim upside down

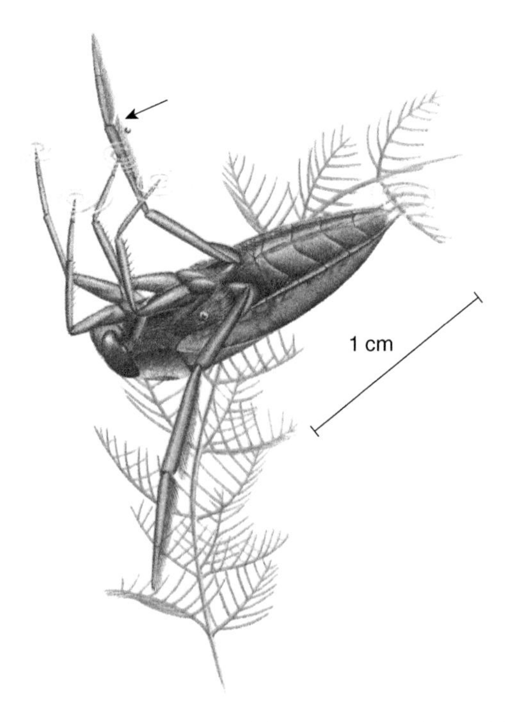

Fig. 7.10 Greater water boatman (*Notonecta glauca*) (from Dipper and Powell 1984).

and often spend much time hanging under the water surface. They then immediately react to any vibration of that surface as caused, for example, by an insect falling on to it. Corixidae (lesser water boatmen) swim the usual way up around submerged vegetation. With a tracheal air-breathing respiratory system, water bugs have to be able to continue air breathing under water for lengthy periods. Both Families use plastron respiration when the insect submerges, which is where air is trapped under the closed wings by hairs on the abdomen over the spiracles. As the adherence to the hairs resists shrinking of the air bubble, more oxygen dissolves out of the water across the meniscus of the bubble as the gas is depleted within it. This considerably extends the time that the insects can spend submerged if calculated only on the amount of oxygen in the original bubble.

Water scorpions and allies (Family Nepidae)

The Nepidae (Fig. 7.11), as the name suggests, have predatory raptorial forelegs reminiscent of the chelae of scorpions as well as an obvious 'tail' in the adults. The latter (the ***siphon***) is in effect a snorkel, allowing the insect resting on vegetation just below the water surface to draw air onto the abdominal spiracles at the base of the siphon. The nymphs do not have a functional siphon, and breathe by six pairs of abdominal spiracles. Common genera are the broad and flat *Nepa* and the long thin *Ranatra* (sometimes called water stick insects).

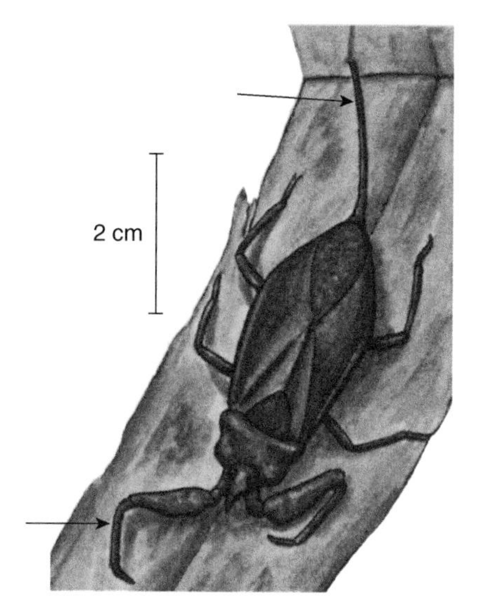

Fig. 7.11 Water scorpion (*Nepa cinerea*) (from Mandahl-Barth 1973, with permission).

Rather larger, as the name suggests, are the giant water bugs (Family Belostomatidae). These are not dissimilar to water scorpions, but the siphon is much shorter. They can be up to 12 cm long and are known to tackle not only other insects, but also small fish, tadpoles and small frogs. These large bugs are food delicacies in parts of Asia. In several genera, the female glues her eggs onto the back of the male.

7.5.1.2 Series Gymnocerata

Gymnocerata means 'free antennae' and, as pointed out above, the members of the Series characteristically have relatively long antennae composed of a few cylindrical segments of equal length arising from a pointed triangular head beyond the margins of which the eyes protrude (e.g. Fig. 7.13). The mouthparts (rostrum) tends to be quite long, projecting past the back of the head and carried between the legs when the insect is walking. They have not been recorded as vectoring plant virus diseases, though some herbivorous species transmit plant pathogenic fungi and species parasitic on vertebrates transmit pathogenic protozoa. There are nearly 40 Families of Gymnocerata, only the more important of which to the reader will be described here.

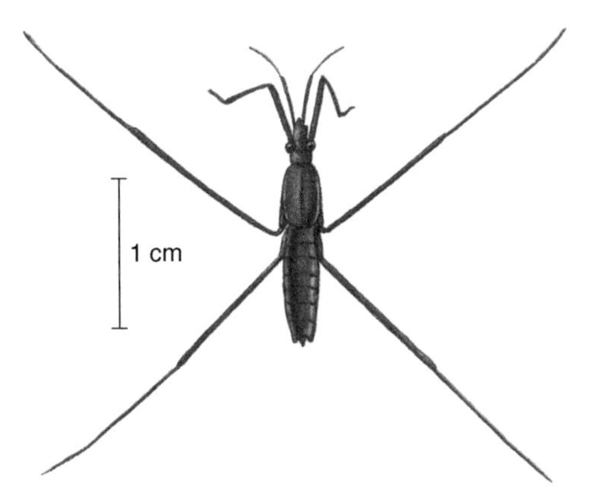

Fig. 7.12 Pond skater (*Gerris najus*) (from Mandahl-Barth 1973, with permission).

Pond skaters (Families Gerridae and Veliidae)

Most Gymnocerata are land bugs, but the pond skaters (Fig. 7.12) live on the surface of water, where they feed on other insects. They have hydrophobic hairs on the pulvilli of the tarsi, and this enables them to skate at the water surface without breaking the surface tension. Being black in colour, they are often most recognisable by the little 'pits' in the water surface where the ends of the legs touch the water. The Veliidae are much shorter and stouter bodied than the Gerridae.

Family Miridae (mirids or capsids)

This is the largest Family in the Series in terms of the number of species. Adult Miridae are easy to recognise by a spot character. Veins project into the membrane part of the hemielytron, but do not extend to the wing margin. They usually form two loops, looking a bit like a letter 'W' (e.g. Fig 7.13).

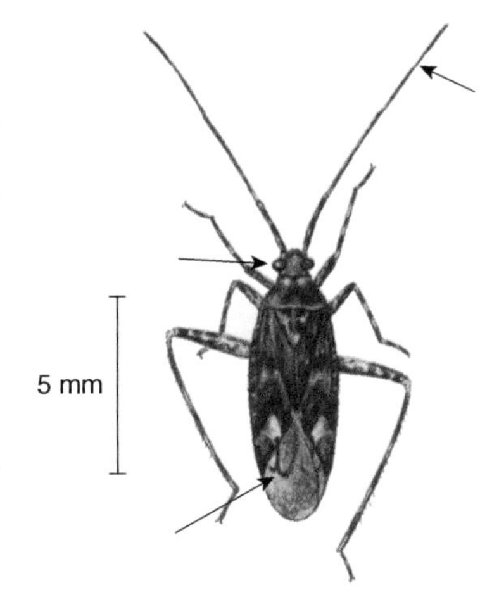

Fig. 7.13 Family Miridae (from Mandahl-Barth 1974, with permission).

The Family was originally called the Capsidae, therefore the common name of 'capsids'. Hence also the apple capsid (*Capsus ater*) which is a minor pest of apples. It causes deformation and reddening of attacked leaves, and the surface of fruits blister around the point of insertion of the stylets. These are reactions to proteins in the saliva of the bugs, and such symptoms are fairly typical of mirids. Although they suck up the contents of mesophyll cells, this is rarely by itself damaging to yield, and most economic damage by mirids is indeed due to the saliva. Gall-like deformations are common, as is 'shot-holing' (Fig. 7.14). 'Shot-holing' is the appearance of clusters of small holes in the leaf. However, the bugs have usually moved on by the time the symptoms appear. The holes are where the expanding leaf has torn at the place where the mirid fed earlier and the bugs are active and prefer young leaf tissue. Thus they often move from one host plant species to another as the season progresses to exploit a sequence of young developing leaves. If one of these plant species is a crop, then there is a potential pest situation.

Eggs are laid singly wherever the mirid happens to be, and are often inserted into the plant tissues. Most mirids are opportunistic carnivores as well as more typically herbivores. They will pierce and suck the fluid out of small caterpillars and other small prey such as aphids that they encounter, and some species (e.g. *Macrolophus caliginosus*) are reared commercially for the control of aphids in glasshouses.

The potato capsid, *Calocoris norvegicus*, causes typical mirid necrotic spots which then tear into holes in the leaves of potato and brassicas in Europe; it is also known as a pest in Canada. Again typically for mirids, the toxic saliva may kill affected shoots. Eggs are inserted into the stems of host plants such as hawthorn on which the bug overwinters, and in the spring the bugs migrate to summer hosts such as potato, usually first spending some time on weed hosts such as nettle. The adult is about 7 mm long, and is green with two small black spots on the pronotum. There are usually two generations a year.

Fig. 7.14 *Lygus* bugs and 'shot-hole' damage to cotton (from Bayer 1968, with permission).

The genus *Lygus* (tarnished plant bugs) contains several species that are pests of soft fruit, apple, cotton and alfalfa, though the insects are generally very polyphagous. Don't be confused by the name *Lygus*. The next Family I shall mention is the Lygaeidae, but paradoxically *Lygus* is a mirid, not a lygaeid. In most crop situations, *Lygus* bugs are normally suppressed by insecticides used against caterpillars, but can become problems when such compounds are replaced by resistant plant varieties (including those produced by genetic modification) or biological control. *Lygus* bugs can then build up large populations, which can even cripple plants by the number of mesophyll cells emptied. The bugs are particularly damaging to cotton (Fig. 7.14), where damaged plants have tattered leaves and leggy growth with shortened side branches. Young flower buds turn brown and are shed. *Lygus* bugs are about 6 mm long and brownish-green. In Europe, they attack potato, sugar beet, alfalfa, apple and other fruit trees. The adults overwinter on trees or in leaf litter until late spring, and eggs are laid in May. There are two generations a year.

Very common in Europe is the green capsid *Lygocoris pabulinus*, sometimes a serious pest on apple and other fruit as well as on potato, sugar beet and beans. It overwinters in the egg stage on perennials such as hawthorn and rose and there are two generations a year. The bug tatters the leaves and callus forms where the insects have sucked the fruit, which renders it unsaleable. Similar damage, as well as blistering of the fruit surface, is done to apples in the UK and some other parts of Europe by another capsid, *Plesiocoris rugicollis*.

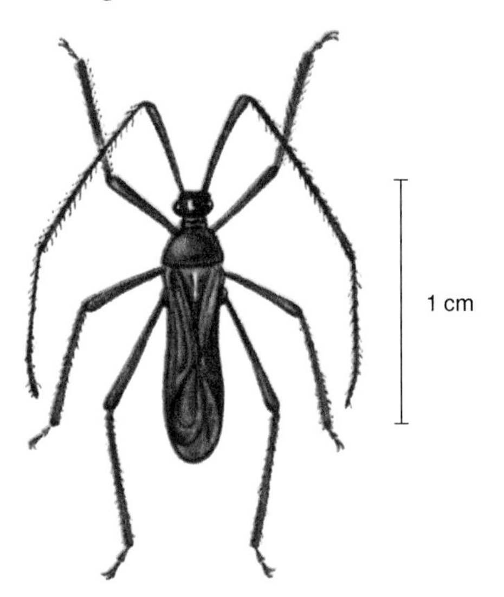

Fig. 7.15 *Helopeltis* sp. (from Bayer 1968, with permission).

Helopeltis spp. (Fig. 7.15) are another genus of polyphagous cotton pests, particularly in Africa. Attacks tend to be sudden and localised; plants become stunted with black lesions on the stems, leaves and bolls. The bugs are characteristically slender for mirids, with longer antennae than is normal. Another mirid pest of cotton is *Pseudatomoscelis seriatus*, the cotton fleahopper (a misnomer, since other hemipteran hoppers are not in the Heteroptera, but in the Auchenorryhncha).

Two species of cocoa capsid, *Sahlbergella singularis* (Fig. 7.16) and *Distantiella theobroma* are very damaging pests of cocoa in West and Central Africa. They feed on the pods and young shoots of mature trees, moving to the foliage of the canopy when the pods have been harvested. Severely attacked trees may die, and control is often essential.

Family Lygaeidae (chinch bugs)

Here the veins are more numerous than in mirids, and extend to the distal margin of the wing as a set of almost parallel lines (Fig. 7.17). An important pest in the USA is the wheat chinch bug, *Blissus leucopterus*. Adults are about 5 mm long (Fig. 7.18), with a black pronotum and wings that are mainly white. Attack and feeding by large numbers of these pests can completely cripple the plants, especially in drought conditions, and even destroy them as seedlings. Drought not only weakens the plants but also speeds up the reproduction of the pest. With chinch bugs a phenomenon is encountered that recurs frequently in the Order Hemiptera, namely 'alary polymorphism' (see also Section 6.5). Adult chinch bugs can occur as two phenotypes, fully winged (macropterous) or short winged (brachypterous). The traditional explanation

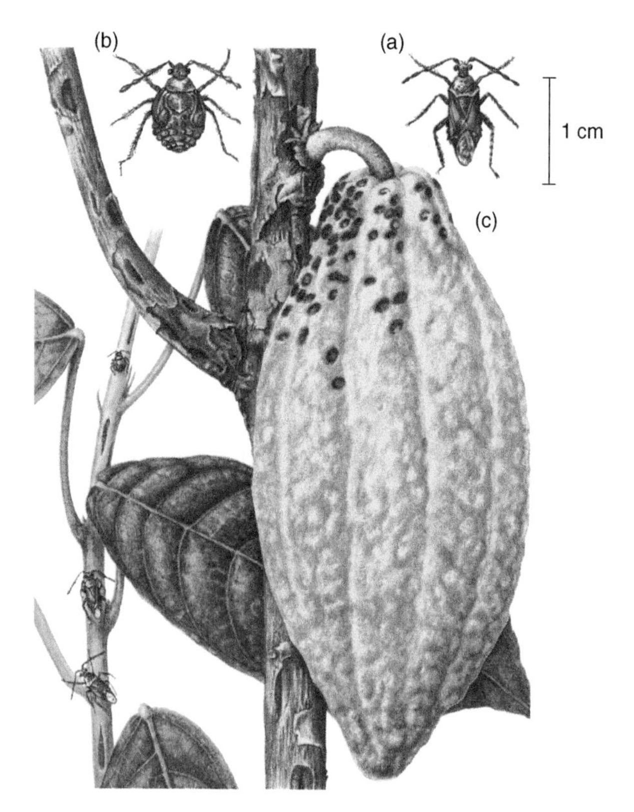

Fig. 7.16 *Sahlbergella singularis*: (a) adult, (b) nymph, and (c) showing feeding damage to stems and pod (from Bayer 1968, with permission).

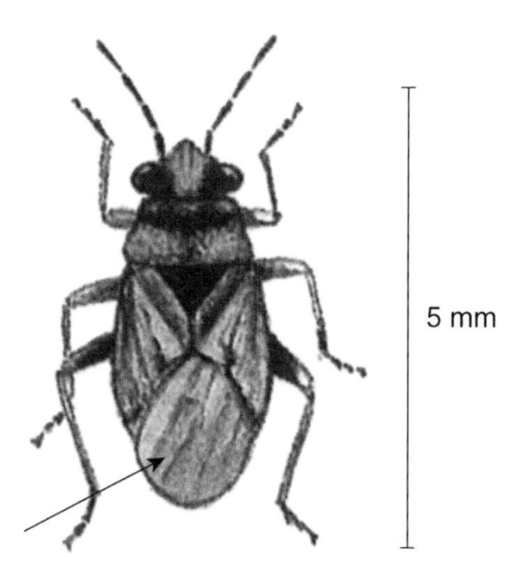

Fig. 7.17 Lygaeidae (from Chinery 1986).

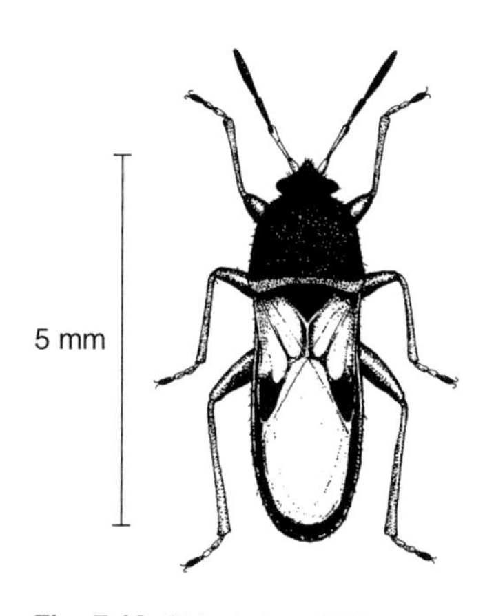

Fig. 7.18 Chinch bug (*Blissus leucopterus*) (from Hill 1987, with permission).

of such polymorphism is that wing muscles are expensive tissues to make, and therefore flightless adults can expend more energy on fecundity. Accordingly, as long as the plants are still providing high-quality food, it pays the species to stay in the environment and not expend energy on wing muscles. As the plants deteriorate, the species will switch to macropterous progeny, since being able to seek a new habitat has priority over maximum reproduction. Thus a second generation happens elsewhere on other cereals such as maize and sorghum. The adults from these sites overwinter outside the crops in hedges and grass tussocks.

Oxycarenus hyalinipennis (cotton seed bug) also attacks other Malvaceae such as okra, but is important on cotton in that it affects the quality of the cotton lint and the seeds. The lint becomes stained and the seeds go brown and shrink; per cent germination is seriously reduced. The eggs are laid singly and loosely between the seeds in the open boll, and the bugs are small with pointed heads and a red abdomen. The cotton seed bug is found throughout Africa and the near and middle East.

Superfamily Pentatomoidea (shield bugs)

Family Pentatomidae (stink bugs)

Stink bugs can exude a predator-repellent odour from ventral 'stink glands' on the sternum between the meso- and the metathoracic legs. The openings to these glands are clearly visible in ventral view and are a spot character for the Family. From the dorsal surface, the large triangular scutellum (shield) is obvious (Figs 7.19 and 7.20), covering about half the length of the abdomen and often with a membranous button-like expansion at the rear point of the triangle. Stink bugs tend to be large, robust and colourful

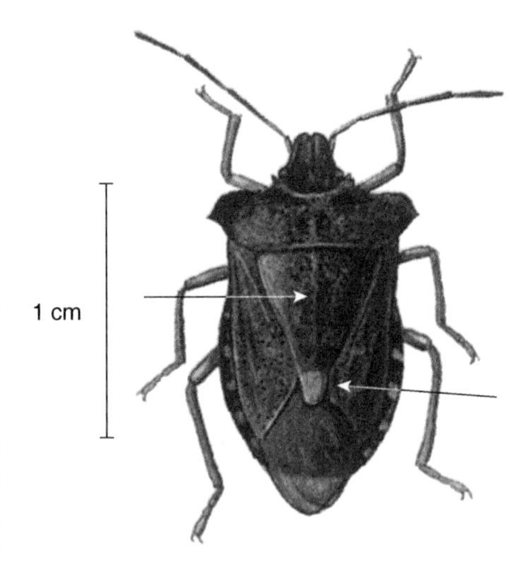

Fig. 7.19 Shield bug (Pentatomidae) (from Mandahl-Barth 1974, with permission).

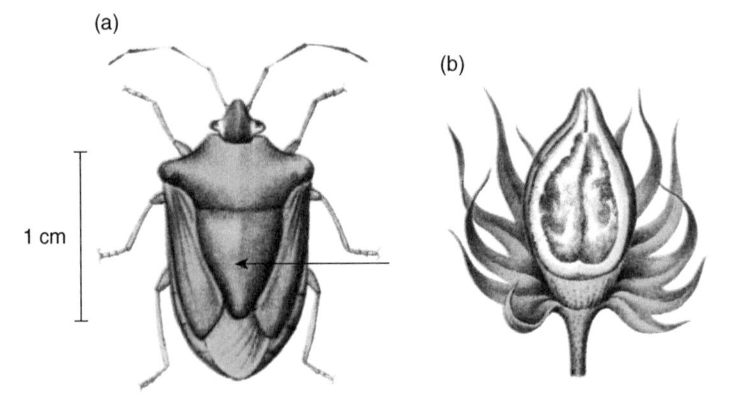

Fig. 7.20 Green vegetable bug (*Nezara viridula*): (a) adult and (b) section through punctured boll showing the stained lint (from Bayer 1968, with permission).

insects. They are frequently found on trees and shrubs, though there are also species attacking wheat.

One of the world's most important pest insects – perhaps even in the top 10 – is a pentatomid, the almost world-wide polyphagous green vegetable bug (*Nezara viridula*). The bug is quite large at 15–18 mm, and is an even green colour above and a lighter green ventrally (Fig. 7.20). Reddish-brown forms are also known. It has a huge host range, including many types of beans, tomatoes, alfalfa, cotton, potato and many cereals. It particularly sucks from fruiting structures, but does not breed on all the hosts that the adults attack. Thus they only visit cotton, yet transmit a *Nematospora* fungus which causes the bolls to rot internally, and only a few bugs can cause a large loss in yield. On other crops it is also the fruiting structures that are attacked, and local necrosis, spotting and deformations results. The eggs are white turning pink and are easily spotted by the rafts of 50–60 stuck to the underside of leaves. At first, the hatchlings stay aggregated without feeding, and only disperse to feed after they have moulted. *Nezara* can feed on parts of the plant other than the fruit and seeds if these are not available; however, when this happens development of the bugs slows or even ceases altogether. Development is slow at the best of times and takes 8 weeks at 26°C, and the adults hibernate at the end of three generations.

One of the main pests of coffee is the antestia bug (*Antestiopsis* spp.; Fig. 7.21). The flower buds may blacken and the young berries rot and fall, but the most important damage is not as visible and results from the introduction into the beans of a fungus, again *Nematospora*. Antestia bug is a problem with as few as two bugs per tree since the infected beans are harvested with the others, and only a few can taint the flavour of a whole stack of otherwise uninfected beans when fermentation takes place. The bug is quire colourful with brown, white and yellow, and it occurs throughout Africa. Keeping the canopy open by appropriate pruning reduces antestia bug populations, but unfortunately also reduces the numbers of its main natural enemy, a parasitoid wasp.

Also in Africa, *Calidea* spp. (e.g. *C. dregii* – which has very many wild and cultivated hosts) feeds on the developing seeds of cotton and can cause bolls to abort. Like *Dysdercus* (see Section 7.5.1.2, Family Pyrrhocoridae) the lint is stained by fungi transmitted by the bugs. The bug visits cotton, but does not breed there. The eggs are

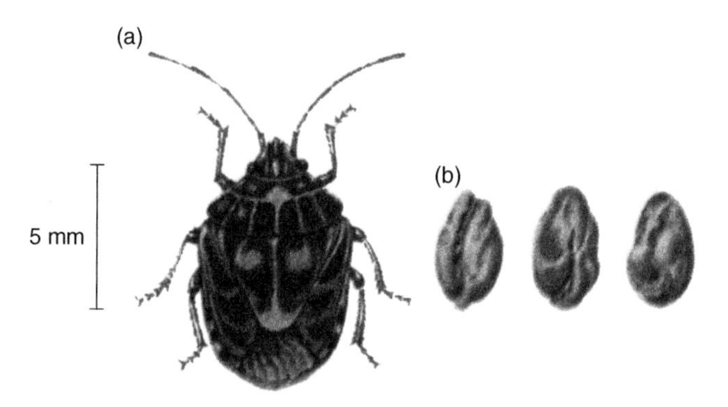

Fig. 7.21 Antestia bug of coffee: (a) adult and (b) shrivelled beans (modified from Bayer 1968, with permission).

laid in a closed spiral of about 40 around a stalk or leaf petiole on one of the many other plant hosts on which the bug does breed (including sorghum and sunflower) and both nymphs and adults (1–2 cm in length) are brightly coloured. The ventral surface is orange or red, and the upper blue or green with large dark spots and stripes. The bug can be so devastating in parts of East Africa that the growing of cotton has been abandoned in some places.

Species of *Rhynchocoris* are quite large at over 20 cm in length and have large spine-like sideways extensions of the pronotum at right angles. The pronotuim is green, and the rest of the dorsum is green or brown. These pentatomids can be quite serious pests of citrus in India, the Philippines and China, feeding on the young fruit and allowing the entry of bacteria and fungi, which then results in the fruit rotting.

Finally among the Pentatomidae, pests of rice worth mentioning are *Scotinophora* spp., the black paddy bugs of rice. All stages feed on the rice stem just above the water level, and inject a toxic saliva; the plants become stunted and the grains do not develop. Very young plants that are attacked may even die. *Scotinophora coarctata* is a stout insect about 8 mm long, brown with faint yellow spots on the thorax; it occurs mainly in India and south-east Asia. *Scotinophora lurida* (Fig. 7.22) is blacker and smaller; it occurs in Japan and China. The life cycles take about 1–1.5 months.

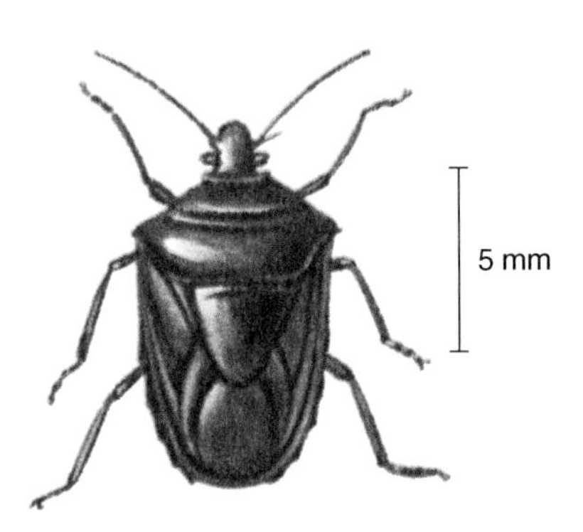

Fig. 7.22 Black paddy bug of rice (*Scotinophora lurida*) (from Bayer 1968, with permission).

Family Acanthosomatidae

These are very similar to the Pentatomidae, but with a long forwardly pointing spine in the mid-line on the ventral surface at the front of the abdomen. Like the pentatomids, they are phytophagous. The scent glands of this Family are on the abdomen.

Family Scutelleridae

These bugs have their shield (scutellum) extending the full length of the abdomen, with the wings pushed to the side and therefore the membrane portions do not overlap as they do in other Heteroptera. A notable pest species is *Eurygaster integriceps* (wheat shield bug or sunn pest) on wheat in Greece, Russia, Asia Minor and Pakistan. The bugs are about 12 mm long, and yellowish (Fig. 7.23). Most damage is done by the bugs sucking the grains. The adults overwinter on weeds and litter on hill-sides, often 10–20 km distant from the wheat crops; at these hillside sites, populations can be very dense with up to 60 bugs per m^2.

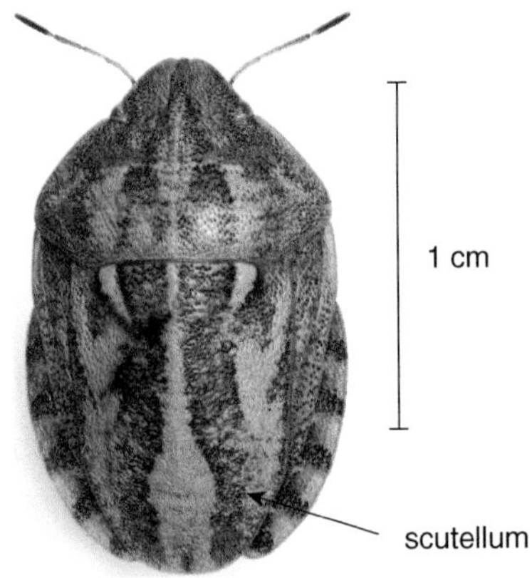

Fig. 7.23 Sunn pest (*Eurygaster integriceps*) (courtesy of V. Neimorovets).

Family Coreidae

These are about the same size as shield bugs, but are usually dark in colour with stout legs and often obvious thorn-like expansions on the ventral surface of the hind femora and at the sides of the pronotum. They include a number of pest species, including genera such as *Anoplocnemis* (Fig. 7.24) and *Acanthomia* (especially *Ac. horrida*), which shrivel and brown the pods of legumes in warmer climates. Much of the damage probably results from the wounding allowing the entry into the pod wall of the fungus *Nematospora*. Coreid bugs have the very low economic threshold of only two per plant reducing seed yield by 40–60% and seed quality by over 90%.

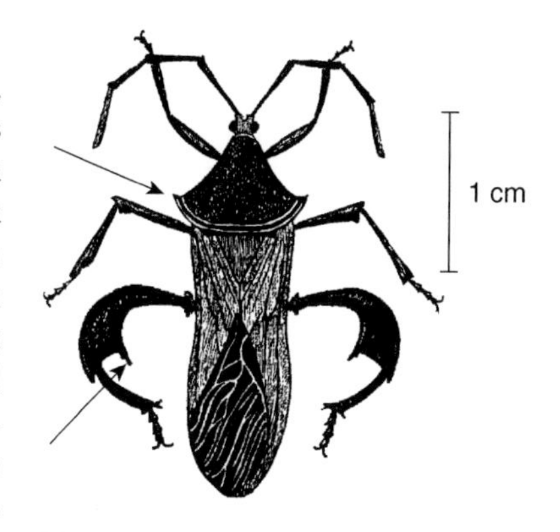

Fig. 7.24 Coreid bug (*Anoplocnemis curvipes*) (from www.tchad.ipm-info.org, with permission).

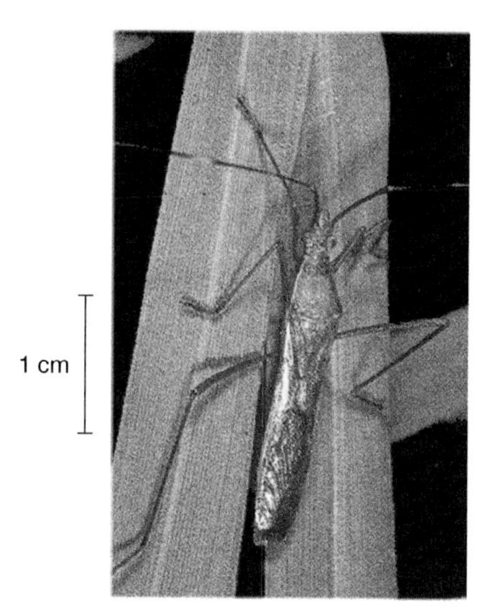

Fig. 7.25 Rice seed bug (*Leptocorisa* sp.) (from International Rice Research Institute).

On rice, *Leptocorisa* spp. (rice seed bug; Fig. 7.25) appear with the rains and suck and shrivel the rice grains. Before the grain is available the bugs feed from shoots and leaves. Yield losses are commonly 15–40%. The bugs are most common in areas with good rainfall or in irrigated crops. They can also survive on wild grasses in the absence of rice, and are found in India, south-east Asia, Indonesia and north Australia. The life cycle takes about a month, giving several generations per year.

The squash bugs (e.g. *Leptoglossus australis*) are pests of cucurbits with a similar distribution to *Leptocorisa* with the addition of central and southern Africa. They are also found on citrus, legumes and – in Malaysia – on oil palm. Fruit are spotted where the bugs have fed and may fall prematurely. Terminal shoots may wither and die. *Leptoglossus australis* has the alternative name of 'leaf-footed plant bug' from the obvious and diagnostic leaf-like flaps at the top of the hind tibiae.

Pseudotheraptus wayi (coconut bug) causes the 'early nutfall' and 'gummosis' symptoms in coconut in east Africa. The toxic saliva causes necrotic spots on the developing nuts and flowers. Coconut naturally sheds up to three-quarters of the young nuts, so the nut fall caused by the coreid may not actually cause a great final yield loss. The bug breeds continuously on coconut and may go through nine generations in a year.

Family Pyrrhocoridae

Pyrrhocorids are smaller than shield bugs and more the size of large mirids. They are conspicuously coloured, especially bright red, as warning colouration of distasteful-

Fig. 7.26 One of the species of cotton-stainer (*Dysdercus voelkeri*) (courtesy of Dr Dieter Mahsberg).

ness to potential predators. In this Family, the major pests are the cotton-stainers in the genus *Dysdercus* (Fig. 7.26), especially *D. fasciatus*. These pests are attracted into the cotton fields by the release from the plant of the volatile gossypol as the bolls open. Here they suck the bolls and transmit a fungus (*Nematospora*) which discolours the lint; hence the name 'cotton-stainer'. Small green bolls may die and go brown, but are not shed. Large bolls show no external signs of attack, but the inner boll wall has distorted growth and/or moist spots at points where the lint is stained yellow by the fungus. The bugs lay their eggs in batches of about 100 in moist litter or soil. The early instars suck fallen seeds, while the later ones seek the bolls. Cotton stainers of one species or another are found wherever cotton is grown.

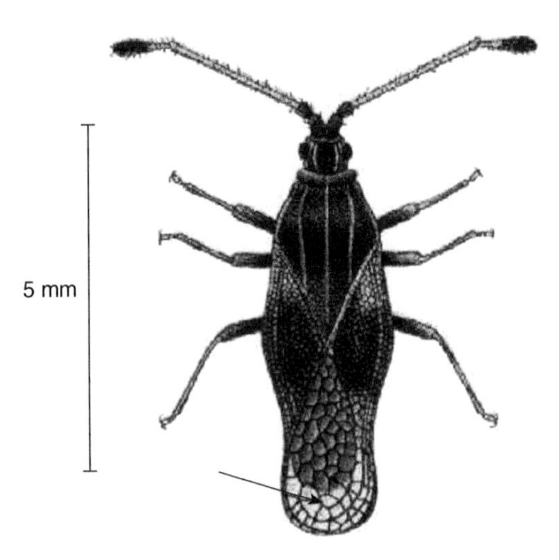

Fig. 7.27 Lace bug (Family Tingidae) (reproduced with the permission of CSIRO, Australia).

Family Tingidae (lace bugs)

The name 'lace bug' refers to a very diagnostic spot character for the Family: the body (often the thorax has small wing-like extensions) and forewings have a raised pattern of anastomosing ridges so that the gaps between them look like the holes in lace (Fig. 7.27). They are relatively small often light-coloured bugs about 5–8 mm in length.

They are common on thistles in the UK, but in India and parts of south-east Asia, the banana lace bug (*Stephanitis typica*) can be a serious problem, distorting the leaves of banana and causing yellow and necrotic spots. There may be many generations a year where it is hot.

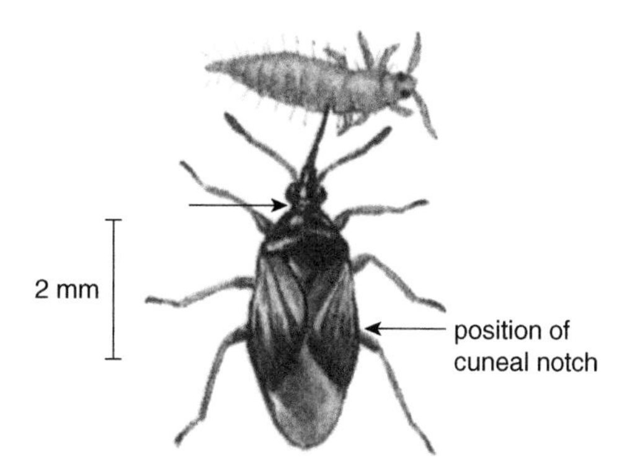

Fig. 7.28 *Anthocoris nemorum* (from Lyneborg 1968, with permission).

Family Anthocoridae (flower bugs)

This and the heteropteran Families that follow are carnivorous, sucking animal rather than plant tissues (Fig. 7.28). The name 'flower bug' is therefore rather misleading in terms of their food preferences, but immature females are frequently found sucking up the nectar at flowers to mature their eggs. Anthocorids are quite small bugs with a number of characteristic features. They have a constricted ring at the front of the pronotum where it abuts the back of the head. This ring is shaped rather like a vicar's 'dog-collar'. The dark and light areas on the hemielytra with the overlapping membranes give a checkerboard pattern, while the body is a characteristic shiny maroon colour. This helps with the recognition of nymphal anthocorids, but probably the most useful diagnostic feature of adults is the forward position of the cuneal notch, so that this indentation divides the leading edge of the horny part of the hemielytron into two subequal lengths. Members of the genus *Anthocoris*, particularly *A. nemorum* and *A. nemoralis* (which has been commercialised), are useful early predators of aphids and small caterpillars on fruit trees. They also fly to crops early enough to predate aphids, arriving there before later predators such as ladybirds and hover flies appear, which also often travel from arboreal habitats.

Anthocorids of the genus *Orius* are much smaller, but are voracious feeders on mites and are bred for sale and release as biological control agents for mite pests by commercial companies.

Family Cimicidae (bed bugs)

These are blood-sucking bugs of birds and mammals. They are brown, dorsoventrally flattened and have small reduced wings more like the wing pads of nymphal Heteroptera (Fig. 7.29). The bed bug attacking humans is *Cimex*

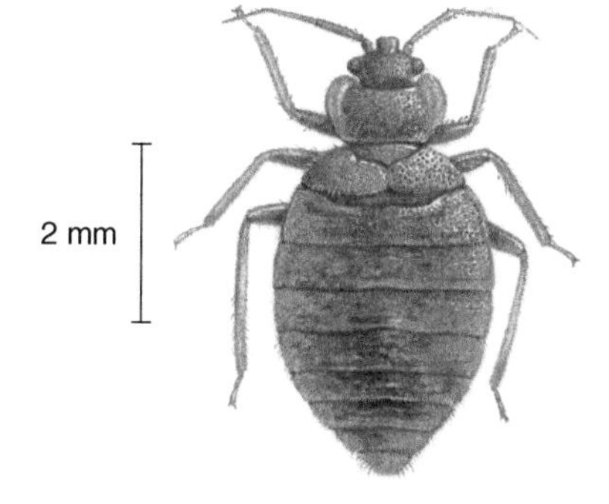

Fig. 7.29 Adult bed bug (*Cimex* sp.) (from Zanetti 1977).

lectularius. The bugs feed at night and hide in cracks and clothing by day. Their bite is irritating and painful, but bed bugs do not appear to transmit any diseases. A strange feature is the aggressive mating of the male, who thrusts his hardened and pointed penis anywhere into the female's body and injects sperm into the body fluids. Because bed bugs thrive in unsanitary conditions, the sperm is accompanied by microorganisms including toxic bacteria, and many females die as a result in spite of having a gland producing an antibiotic specifically to counter such infections.

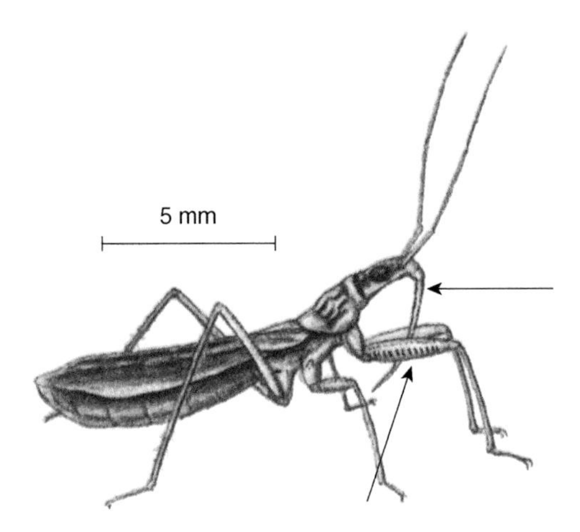

Fig. 7.30 Family Nabidae (*Nabis ferus*) (from Lyneborg 1968, with permission).

Family Nabidae (damsel bugs)

Damsel bugs seems a strangely decorous name for these rather uniformly brown predators of aphids and caterpillars. They have a vicious hardened rostrum curved like a billhook and which cannot be carried flat under the body (Fig. 7.30). The front legs are clearly raptorial (i.e. for capturing and holding prey) with enlarged femora equipped with 'teeth' on the ventral edge. Nabids can commonly be caught in large numbers by sweeping ungrazed/uncut grassland by day, but they are predominantly night active and numbers will be badly underestimated.

Important species are *Nabis ferus* and *N. flavomarginatus*. Like anthocorids, nabids have the utility as biological control agents of arriving early in crops. Alary polymorphism, as for example in *N. flaveomarginatus*, results in brachypterous adults of some species being the phenotype most frequently encountered for much of the early summer.

Family Reduviidae (assassin bugs)

The reduviids are the predators equivalent to the nabids in the tropics. They tend to be larger and more colourful. *Phonoctonus fasciatus* is an interesting species in that it preys on cotton-stainer bugs and closely mimics their red and white pattern of warning colouration.

The Family also includes species that are parasitic on mammals, including piercing the skin and sucking the blood of man. In doing this, they may transmit diseases of humans such as Chagas' disease (a sleeping sickness caused by *Trypanosoma cruzi*) transmitted by bugs of the genus *Triatoma* (Fig. 7.31). It is thought that Charles Darwin may have suffered from Chagas' disease, having become infected during his voyage on the Beagle.

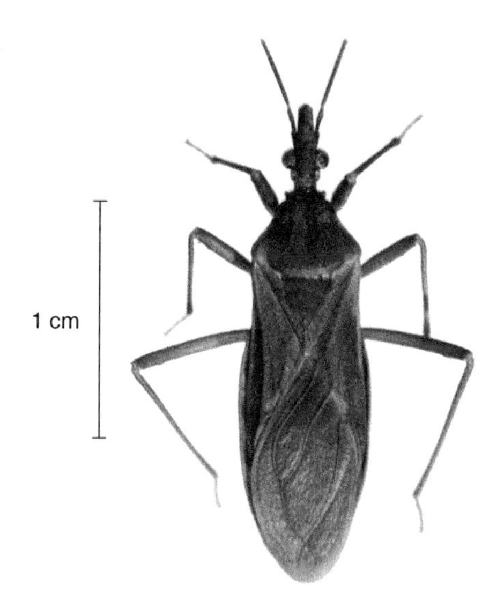

Fig. 7.31 *Triatoma rubida* (University of Arizona, with permission).

7.5.2 Suborder Auchenorryncha (hoppers)

The Auchenorryncha (formerly grouped with the Sternorryncha as the Suborder Homoptera) do not have the hemielytron of the Heteroptera, and the front wing can either be entirely horny (e.g. Fig. 7.34) and often brightly coloured or entirely membranous (e.g. Fig. 7.39a). Also, although nearly all the Heteroptera fold their forewings flat over the abdomen with the distal areas (their membranes) lying over each other, the Auchenorryncha hold their forewings at an angle along the sides of the abdomen to form a pitched roof.

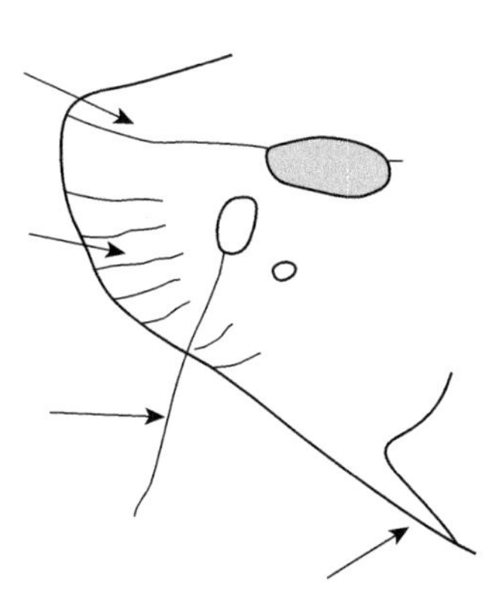

Fig. 7.32 Cartoon of the head of a hopper (Auchenorryncha).

The head of a hopper also has a characteristic appearance (Fig. 7.32). The overall shape is a bit like the prow of a boat, the eyes sunk into the smooth outline of the head and the front sloping somewhat backwards from dorsal to ventral with, in side view, the short rostrum in line with this angle. Viewed face on, the front of the head bulges a little forwards with a pattern of slightly oblique stripes to accommodate the muscles working the ***cibarial pump*** above the origin of the mouthparts. That the rostrum continues in a straight line from the front of the head is the characteristic giving rise to the designation Auchenorryncha (straight mouthparts). In stark contrast to the Heteroptera, the antennae are rather insignificant and consist of a short bristle arising from the tip of a short articulated basal segment.

Hoppers are all herbivorous, and have much more specific plant associations than the Heteroptera. The eggs are laid in batches, and overwintering is on a host plant as egg or adult. Often, specific wild host plants are important reservoirs of the pest over winter. The Suborder, again unlike the Heteroptera, vectors some important plant virus diseases, which may overwinter in the vector rather than in the plant (if an annual).

7.5.2.1 Superfamily Fulgoroidea

Family Fulgoridae (lantern flies)

This Family of hoppers includes the remarkable tropical lantern flies, which have grotesquely enlarged heads with patches that glow in the dark.

Family Delphacidae (planthoppers)

This Family of phloem feeders is easily recognised by the thick articulated spur at the end of the hind tibia (Fig. 7.33). In some species this may be almost as long as the tarsus, giving the impression of a forked structure at the end of the tibia. Another delphacid feature is the relatively long (compared with other Auchenorrhncha) basal segment of the antenna carrying the short bristle. There is much alary polymorphism with macropterous and brachypterous phenotypes in this Family.

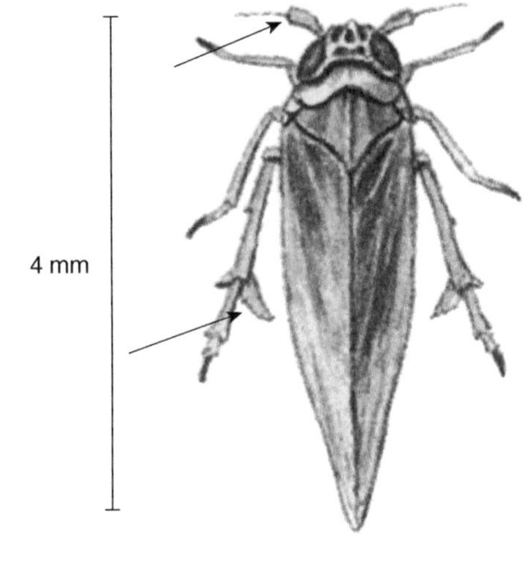

Fig. 7.33 A planthopper (Delphacidae) (from Lyneborg 1968, with permission).

There are some really important pests in this Family. Pre-eminent, and one of the world's top ten pests, is the brown planthopper of rice, *Nilaparvata lugens* (Fig. 7.34). This insect is referred to so often that its common name is usually replaced by the acronym BPH. As a major pest of rice, BPH is a threat to the subsistence cereal of a large proportion of the world population. Small white eggs are laid in groups inside the leaf sheath and on the midrib. Each female can lay 200 eggs, and the generation time can be just over 2 weeks. Adults and nymphs feed at the base of the rice stems in dense colonies, just above the water level in the rice paddies. Their feeding weakens the plants, but the pest also transmits the debilitating *Tungro virus*. This initially shows as yellow patches of plants where the initial fully winged colonising hoppers have multiplied and spread. Before the 1960s, BPH was not a serious hopper problem compared with the green rice leafhopper (see Section 7.5.2.2, Family Cicadellidae). However, the new high-yielding rice varieties of the 'green revolution' proved highly susceptible to BPH, and the species soon became the predominant rice hopper problem. As the common name suggests, the insects are brown; there are many brachypterous adults and males at 2.5 mm long are smaller than females (3 mm). Japan is re-invaded each year from China, for the winter in Japan is too cold for the pest to perennate. BPH-resistant rice varieties have played a major part in controlling the pest but, like insecticides, they may fail due to their selection for tolerant strains of the pest.

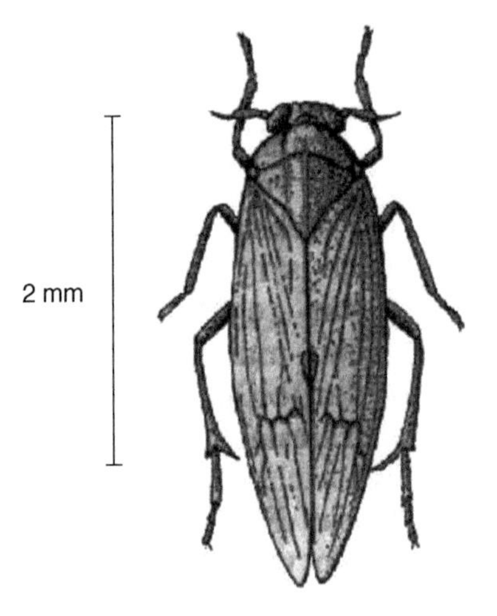

Fig. 7.34 Brown planthopper (*Nilaparvata lugens*) (reproduced with the permission of CSIRO, Australia).

Sogatella furcifera also attacks rice, mainly in the Far East. The insect appears early in the life of the crop and can quickly reach high populations. Damage is usually direct, and 'hopper-burn' (reddening and desiccation of the leaf) is a common symptom. Surviving plants are stunted and show delayed tillering. However, the importance of the pest rises where it transmits two rice virus diseases (*Rice yellows virus* and *Rice stunt virus*). The forewings are clear with dark veins, with a dark spot on the trailing edge. The body is black, but the pronotum is pale yellow, and adults may again be macropterous or brachypterous. Generation time is 2–3 weeks. This species also attacks and transmits virus to maize, oats and wheat in Europe and northern Asia.

Another important delphacid pest is the sugar cane planthopper, *Perkinsiella saccharicida*, in parts of South America and South Africa, tropical parts of Asia and in Australia and Hawaii. Damage is to the leaves from a combination of sap sucking and wounding with the ovipositor. The pest also transmits Fiji disease of sugar cane. The eggs are laid at night in the leaf midrib (up to a dozen eggs per incision); each female can lay 200 eggs during its 8 weeks of adult life. Excretion of honeydew can result in sooty moulds developing. The eggs may not hatch for 2–6 weeks, and the nymphal stage then lasts about 20–40 days, and there are five to six generations per year. Both sexes can be fully winged or brachypterous as adults. The pest is one of the success stories of biological control, with successful introduction into Hawaii in 1923 of the egg predator *Tytthus mundulus* (Miridae).

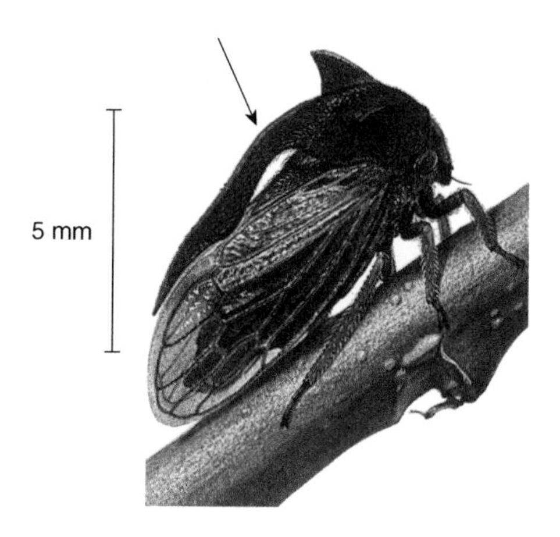

Fig. 7.35 Treehopper (Membracidae) (from Zanetti 1977).

7.5.2.2 Superfamily Cicadelloidea

Family Membracidae (treehoppers)

These attractive insects have the dorsal surface of the pronotum pulled out into a backwardly directed ridge ending in a point. Sometimes this character is so large that it dwarfs the rest of the insect (Fig. 7.35).

Family Cicadellidae (leafhoppers)

These are the most commonly encountered Auchenorryncha, and – apart from aphids – are probably the most abundant Hemiptera. Leafhoppers feed on the phloem and mesophyll and many viruses diseases of plants are transmitted by them. They can be distinguished from the Delphacidae and Cercopidae by the parallel rows of long spines on both sides of the rear tibia. It looks rather like fish bones projecting from the fish vertebral column (Fig. 7.36). The forewings are not infrequently brightly and even multicoloured.

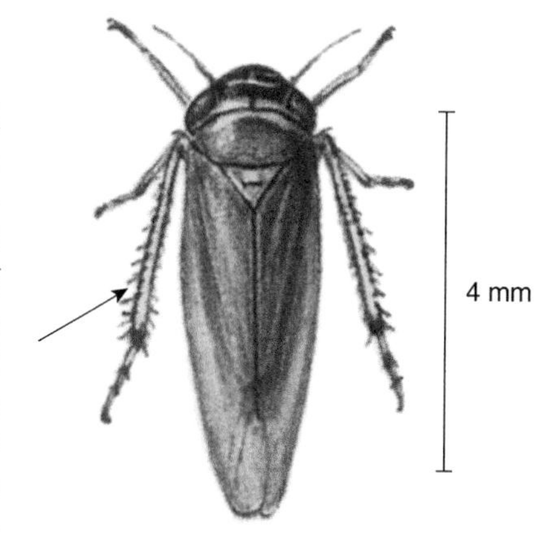

Fig. 7.36 Planthopper (Cicadellidae) (from Lyneborg 1968, with permission).

The genus *Empoasca* is particularly widespread and contains pests of legumes (e.g. *E. kraemeri* and *E. dolichi*), cotton (*E. facialis* and *E. libyca*), potato (*E. fabae*) and many other crops (e.g. *E. vitis* on grape vines). Pest species of *Empoasca* occur across sub-Saharan Africa, but also around the Mediterranean (e.g. Morocco, Tunisia and Spain). Hairy cottons are resistant, but unfortunately extrasusceptible to whitefly. Attacks often result in hopper-burn as a result of the toxic saliva (see Section 7.5.2.1, Family Delphacidae), beginning at the leaf edges which curl under. The eggs are green and curved like a banana, and they are laid into plant tissues, especially thicker ones such as petioles and main leaf veins. The adults are an even pale green, and the insects are 2–3 mm long and narrow, especially to the rear, so that they look from above like a narrow blunt triangle. Generation time is 3–4 weeks, and 10 generations can occur in warm regions where breeding is continuous.

On fruit trees and bushes, *Typhlocyba* spp. are the widely distributed polyphagous leafhoppers. They are pests of apple, plum and nearly all soft fruit (especially strawberries). They are also pests of potato and many ornamentals. *Typhlocyba rosae* damages ornamental roses. Feeding damage does not involve hopper-burn, and begins as silvery flecks on the leaves. Widespread mottling and curling of the leaves then follows, with stunting of fruit. There are two generations during the season, and these leafhoppers overwinter in the egg stage.

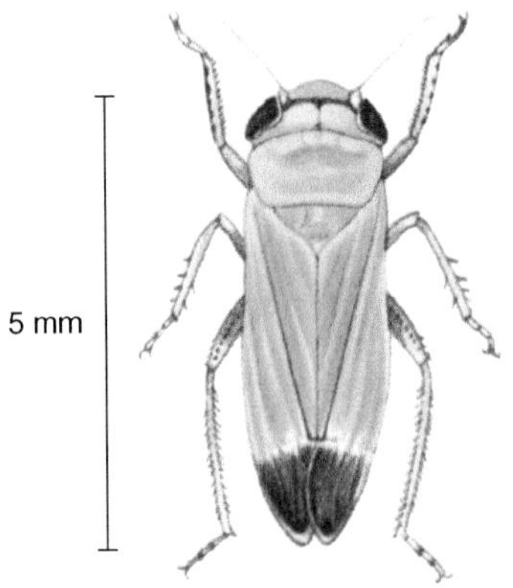

5 mm

Fig. 7.37 Green rice leafhopper (*Nephotettix cincticeps*) (from Bayer 1968, with permission).

As mentioned above, the green rice leafhopper *Nephotettix nigropictus* and the related *N. cincticeps* (Fig. 7.37) can be a problem at high numbers, though taking second place on rice to the brown planthopper (BPH). Its sap sucking causes direct damage, and the insects also transmit several plant viruses, especially *Yellow dwarf virus*. Eggs are laid in the leaf sheaths, and generations are as short as 22 days. Adult *N. nigropictus* are 3–5 mm long, and a bright green with black spots on the wing and at the wing tip. The green rice leafhopper is found in all the tropical and subtropical parts of Asia.

Important primarily for its transmission of *Curly top virus* of beet and tomatoes (especially in the USA) is the beet leafhopper, *Circulifer tenellus*. The virus causes leaves to turn yellow, curl and often develop a purple tinge. Plants remain stunted and fruit is deformed and ripens early. Direct feeding of large numbers leads additionally to hopper-burn. The hopper has a very wide host range and indeed not only overwinters among weeds, but feeds on weeds for its first generation. The adults then move to other weeds and also crops. Eggs are laid into stems and leaves; each female may lay up to 400 eggs. A generation takes 2–3 months according to temperature so that, in the USA, there are only three generations in the north, but five or more overlapping generations in California. The adult hopper is about 3 mm long; it is pale green with some darker blotches. The importance of *Curly top virus* in the USA tends to make one forget that the insect is very widely distributed elsewhere, from Spain through Africa, the Caribbean and Hawaii.

However, if we are talking pests, then here we have another in the world top 10. This is *Cicadulina mbila* (Fig. 7.38), the maize leafhopper in Africa. A pest of another important world staple, this insect overwinters on wild grasses and flies huge distances in the spring to re-infect the new crop over vast areas, not only with itself but also with the crippling *Maize streak virus*. This can be spread by remarkably few hoppers; one per 20 maize plants is considered a high infestation! Feeding by the tiny (2–3 mm) insects therefore does little direct damage, and moreover not all races transmit the disease. The adult has a diagnostic brown stripe along the length of the wing, while the whole body is yellow with some dorsal brown markings.

Fig. 7.38 Maize leafhopper (*Cicadulina mbila*) (courtesy of M.F. Claridge).

In 1994, there was consternation in the Californian wine industry when the glassy winged sharpshooter, *Homalodisca vitripennis*, was first reported there. It was already native in the south-eastern USA, and probably had already reached California in the 1960s. This is a leafhopper that atypically feeds not from the phloem, but from the xylem. It transmits the serious bacterial Pierce's disease to vines in California and unfortunately control is made difficult by the many wild and garden alternative host plants of both the leafhopper and the disease.

7.5.2.3 Superfamily Cicadoidea (Family Cicadidae – cicadas)

Cicadas are large bugs (Fig. 7.39a) with the males (only) possessing amazing sound-producing ability. Indeed, in the tropics their monotonous 'singing', which has been likened to knives being sharpened on a grinding wheel, can be close on deafening in the evenings. The sound is produced by the vibration of shell-like 'drums' located at the end of the abdomen.

The nymphs are subterranean, with enlarged front femora for digging (Fig. 7.39b). They suck the juices from the smaller plant roots, and both adults and nymphs (like froghoppers, see Section 7.5.2.4, Cercopidae), feed from the xylem and not, as do most Auchenorryncha, the phloem. The Family includes the bizarre 'periodical cicadas' (*Magicicada* spp.) of the USA, which appear in large numbers at the end of extremely long life cycles of 13 (in the south) and 17 (in the north) years.

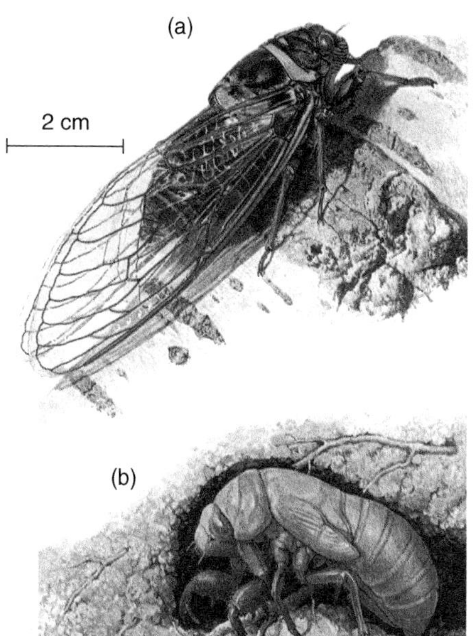

Fig. 7.39 Cicada: (a) adult and (b) nymph (from Zanetti 1977).

7.5.2.4 Superfamily Cercopoidea (Family Cercopidae – froghoppers)

The froghoppers have horny forewings and can again be distinguished by characteristics of the hind tibia (Fig. 7.40). This has just a very few, usually black-tipped spurs like thorns, and similar spurs are found in a ring around the bell-bottomed tip of the tibia. To a lesser extent, the tips of the tarsal segments are miniature replicas of the ring of spurs at the tip of the tibia.

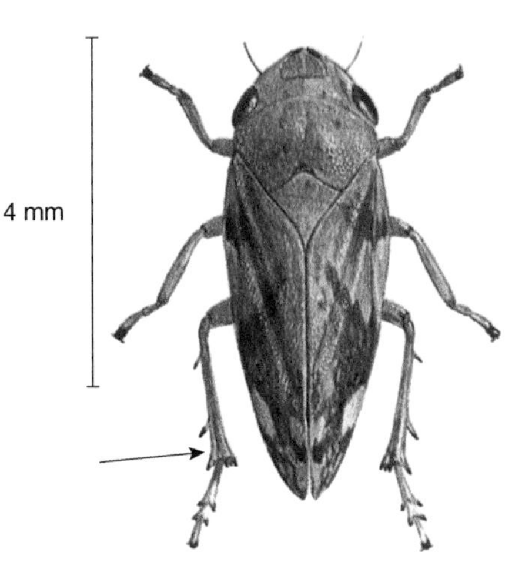

Fig. 7.40 Froghopper (Cercopidae) (from Zanetti 1977).

The nymphs of froghoppers develop singly under the protection from desiccation of a mass of viscous bubbles produced by the mother blowing air from a special chamber with spiracles under the anus into excreted xylem when one egg at a time is laid. This frothy mass looks like spit – hence the name 'cuckoo spit' for these phenomena. Indeed, froghoppers have the alternative name of 'cuckoo-spit insects'. In the UK, the genus *Philaenus* has several common species and is both common and widely distributed in grassland. Some cercopids live underground during the nymphal stage, yet even so are enveloped in 'cuckoo-spit'.

By and large, froghoppers are not pests of economically important plants. However, in South America and the West Indies, the sugar cane froghoppers *Tomaspis saccharina* (a striking red and black insect) and *Aeneolamia* spp. can be serious problems, and *Tomaspis* has been a target for biological control. Eggs are laid into plant tissues or in the soil (the latter especially by *Aeneolamia*). The saliva, injected by the adults and nymphs feeding concealed in spittle, results first in necrotic feeding spots from which streaks spread across the leaf, which eventually goes brown, wilts and dies. Some nymphs are found feeding on the roots, but little damage results. With a life cycle lasting about 2 months, there are four to five generations a year.

More interesting for entomologists than pest status is that many froghoppers have evolved to feed solely on the watery contents of the xylem vessels. This has been called 'the art of the impossible', since the need to suck up large quantities of the fluid to compensate for its low concentration of nutrients is exacerbated by the fact that the evapotranspiration of the leaf canopy maintains a strong suck in the xylem vessels in exactly the opposite direction! The hoppers might succeed if they only fed at night when evapotranspiration is lower; they may well do so, but they also feed during the day, apparently with little difficulty. This is illustrated by the tree species, particularly in Africa, called 'rain trees' because the throughput of watery xylem by froghoppers is so vast that it appears to be raining under the tree. The output of 20 individuals in 1 hour has been measured as over 18 litres.

If the froghopper indeed has to 'out-pull' the xylem vessel, then it would help to be able to evaporate water from the body. Yet this is prevented in nymphs by the surrounding liquid cuckoo-spit.

7.5.3 Suborder Sternorryncha (plant lice)

'Sternorryncha' means 'belly mouthparts' and refers to the fact that the mouthparts appear to arise clearly behind the front of the head and often appear to leave the head close to the base of the front legs (Fig. 7.41). This is usually pretty obvious when the insect is viewed from the side. Wings are membranous and usually held along the sides of the abdomen at rest. Antennae are variable; they are usually of obvious length and thread like, but the segments are not roughly equal in length as in the Heteroptera but get shorter towards the tip.

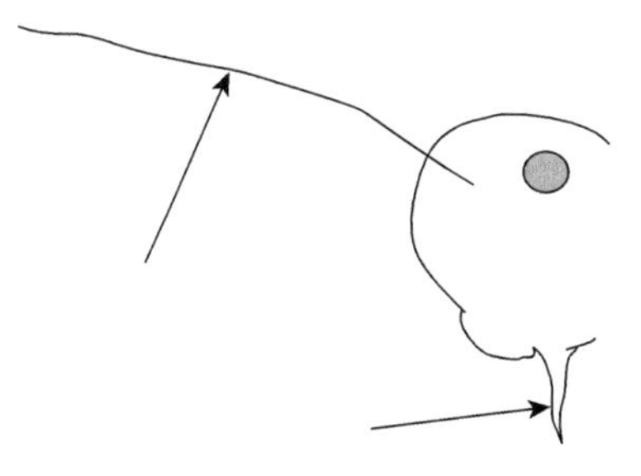

Fig. 7.41 Cartoon of the head of Sternorryncha.

With this Suborder, we see a continuing progression from the Auchenorryncha to closer adaptation to the plant from the relatively solitary active psyllids to the sessile scale insects, which become plumbed into the plant. There is also increasing dependence on phloem feeding. This has a higher nutrient concentration than mesophyll, but is forced into the insects under pressure. This makes it especially necessary for Sternorryncha to rid themselves of copious sugary excess phloem by excretion from the anus. This sticky 'honeydew' is itself a cause of damage to plants. Sternorrhyncha mainly feed on the underside of the leaves, so that the honeydew falls on the leaf below. Here it proves an ideal food source for sooty moulds, which are black and obscure light from the photosynthetic tissue. Moreover, as the honeydew becomes an increasingly concentrated deposit of sugar through evaporation, osmosis pulls water out of the leaf and the cells there eventually plasmolyse.

7.5.3.1 Superfamily Psylloidea (Family Psyllidae – suckers or jumping plant lice)

These live singly or in small groups on the underside of leaves, which are often at least partly rolled as a result of their feeding. The shape of the head is not dissimilar from that of hoppers (Fig. 7.43a), in fact they look rather like miniature cicadas, with very obvious wing pads (Fig. 7.43b). Adults can be recognised by a distinctive wing venation (Fig. 7.42) where a single thick 'vein', made up of the radial, medial and cubital sectors, crosses the wing membrane and does not fork for some considerable distance.

In temperate countries, they are only known as pests where top fruit is concerned. The apple sucker (*Psylla mali*) (Fig. 7.43) is an occasional pest of apples in Europe. Overwintering eggs are laid on the twigs, and the nymphs suck sap from the flower trusses and leaf buds from early spring; infested flowers go brown and leaves curl.

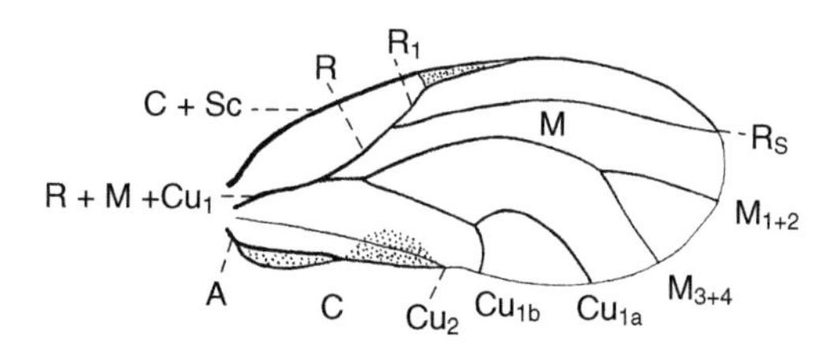

Fig. 7.42 Forewing venation of a psyllid. C, costa; Sc, subcosta; R, radial sector; M, medial sector; Cu, cubital sector; A, anal vein (from Richards and Davies 1977).

Fig. 7.43 Apple sucker (*Psylla mali*): (a) adult (from Bayer 1968, with permission); (b) nymph (from Zahradník and Chvála 1989); (c) leaf curling damage (from Bayer 1968, with permission).

There is one generation a year. The pear sucker (*P. pyricola*) overwinters as adults and has three generations. The first generation damages pear flowers in the same way as *P. mali* on apples. The second generation feeds on the leaves, with the third generation in the autumn feeding on the fruit buds; this can cause loss of crop the following year.

In the subtropics and tropics, *Trioza erytreae* (African citrus psyllid) can be a problem on citrus, forming pimple-like galls on the young leaves. The galls are open on the lower leaf surface. The psyllids are especially serious on young plants in the nursery, and it has been suggested that they also vector *Citrus tristeza virus*, but the evidence for this is far from conclusive. The eggs are elongate and pear shaped, and are anchored to the leaf surface. The tiny yellow nymph with two obvious red eyes walks for a while and then starts to feed on the leaf underside; it will not move again unless prodded. As the leaf grows, the characteristic gall begins to form as an inverted pit at the feeding site, but the nymph never moves into the gall. Generation time is 3–4 weeks.

A recent addition to the psyllid rogues' gallery is the Asian citrus psyllid, *Diaphorina citri*, which has become a threat to the citrus industry in Florida as a vector of citrus greening disease, caused by a phloem-limited bacterium. In attempts to halt its spread, infected trees are felled and burnt on the spot, but unfortunately the bacterium has other hosts.

Any herbivorous insect is a candidate biological control agent if its host plant is a noxious weed and especially if the insect is highly specific to that one plant. Japanese knotweed (*Fallopia japonica* – Family Polygonaceae) was purposely introduced by the Victorians as a garden ornamental, but has become a major weed problem in the UK since it is very hard to kill with herbicides, which have to be injected into individual cut stems of plants. There is currently hope that the psyllid *Aphalara itadori* from Japan may prove an effective biological control.

7.5.3.2 Superfamily Aphidoidea (aphids)

Taking the world crops as a whole, aphids are probably the most serious group of pests in the Class Insecta, especially if one includes the yield loss caused by plant viruses that they vector.

The diagnostic feature of aphids is found only in the adult (Figs 7.44 and 7.50); it is the possession of a lobe (the ***cauda***) projecting rearwards at the dorsal edge of the anus. Most species in the Family Aphididae also have a pair of barrel-shaped organs (the ***cornicles*** = ***siphunculi***) projecting from the dorsal surface towards the rear of the abdomen. These, unlike the cauda, are also present on nymphs. However, not all species of aphids have them; they may be represented by pores at the surface of the cuticle or be completely absent.

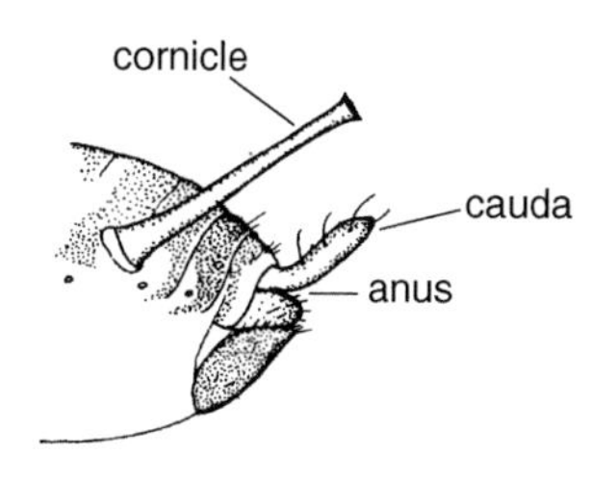

Fig. 7.44 End of the abdomen of an aphid (from Blackman and Eastop 2000, with permission).

Family Lachnidae

This primitive group of aphids is restricted to coniferous trees, and large populations can cause damage. The lachnids are parenchyma and not phloem feeders, so honeydew is not an issue. Many lachnids secrete a white wax from glands in the cuticle, and this is often more obvious among the needles than the aphids themselves. *Adelges* and *Greenidea* are important genera. Lachnids are not virus vectors.

Family Aphididae (aphids, greenfly, blackfly)

The Aphididae are phloem feeders and cause damage to plants in a variety of ways. They deplete the plant's food reserves by imbibing large quantities of plant sap, but perhaps even more damaging is the injection of saliva. Plants react to the foreign proteins in the saliva by a long-term increase in respiration rate; this burns up carbohydrate in addition to that removed by the feeding of the aphids. The excretion by aphids of excess sap ('honeydew' – see Section 7.5.3) causes plasmolysis and obstructs photosynthesis, especially as the honeydew gets colonised by sooty moulds (usually a species of *Cladosporium*). All these demands on the photosynthates of the plant result in the first symptoms (although of course rarely observed) of aphid attack being a dramatic reduction in the amount of root. This, rather than removal of sap, is probably the main reason why heavily infested plants wilt.

The toxic saliva of aphids causes reddish colours to appear in plant tissue, leaves to curl by the removal of plant growth substances from just the lower leaf surface where the aphids feed, and even the proliferation of plant tissue to form galls. Though such symptoms are usually a problem, there is at least one example where they fuel an economy. In some eastern parts of China, a main source of income is the extraction and sale of pigments from galls on the tree *Ailanthus*.

However, many species of aphids are pests more because of the virus diseases they transmit than because of the depletion of photosynthates. Thus even quite low populations of some aphid species can present a problem that needs controlling. Piercing and sucking mouthparts are ideal 'hypodermic needles' for injecting the plant with disease, to the extent that many viruses can equally be transmitted from an infected to a healthy plant with a pin. Aphid transmission of such viruses involves rather more biology than simple contamination of the stylets with virus particles, but they are classified as 'stylet-borne' in contrast to the 'circulative' viruses. The latter have to undergo a cycle in the insect before a healthy plant can be infected by virus particles that have passed through the gut wall into the haemolymph and then accumulated in the salivary glands. This difference between the fate of the virus in the aphid lies behind many of

Criterion	Stylet-borne	Circulative
Is transmission reproducible with a pin?	Usually	No
Plant tissue where virus is acquired	Epidermis	Mesophyll/phloem
If vector is starved, will acquisition of virus be quicker?	Yes (immediate probing)	No
Feeding time needed for vector to acquire virus	Seconds	Hours/days
Interval between acquisition and ability to transmit	None	There is a 'latent period' of days/weeks
Duration over which **feeding** vector remains infective	Minutes/hours	Days/life/future generation(s)
Duration over which **fasting** vector remains infective	Hours	As above
Number of plants that can be infected from one virus acquisition	A few (4?)	Many
Is virus retained when vector moults?	No	Yes
Feeding time needed for vector to transmit virus	Seconds	Longer
Specificity of virus to a particular species of vector	Low	High
Control of virus by insecticide against vector	Difficult (because of time taken to die)	Relatively easy (because of latent period)

Table 7.1 Criteria for distinguishing between insect-vectored stylet-borne and circulative viruses.

the criteria that distinguish the two groups (Table 7.1). 'Stylet borne' and 'circulative' fairly closely equate, respectively, to the older terms 'non-persistent' and 'persistent', though there is an additional third intermediate category of 'semipersistent'. This older classification is still the more widely used.

Note that the terms in both classifications are properties of the virus, not the vector. Any one virus will always be transmitted in the same manner.

One thing that clearly happens in at least many circulative viruses is that the virus multiplies in the vector. This has been shown in tissue cultures from vectors, but was also demonstrated more than 50 years ago by some elegant experiments with leafhoppers. Vectors were allowed to transmit virus, but the plants were continually changed for virus-uninfected plants so that no new virus could be acquired. Transmission continued for generation after generation – after 21 generations the volume of virus the first generation would have had to acquire (had there been no virus multiplication since) would have far exceeded the volume of the insect!

An important feature of aphids is their impressive powers of reproduction. Most of the time, all the individuals in a population are female and reproduce asexually (by a process known as parthenogenesis) and with live births (viviparity). This enables an amazing telescoping of generations whereby individuals begin their development within their mothers and before these are even born. In fact, an adult aphid already contains the growing embryos of her great-grandchildren. It is not unlike a 'Russian doll'. Generations can be less than a week. With fecundities of 80 per mother in some species, populations can double every few days and generations soon completely overlap. I do not vouch for the truth of the space-fiction statistics that others have calculated, but one calculation is that the cabbage aphids arising from just one mother at the end of 1 year would weigh 250 million tons and would, sitting nose to tail, encircle the earth at the equator a million times! Aphids are soft bodied and have relatively little defence against predators, of which there are many. Breeding faster than they can be eaten usually ensures survival of an aphid population. Quite a number of species in two plant gall-forming Subfamilies (the Hormaphidinae and the Pemphiginae) actually have first instar individuals, which are sterile and never progress to later instars, but which have sclerotised front legs and act as 'soldiers'. They are often prolific, and are highly effective at protecting the entrance to the gall against a range of natural enemies; they even repair galls that have been damaged by chewing insect herbivores such as caterpillars.

Honeydew has already been mentioned several times in connection with the pest status of phloem-feeding Hemiptera. Its accumulation can be a problem for sessile insects. Many aphids merely kick honeydew droplets that form at the anus away from themselves to land elsewhere on the plant or fall to the ground. I recall a litigation where I was called as an expert witness to confirm that the far too frequent and very expensive re-thatching of a cottage was caused by the moulds growing on honeydew produced from aphids on a neighbour's overhanging lime tree. Quite apart from the mould attacking the thatch, the honeydew itself caused the most unsightly and destructive damage to the paintwork of a car normally parked outside the cottage.

Many aphid species have formed an association with ants, and are said to be 'ant-attended'. The ants remove the honeydew for use in their nests as food and as substrates for 'fungus gardens' – like termites (see Section 6.11) they utilise the ability of fungi to digest cellulose. Ants will approach aphids from the rear and with a front leg pull on a cornicle, rather like a barman pumping a pint of beer. This movement of the cornicle stimulates the aphid to release the honeydew it has been accumulating in its rectum. The ants also protect the aphid colony by killing or driving off insect predators and parasitoids. Colonies of ant-attended aphid species thrive compared with ones to which access to ants is prevented; ant-attendance is really 'farming'.

Some non-ant-attended species have problems ridding themselves of their honeydew. One such is the cabbage aphid (*Brevicoryne brassicae*) whose honeydew simply rolls down the shiny leaves and accumulates in the heart of the plant and in the leaf axils. Other aphids (e.g. members of the genus *Pemphigus*) live in galls. Such aphids, which cannot escape their own honeydew, escape the danger of drowning in it by covering the honeydew droplets in wax. These wax-coated drops do not coalesce and the aphids can even move among them and push them aside as if they were footballs.

Phloem is high in carbohydrates and low in nitrogen; hence the need of aphids to excrete surplus carbohydrate with maximum assimilation of soluble nitrogen. Aphids are unusual in that much important nitrogen metabolism and conversion to essential amino acids is accomplished, not by the aphid itself, but by obligate symbionts (especially the bacterium *Buchnera aphidicola*) passed down through the generations and housed in special cells (***mycetocytes***) in special tissues (***mycetomes***) on the outside of the gut wall. The aphids cannot survive without these symbionts (e.g. if fed antibiotics in a sterile artificial diet). In the 1970s, it was reported that the fungicide benomyl was also an effective aphicide. It turned out later that benomyl disrupted the matrix material of the mycetome, which killed the symbionts before the then inevitable death of the aphids.

The cornicles have been mentioned above in relation to ant-attendance. However, they also have a defensive function, and can be used to blind predators like ladybird larvae with a quick-setting opaque wax as they approach from the rear. Moreover, the cornicles of an aphid attacked by a predator may also release 'alarm pheromone', which causes other aphids nearby to up-anchor and rapidly move away from the area.

Aphids have a variety of life histories, often involving polymorphism. One variation is whether the aphids have a sexual cycle in the autumn; the other main variation is whether two hosts are involved. Aphids that have a sexual cycle are termed 'holocyclic'; those that remain asexual all year round are 'anholocyclic'. Species differ in this respect, but so may genotypes of the same species as in the peach–potato aphid (*Myzus persicae*) where even a third life history has been recorded (androcycly, where males are produced but never females). Anholocyclic aphids need only a single host plant species ('monoecious' aphids), but many holocyclic aphids must alternate between a summer and winter host ('dioecious' aphids). The summer (secondary) host is usually herbaceous, and the winter (primary) host is a woody perennial shrub or tree on which the eggs can pass the winter. There are a number of variations as to whether sexual or asexual aphids disperse to the winter host in autumn, but the life cycle of the black bean aphid (*Aphis fabae*) is illustrated in Fig. 7.45. This species alternates between spindle tree (*Euonymus europaeus*) as the primary host and various herbaceous plants (especially broad bean, *Vicia faba*, and sugar beet, *Beta vulgaris*) as secondary hosts.

The morphs of adult aphids that may occur are: ***fundatrix***, parthenogenetic female that hatches from the egg; ***fundatrigenia***, parthenogenetic female on the primary host; ***virginopara***, parthenogenetic female on the secondary host and bearing more parthenogenetic females; ***sexupara***, parthenogenetic female giving rise to sexual morphs; ***gynopara***, parthenogenetic female giving rise to female morph only; and male and female (***ovipara***).

Aphis fabae is typical of the extensive host specificity of many aphids, so common names often refer to a plant on which the species is often found. With economically important species we usually use the name of the crop plant affected. So in naming *A. fabae* the black bean aphid we ignore the fact that it could equally be named after its winter host, the spindle tree. However, rosy apple aphid (*Dysaphis plantaginea*) is named after its winter host. It could equally be called the 'hedge woundwort aphid' after its weed summer host *Stachys sylvatica*. Sometimes, both summer and winter hosts are crops; then both figure in the common name (e.g. *M. persicae* referred to above in this Section).

Dispersal between host plants (whether summer and winter or new summer hosts in the summer) is passive rather than a directed migration. Most of the time, aphid reproduction is by wingless adults (apterae). The switch to winged adults (alatae) in summer dispersal is triggered by overcrowding and deterioration of the condition of the plant through age or the effects of the aphid infestation. The principle factor inducing production of alate adults in the autumn is shortening day length. Winged and wingless adults, together with parthenogenetic or sexual reproduction, combine to produce polymorphism where a number of phenotypes can be distinguished (see Fig. 7.45).

Long-range dispersal itself is passive once the winged adult has taken off from the plant. This usually is synchronised for the early morning when the temperature has risen above a threshold and the warming of the ground by the sun has created thermal up-currents. Once in a thermal, the aphids merely parascend with outstretched wings, and may be lifted up to over 600 metres. Here horizontal wind speeds can be quite high, and the aphids may be blown a long way laterally before the thermals cease and

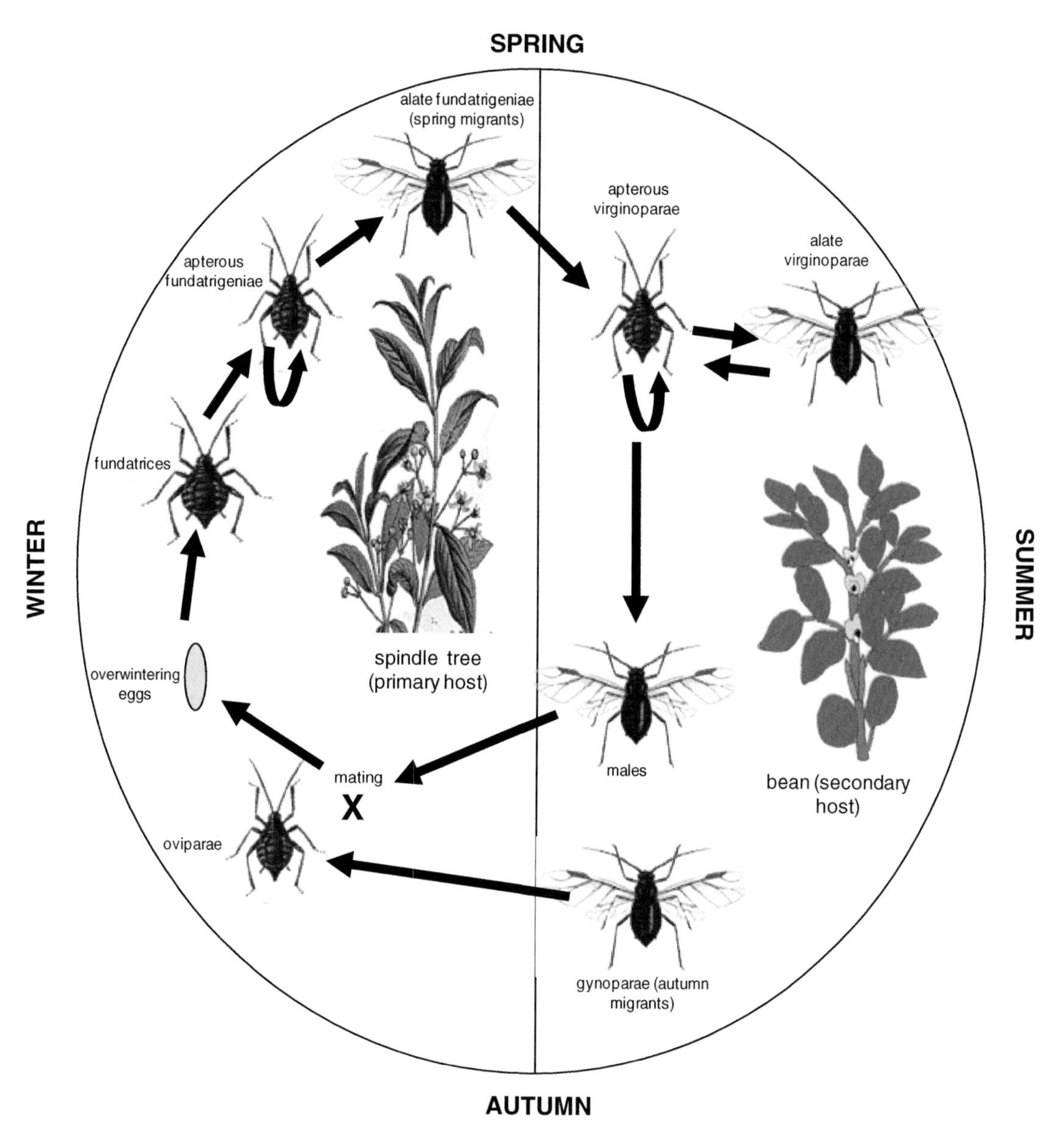

Fig. 7.45 Life cycle of the black bean aphid (*Aphis fabae*). The U-shaped arrows denote when there are several apterous generations before the subsequent alate generation.

the aphids descend to wherever gravity takes them, be it a road, a lake or the shirt of a cricketer in the outfield. A few may land on plants, but usually not an acceptable one. Host selection is therefore a decision to stay; the arrival is by chance. This is rather important in relation to the transmission of stylet-borne viruses, as it is sometimes possible for aphid species for which the crop is a non-host to transmit these viruses while testing plants before taking off again the next morning.

World-wide, the most important crop pest aphid species are probably:

Acyrthosiphon pisum. The pea aphid is a rather large green or pink aphid with long slender legs found on the shoots and pods of many cultivated and wild leguminous plants. It transmits more than 30 virus diseases, including the persistent *Pea enation mosaic virus* and *Bean leafroll virus*. It has an almost world-wide distribution and, in cold temperate regions, is holocyclic but without host alternation.

Amphorophora idaei is often referred to as *A. rubi* (a different species found on blackberry). The large European raspberry aphid is a quite large pale green aphid with pale cornicles. It is monoecious and holocyclic on raspberries, rarely numerous but nonetheless an important pest through transmission of *Raspberry mosaic virus*. It occurs in Europe.

Aphis craccivora. The cowpea or groundnut aphid is a small, dark brown aphid with a shiny, black dorsal shield. Although found mostly on legumes, it is more polyphagous than *A. pisum*, attacking some 50 crops in 19 different Families. Of the about 30 plant virus diseases it transmits, many are non-persistent. However, it transmits three potentially important persistent viruses: *Subterranean clover stunt virus*, *Peanut mottle virus* and *Groundnut rosette virus*. The aphid is now more or less world-wide, though most serious as a pest in warmer climates. It does not host alternate and is predominantly anholocyclic.

Aphis fabae. The black bean aphid (Fig. 7.46) is a major pest of beans in Europe, though it also infests sugar beet and there transmits yellows viruses. It is a dull black colour, but individuals in large colonies often develop waxy white stripes. Its wide polyphagy suggests it is probably a mixture of species. The member of this complex that attacks beans and beet is holocyclic and host alternating, with spindle (*E. europaeus*) as its winter host. The species occurs world-wide in temperate regions in the northern hemisphere, South America and Africa.

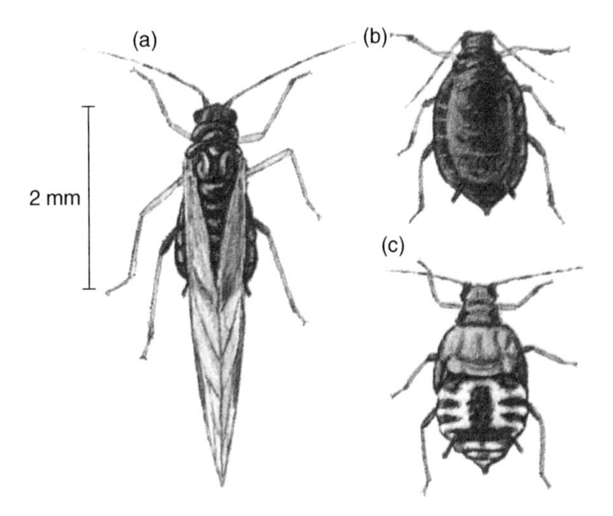

Fig. 7.46 Black bean aphid (*Aphis fabae*): (a) alate adult; (b) apterous adult; (c) alate nymph (all from Lyneborg 1968, with permission).

Aphis gossypii. The cotton or melon aphid (Fig. 7.47) is probably another polyphagous species complex. It is a small aphid which is bluish-black in colder climates, but varies from blue-green through yellow-green to pale yellow as temperatures increase.

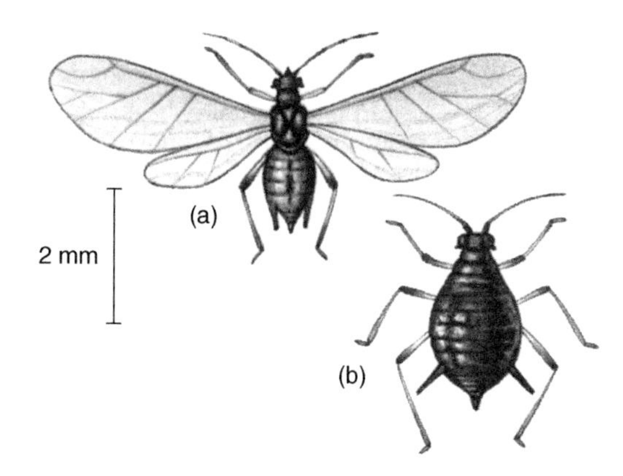

Fig. 7.47 Cotton/melon aphid (*Aphis gossypii*): (a) alate adult and (b) apterous adult (from Bayer 1968, with permission).

Its cauda is usually paler than the cornicles. It has been recorded on over 100 crop plants and transmits more than 50 plant viruses. Many are non-persistent and these are especially serious in cucurbits; important persistent viruses are *Cotton anthocyanosis virus* and *Pea enation mosaic virus*. It is anholocyclic, though in Japan and parts of the USA indistinguishable populations are holocyclic and overwinter on a number of shrubs. The cotton or melon aphid, as its name suggests, is a major pest of cotton and cucurbits but is also an important pest of glasshouse chrysanthemums in temperate regions. It is almost world-wide as an outdoor species in warmer climates.

Aphis spiraecola. The spiraea or green citrus aphid is a small yellow or yellowish-green aphid; the cornicles and cauda are black. It is primarily a pest on citrus. Its ant-attended colonies are in curled leaves near the tip of shoots of plants in more than 20 Families, particularly shrubs. Its sheer numbers cause damage, but it also vectors several viruses including *Citrus tristeza virus*, *Cucumber mosaic virus*, *Water melon mosaic 2 virus* and *Zucchini yellow mosaic virus*. It is almost world-wide and anholocyclic on secondary hosts, except for East Asia and North America where it has a sexual phase on *Spiraea*.

Brevicoryne brassicae. The cabbage aphid (Fig. 7.48) is a medium-sized dark green aphid which appears grey because it is covered in a whitish wax. Because of this, it is sometimes called the 'mealy cabbage aphid'. It is restricted to the Brassicaceae and causes damage by sheer weight of numbers feeding in the heart leaves and on the undersides of the older leaves. It shows a sharp increase in numbers in late summer, when it may invade the growing buttons on Brussels sprout plants. Inside the sprouts, it is exceedingly difficult to control. The aphid shows no host alternation and is mainly anholocyclic, though oviparae and eggs are found in very cold spells

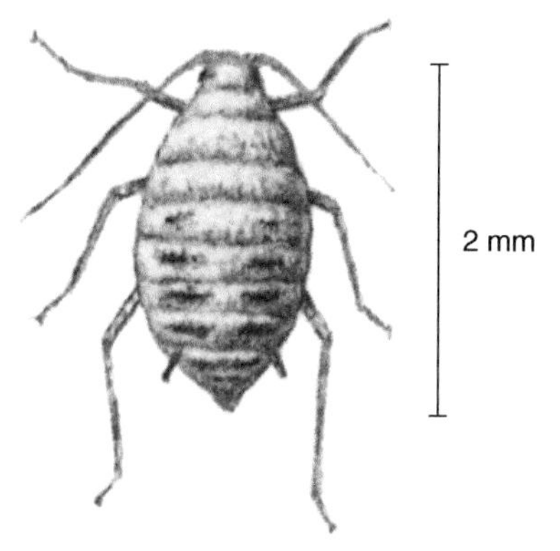

Fig. 7.48 Cabbage aphid (*Brevicoryne brassicae*) apterous adult (from Lyneborg 1968, with permission).

during the winter. The aphid prefers colder climates and so is common in Scandinavia, but progressively decreases in abundance further south.

Diuraphis noxia. The Russian wheat aphid is a slender yellowish-green aphid which was little known outside southern Russia until the late 1970s. It has since spread to much of Europe, South Africa and both North and South America. It is limited to the Poaceae and is a pest particularly on wheat and barley. Its saliva seems very toxic to plants and it also transmits the persistent *Barley yellow dwarf virus*. It has an anholocyclic life history.

Dysaphis devecta. The rosy leaf curling aphid 'does what it says on the tin'; that is in spring the aphid rolls the edges of apple leaves downwards and these rolls turn an obvious red. Inside these rolls are the wax-covered bluish-grey apterae and dark green apterae destined to become alatae. Later the young shoots are colonised. The species is monoecious and holocyclic, but is unusual in that the sexual morphs are produced very early in the summer, and after only three parthenogenetic generations. Attacks causing yield loss tend to be on mature trees and local in appearance. It occurs only in Europe.

Dysaphis plantaginea. The rosy apple aphid is also holocyclic, but in contrast to *D. devecta* is heteroecious, with plantain (especially *Plantago lanceolata*) as the herbaceous summer host. Here the aphid feeds on the undersides of the leaves, causing severe curling, yellowing of the leaves and much distortion. Fruits are stunted and malformed. Some aphids may, however, stay on apple throughout the summer. On apple, the aphids live on the leaf underside in blister-like galls; the aphids are a purple-grey colour, but wax covered. The aphid is a serious pest of apples. The species has a much wider distribution than *D. devecta*, and is found in Europe, the Middle East, Central Asia, Africa and South America.

Lipaphis pseudobrassicae. The turnip or mustard aphid (or false cabbage aphid) has a mealy grey waxy bloom hiding the true yellow, grey or dark green colour. The aphid forms dense colonies on the underside of leaves or flowering stems of many Brassicaceae, on which the leaves curl and become yellow. It vectors about 10 non-persistent viruses, including *Turnip mosaic virus* and *Cauliflower mosaic virus*. It is a pest in warmer climates, and is anholocyclic although a holocycle without host alternation has been recorded in Japan.

Eriosoma lanigerum. The woolly apple aphid is so called because it covers itself in long waxy filaments (Fig. 7.49) so that colonies of the aphid look like cotton wool. It feeds on the stems of apple trees, especially at pruning cuts. The aphid is found wherever apple is grown. Damage is mainly caused by fungal canker which can enter the tree where the aphid has fed. The aphid does not host alternate but is holocyclic with the peculiarity that each ovipara lays only one egg. Because this makes the insects hard to find in winter, it was long assumed that they migrated to the roots. However, the aphid there is a different species.

Macrosiphum euphorbiae. The potato aphid is a medium to large fusiform aphid. It is usually a deep green but can also be pink. Adults are shiny, but nymphs are lightly dusted with a grey wax. The species is holocyclic in the north-east USA with rose as the winter host, but elsewhere it is mainly anholocyclic though sexual morphs are sometimes observed. It is an important pest of potato, though it is highly polyphagous with over 200 host plants in more than 20 Families. In keeping with its polyphagy, it vectors over 40 non-persistent and five persistent viruses. Most important are *Beet yellow net virus, Pea enation mosaic virus, Bean leaf roll virus, Zucchini*

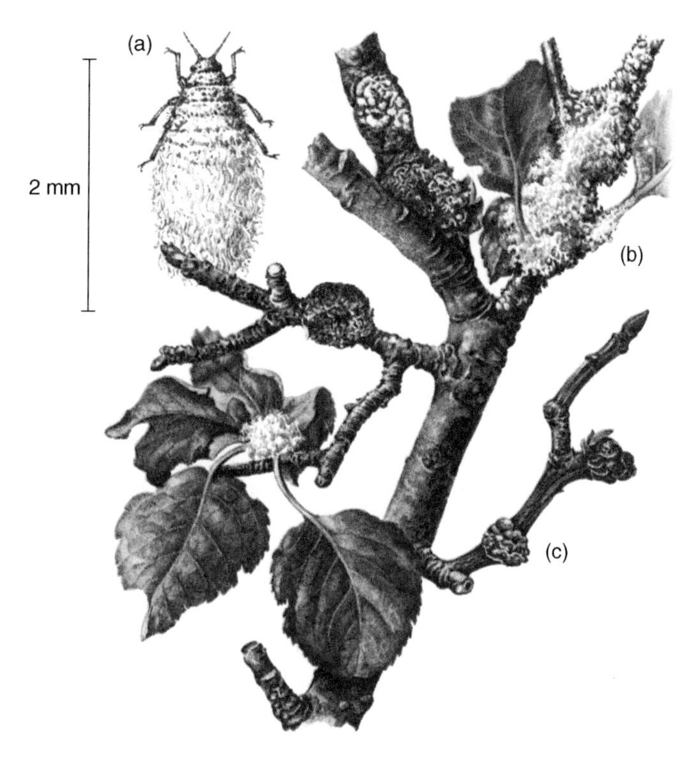

Fig. 7.49 Woolly apple aphid (*Eriosoma lanigerum*): (a) apterous adult; (b) wax- covered colony; (c) cankerous galls (all from Bayer 1968, with permission).

yellow mosaic virus and *Potato leaf roll virus*. The species is almost world-wide.

Myzus persicae. Peach–potato aphid or green peach aphid (Fig. 7.50) is a highly efficient virus vector, highly polyphagous and shows much variation in life history (see earlier in this Section). Adult apterae are mostly small and pale green, but can vary greatly with shades of pink or red. Alatae of the genus *Myzus* have a shiny black area on the back of the abdomen. Holocyclic clones host-alternate with peach as the winter host in most places, and such host alternation is found wherever there are peaches and low winter temperatures, which are required for sexuals to be produced. In the spring, dense colonies form on peach and cause leaf curling. However, even in temperate regions, most clones of *M. persicae* are anholocyclic and overwinter as aphids on herbaceous hosts. The high host specificity for the winter host is not shown with the herbaceous summer host, which can be in more than 40 Families and include too many economically important plants to list. The persistent viruses *Bean leaf roll virus*, *Beet western*

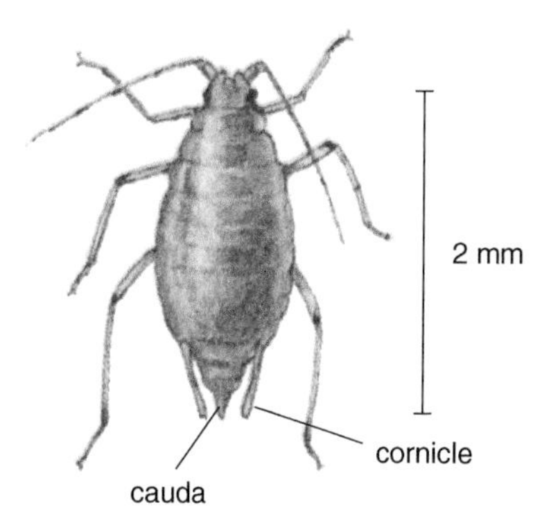

Fig. 7.50 Peach–potato aphid (*Myzus persicae*) apterous adult (from Lyneborg 1968, with permission).

yellows virus, Beet mild yellowing virus, Beet yellow net virus, Pea enation mosaic virus, Potato leaf roll virus, Tobacco vein-distorting virus, Tobacco yellow net virus and *Tobacco leaf vein virus* are vectored by this one species. The species is world-wide in its distribution.

Nasonovia ribisnigri. The currant–lettuce aphid is holocyclic and, as its name suggests, is heteroecious between various species of currant (*Ribes* spp.) as the winter host and lettuce (but also many other species in the Asteraceae as well as in the Scrophulariaceae and Solanaceae) in the summer. The species is a medium-sized aphid, rather spindle-shaped, and varying in colour from a shiny pale green on *Ribes* to a much more yellowish-green on its summer hosts, where the abdomen also has dark markings and dark tips to the cornicles. Individuals are not usually at high density, and damage caused is mainly on gooseberries from the transmission of *Gooseberry vein banding virus*. On lettuce foliage, economic damage mainly occurs when the aphid infests the heart of the plant. The species is found in Europe, much of Central Asia, the Middle East and North and South America.

Pemphigus bursarius. The lettuce root aphid is a holocyclic species alternating between Asteraceae (including lettuce) in the summer and poplar as the winter host, where the aphids live in purse-shaped galls on the leaf petioles. The aphids are yellowish with a fair amount of wax. Economic damage is on lettuces, where the aphids feed on the roots and can cause complete collapse of the plants; they also transmit *Lettuce mosaic virus*. Somewhat unusually for root aphids, colonies of *P. bursarius* are not ant-attended. The species is very widely distributed, except for South America.

Phorodon humuli. The damson–hop aphid is holocyclic and heteroecious between *Prunus* spp. in winter and hops in the summer, where it transmits a number of virus diseases including *Hop mosaic virus*. The aphids are small and rather yellowish, and can cause damage to hops by sheer numbers on the leaves, flowers and fruit. On plum they transmit *Plum pox virus* ('Sharka' disease). This aphid appears to have some innate resistance to organophosphate insecticides, even before any selection resulting from use of such materials. The species is widely distributed, occurring in Europe, south-west Asia, North Africa, North America and New Zealand.

Rhopalosiphum maidis. Apterae of the corn leaf aphid are small to medium bluish-green aphids with short antennae and dark legs, cornicles and cauda. They are especially pests on the young leaves of maize, sorghum and barley, but they also live on many other wild and cultivated Poaceae in more than 30 genera. The species is a major vector of the persistent *Barley yellow dwarf virus, Sugar cane mosaic virus* and *Maize dwarf mosaic virus. Rhopalosiphum maidis* is probably the most important aphid on cereals, and pest populations are anholocyclic and unable to survive severe winters; apart from this restriction the species is cosmopolitan.

Rhopalosiphum padi. The bird cherry–oat aphid is of a variable green colour between yellowish and dark green and often has rust-coloured areas around the cornicles. It feeds on many Poaceae as well as on other monocotyledonous and some dicotyledonous plants. It vectors several viruses; probably the main economic impact this aphid has is to transmit persistent *Barley yellow dwarf virus*, which is present in a symptomless condition in grassland and is brought to cereal crops by the aphid. As its name suggests, the species is holocyclic and host alternating with bird cherry as the winter host. The species is also cosmopolitan.

Schizaphis graminum. The greenbug is a major pest in the winter wheat growing areas of North America, where it is anholocyclic. However, further south in the USA and in Asia sexuals are produced, though there is no host alternation. It is a small bluish-green aphid with pale cornicles with dark tips, which is restricted to the Poaceae. Wheat reacts to its saliva with yellowing and severe stunting, but the aphid also transmits *Barley yellow dwarf virus*, *Sugar cane mosaic virus* and *Maize dwarf mosaic* among others. Greenbug is very widely distributed, although it is absent in northern Europe and probably also Australasia.

Sitobion avenae. The grain aphid is a medium-sized light green or reddish aphid with a reticulated pattern on the cornicles, black antennae and a pale cauda. It colonises many wild and cultivated Poaceae and is holocyclic without host alternation, though it may also show an anholocyclic pattern when the winter is mild. Direct damage is done by curling the flag leaf of wheat and later when feeding on the developing grains; the aphid transfers to the ear as the tiller extends past the flag leaf. Additional damage is done by the honeydew on which black moulds develop and these and the stickiness interfere with the milling process. The aphid is also an important vector of *Barley yellow dwarf virus*, but since it transmits this virus to the wheat in spring, it is less serious than *R. padi*, which brings in the virus much earlier in the previous autumn. The species is very widely distributed, but is not found in the Far East or Australasia.

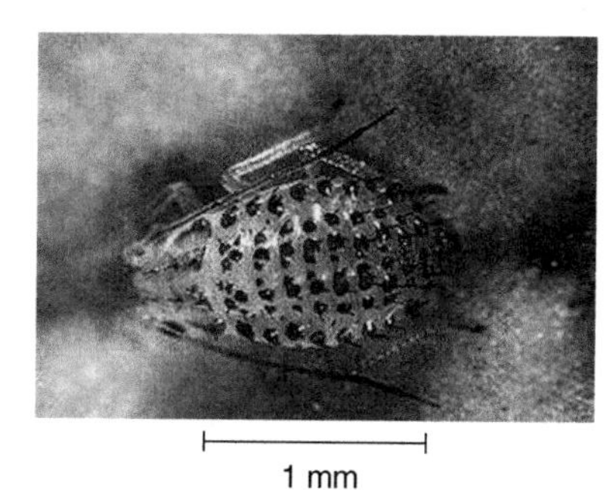

Fig. 7.51 Spotted alfalfa aphid (*Therioaphis trifolii maculata*) apterous adult (courtesy of Ali Zarrabi).

Therioaphis trifolii maculata. The spotted alfalfa aphid (Fig. 7.51) has a knobbed cauda and a bilobed anal plate. It is a shiny yellowish-green with four to six longitudinal rows of darker dorsal tubercles. The aphid was introduced to the USA from Europe and became a major pest of the alfalfas forage crop. The species also occurs in North Africa and the Middle East. Again the plant's reaction to the aphid's saliva is the main problem, and plants can be killed quite quickly. The species is anholocylic and there is no host alternation.

Family Phylloxeridae

The grape phylloxera (*Viteus vitifolii*) attacks both the roots and leaves of vines (Fig. 7.52). At both sites the tiny aphids live within galls and cause severe stunting and failure of the grapes to form. The pest wiped out the French vineyards between 1875 and 1892 and wine production across Europe was only saved because some old discarded varieties used to make Communion wine had been taken by religious settlers to California and turned out to be resistant to the pest. This resistance was graft transmissible and, by grafting German and French varieties onto these Californian rootstocks, wine production in Europe was saved. However, in 1993, a strain of the aphid able to break this resistance appeared in Germany.

7.5.3.3 Superfamily Aleyrodoidea (whiteflies) – Family Aleyrodidae

Whitefly adults (Fig. 7.53a) are small insects of moth-like appearance with triangular forewings and covered in wax, which is usually white (hence white fly). However, *Aleurocanthus woglumi* (Fig. 7.53c) is black. The term 'blackfly' is already used for

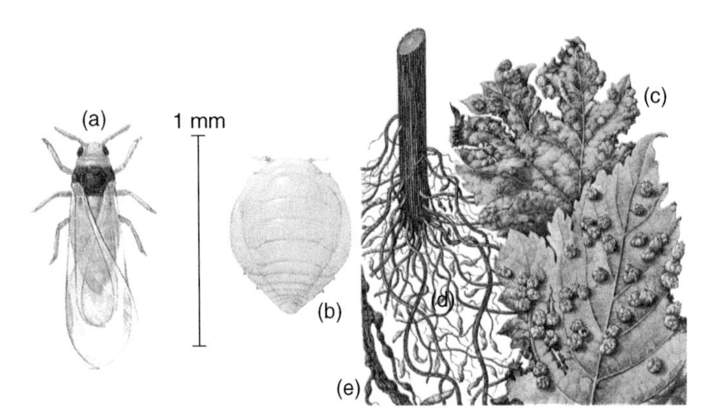

Fig. 7.52 Grape phylloxera (*Viteus vitifolii*): (a) winged sexual aphid; (b) fundatrix; (c) leaf galls; (d) galls on younger roots; (e) galls on old roots ((a,b) from Zanetti 1977; (c,d,e) from Bayer 1968, with permission).

black aphids (especially the black bean aphid), and 'black fly' is a biting fly. So, what do we call *A. woglumi*? Rather whimsically, it is known as 'the black whitefly of citrus'!

Adults are of both sexes, but even so reproduction is often parthenogenetic. The adult female lays her eggs in a circle by rotating on her stylets. The eggs hatch into tiny wingless nymphs called 'crawlers', because it is this first instar that disperses from the oviposition site and settles to moult into a legless sedentary scale-like second instar (Fig. 7.53b). Here the insect remains throughout subsequent moults. Unlike aphids, they therefore cannot move to change their feeding site; instead they have to re-position the tips of their stylets. Whitefly nymphs therefore have very long stylets, which can be retracted and pushed out again from 'reels' in the left and right sides of the head. This eventually leaves many long branched stylet tracks arising from each scale, and the process causes much disruption of the leaf tissue, with breakdown of permeability barriers and much chlorosis. The last immature moult gives rise to a scale of rather different appearance, considered by some to be almost a 'pupa' and to be the link with the complete metamorphosis of the Endopterygota.

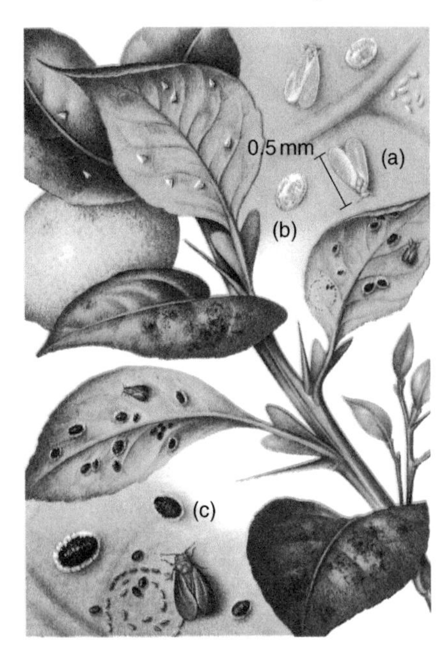

Fig. 7.53 Two whitefly pests of citrus: (a) adult and (b) nymph of *Dialeurodes citri*; (c) life stages of *Aleurocanthus woglumi* (all from Bayer 1968, with permission).

In warmer climates, *Bemisia tabaci* is a serious cosmopolitan pest of cotton, sweet potato, outdoor tomato, tobacco and many other crops. It is extremely polyphagous, having been recorded from over 200 host plants in more than 60 plant Families. Plants are weakened and stunted, and a number of important virus diseases (including *Cassava mosaic virus, Cotton leaf curl virus* and *Tobacco leaf curl virus*) are transmitted. The eggs are anchored into stomata, and hatch in about a week; generation time is only 3–5 weeks leading to the potential for population explosion. Breeding can be continuous, with many generations a year. Reproduction is stimulated by high nitrogen fertilisation to the plants, and the high nitrogen plots in cotton experiments can often be identified by the clouds of whitefly which fly from the plants when someone walks through the experiment.

In Europe, whitefly is known mainly as a pest of glasshouse crops such as tomato and cucumber. Here the main species are the glasshouse whitefly *Trialeurodes vaporariorum* and the sweet potato or tobacco whitefly *B. tabaci* (see previous paragraph). More recently, the silver leaf whitefly (also *B. tabaci* but known as *B. argentifolii*) has been accidentally introduced into Europe, and has become a serious problem on a range of vegetables and tomato, especially in glasshouses.

Outdoors in Europe and other regions including Russia, part of Africa, Brazil and New Zealand, mild winters can see outbreaks of *Aleyrodes proletella* on brassica crops. The insect has several generations a year but only seems to become a problem overwinter; it can even breed at subzero temperatures. The nymphs are covered in wax and heavy populations stunt plant growth with sooty moulds developing on the copious honeydew.

Aleurocanthus woglumi (Fig. 7.53c) has already been mentioned. In the New World the species is also a pest of coffee. The black scales with a white fringe are on the undersides of leaves, while the upper leaf surfaces are contaminated with honeydew and sooty moulds. The batches of 30 or more eggs, yellow at first before becoming black, are anchored on the leaf in a characteristically spiral arrangement.

The white *Dialeurodes citri* (the citrus whitefly) (Fig. 7.53a,b) is another pest of citrus. It is also found on coffee and ornamental trees and shrubs. The honeydew with associated sooty moulds may cause cosmetic damage to fruit, and this can often be the main problem. Each female can lay over 100 eggs, and the length of the life cycle shows a huge range (1–10 months) depending on temperature; much of the time (as much as 300 days) can be spent in the quiescent pupal stage.

7.5.3.4 Superfamily Coccoidea (scale insects and mealybugs)

These are in some ways the 'aphids' of the tropics in that they are also phloem feeders and honeydew excreters. They become progressively more common as latitudes get closer to the equator while aphids (dominant in temperate zones) become less common. Honeydew production can be so large that many species depend on attending ants to remove the liquid. An early and effective control of scale insects and mealybugs on trees was to grease-band the trunk and so prevent ants from reaching the coccoids, leaving them to drown in their own honeydew.

All stages other than the 'crawler' first instar and the small and rarely seen males are sessile, so that the 'scale' facies of the whiteflies extends to the adult females. Most reproduction is parthenogenetic and the eggs are laid under the protection of the body or hard scale covering of the mother; they hatch from that position as the mother shrinks and dies. The tiny winged males are rare; indeed they are unknown for many species of scale insect. The males (Fig. 7.54) are unusual in having only functional forewings, the rear wings are reduced to 'halteres' (as in the Diptera, see Section 10.1).

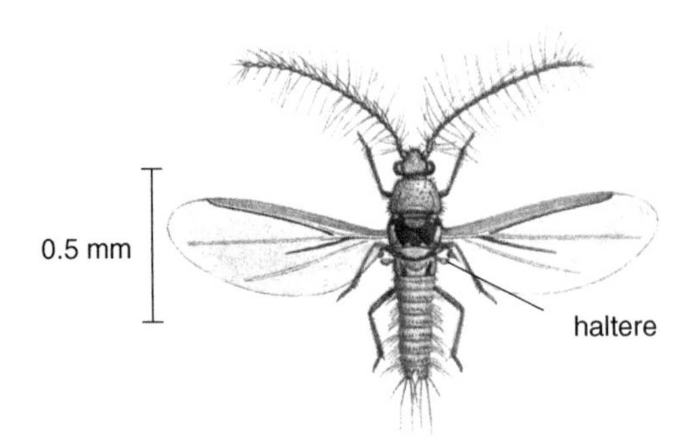

Fig. 7.54 A male coccid (from Zanetti 1977).

There are many families; only those of some economic importance will be listed below.

Family Pseudococcidae (mealybugs)

These often extrude long white waxy filaments from the abdomen which gives them a 'woolly' appearance. They are regular pests on ornamentals in glasshouses in temperate countries, but in the tropics they can also be pests outdoors, usually on woody rather than herbaceous plants. Especially important is *Phenacoccus manihoti* (the cassava mealybug) because of the importance of cassava as a subsistence crop in Africa. The mealybug was introduced to Africa from South America in the late 1970s and spread rapidly in Africa to become an extremely serious problem. Since the successful introduction of a parasitoid wasp (*Anagyrus lopezi*), it has declined in importance. The mealybug sucks on the young shoots with injection of toxic saliva, and leaves become distorted and the shoot internodes shortened; the shoot may die. The shortening of the shoots leads to the diagnostic 'bunchy top' symptom. The mealybug is about 3 mm long and covered with white wax. The life cycle is only about 3 weeks, so many generations per year occur.

The citrus mealybug (*Planococcus citri*; Fig. 7.55b) is found as white insects under a crust of pale green *Polyporus* fungus on the roots of citrus, coffee and cocoa. The mealybugs are also found on leaves, shoots and fruit. The plant looks drought stressed with yellow leaves. The pest is a vector of *Swollen shoot virus* of cocoa. The related Kenya mealybug (*P. kenyae*) is found aerially on coffee between the flowers or berries; it also attacks many other cultivated and wild plants. The mealybug is a prolific honeydew producer and although the eggs are covered in wax the nymphs are bare. It has been under successful biological control with the pteromalid parasitoid *Pachyneuron* since the late 1930s.

The pink sugar cane mealybug (*Saccharicoccus sacchari*) is the most important mealybug on sugar cane. It is found low down on the stem, even just below ground level, as well as on the lower surface of the leaves. It also produces copious honeydew, and so is associated with sooty moulds as well as being ant-attended.

Family Margarodidae (cushion scales)

These have patches rather than filaments of white wax. An important pest is the cottony cushion scale of citrus (*Icerya purchasi*; Fig. 7.55a), which has the additional

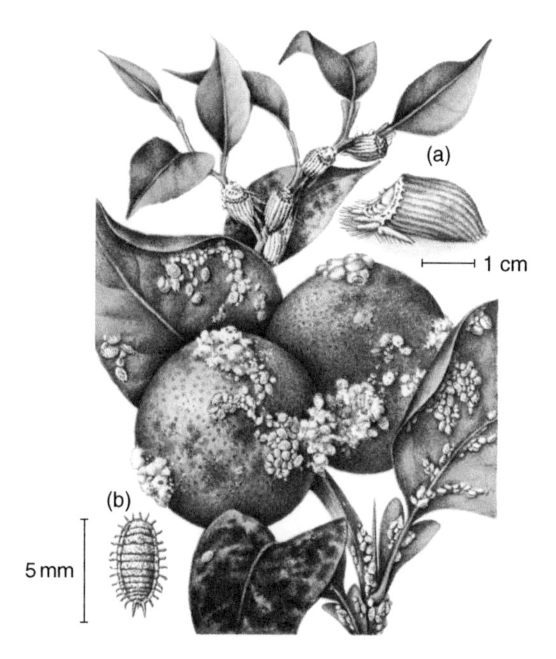

Fig. 7.55 (a) cottony cushion scale (*Icerya purchasi*) and (b) mealybug (*Planococcus citri*) (from Bayer 1968, with permission).

distinction of being the target in the late 1880s for what is regarded as the first example and foundation of modern biological control. It was introduced to California from Australia in the 1860s and is now found in most subtropical and tropical citrus-growing areas in the world. This scale insect is polyphagous, but its main host is citrus. The large fluted scales are very obvious on the leaves and twigs and there is plenty of honeydew around. Leaves fall prematurely, and young shoots (and therefore nursery stock) can be killed. Also, much honeydew is produced. Males are rarely produced; the females are hermaphrodite. The adult female is over 3 mm and is brown in colour dusted with white wax. The most conspicuous feature is the large fluted egg sac, which is often more than twice the 3.5 mm length of the body and contains about 1000 red eggs. Generation time varies with temperature, but 2–3 months is normal.

Family Kerridae

I mention this because it reminds us that insects can be economically beneficial. Even in groups of major world pests such as the Coccoidea is the shellac scale (*Kerria lacca*). These insects are covered with clear hard scales that can be harvested and dissolved in alcohol to yield the quality varnish known as 'shellac', and the insects are purposefully farmed for their scales. As you can imagine, genuine scale insect-derived shellac is not cheap.

Family Asterolecaniidae

In this Family, the principal pest species is *Asterolecanium coffeae*, the star scale. This changes the shape of the tip branches of coffee in a characteristic way; they are sharply angled at the nodes bearing drooping dead leaves and new branch growth is thin and whippy. The red and yellow scales are covered with a hard but transparent lid, and have a fringe of projecting hairs. The life cycle takes about 6–7 months. The pest occurs in Africa, particularly in the east.

Fig. 7.56 Soft scales (Coccidae) on citrus: (a) *Coccus hesperidum* and (b) olive scale (*Saissetia oleae*) (from Bayer 1968, with permission).

Family Coccidae (soft scales)

When you look at a 'soft' scale from above, you see the dorsal surface of the actual insect (see Diaspididae below for the contrast, Section 7.5.3.4, Family Diaspididae). There are many soft scales on citrus, many as minor pests and the result of the destruction of natural enemies by insecticides. It is said that attempts to control the cottony cushion scale (see Section 7.5.3.4, Family Margarodidae) with insecticide raised the number of scale pests of citrus from one to 97! The soft green scale *Coccus viridis* and soft brown scale *C. hesperidum* (Fig. 7.56a) can be damaging to young trees and occur throughout the tropics as flat green or brown scales at the shoot tips and along the main leaf veins. Each generation takes 1–2 months. Honeydew and sooty moulds are prevalent.

The olive scale, *Saissetia oleae* (Fig. 7.56b), is a cosmopolitan polyphagous pest, the main hosts being olives, citrus and fig. The large jet black scales are found encrusting the twigs and shoots with sooty moulds growing on the honeydew. Heavy infestations cause the leaves and shoots to wither. Males are rarely found; reproduction is mainly by parthenogenesis. The life cycle takes 3–4 months, but the nymphs can enter diapause to bridge unfavourable conditions. The related helmet scale (*S. coffeae*) of coffee has many alternative hosts, both wild and cultivated. Severe outbreaks can occur, especially when the crop is suffering other stresses. The adult female scales are shaped like a flat-topped helmet, which has a diagnostic yellow mark rather like the letter H. The scales begin green but darken to brown in later instars; they are often lined up at the leaf edge.

The brown scale (*Parthenolecanium corni*) is one of the few scale insect pests found in the UK and occurs under glass but also outside on the stems of soft fruit bushes. It is found throughout Europe on stone fruits, apple and grapevines. The small brown scales are obvious on the twigs, and often have egg masses under the rear end of the scale. The eggs hatch in early summer and it is the second instars that overwinter. Feeding resumes in spring, and the adult stage is reached in April.

Family Diaspididae (armoured scales)

When you view an armoured scale from above, you do not see the dorsal surface of the living insect, but an inanimate hard shell secreted by the insect and enlarged as it grows beneath it (Fig. 7.57). These instar layers are usually discernible as rings.

The San José scale (*Diaspidiotus perniciosus*) is a widely distributed and polyphagous armoured scale. It is a major pest of yam, orchard fruits and currants in the subtropics. It is rarely found in the tropics. The circular grey scales have a raised central pimple and are so small as to be barely visible, yet the tree bark may be virtually covered in heavy infestations. As well as on the bark, the scales are also found on the fruit, which

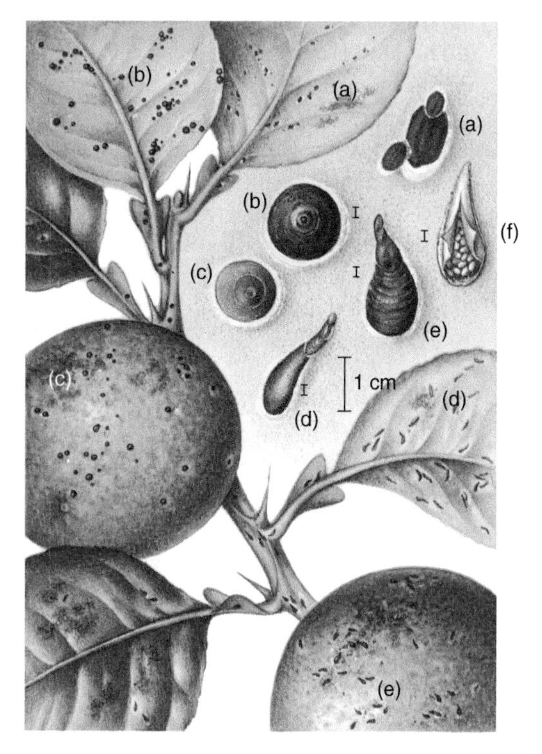

Fig. 7.57 Armoured scales (Diaspididae) on citrus (from Bayer 1968, with permission): (a) *Parlatoria ziziphi*; (b) *Chrysomphalus aonidum*; (c) *Aonidiella aurantii*; (d) *Lepidosaphes gloverii*; (e) *L. beckii*; (f) *L. beckii* showing space under the scale with eggs. The small scale bars on the images in the top right corner are natural size relative to the 1-cm bar.

makes them unmarketable, and in general trees may be badly weakened, or even killed, with the bark splitting and oozing gum. The life cycle takes 18–20 weeks, with four to five generations a year in warmer climates.

Mussel scales (*Lepidosaphes* species, especially *L. beckii*; Fig. 7.57e) are pests of several crops including citrus, where they often are clearly visible in groups of small purple mussel-shaped scales (about 2 mm in length) on the surface of the fruit, leaves and twigs. Symptoms of heavy attack are leaf fall and die back. The pest is of oriental origin, but now occurs throughout the tropics and subtropics. The life cycle takes 2–4 weeks. On apple, pear and soft fruit bushes in temperate climates, the mussel scale *L. ulmi* is a frequent pest. Incrustations of scales on the twigs stunt the growth of the plant, and there is one generation a year in northern Europe. Eggs are laid in late summer, but the female covering them dies and they do not hatch till late May and the nymphs reach the adult stage late in July. There are both sexual and parthenogenetic genotypes.

The most serious scale insect attacking citrus in the USA is the California red scale, *Aonidiella aurantii* (Fig. 7.57c). The scales have a central pimple and look a bit like tiny Mexican hats, and infestation can be so bad that the crawlers cannot find a clear space on the leaf undersides and, as the scales grow, they are forced to sit partly on top of each other. Branches of infested trees may die back, and the scales on the fruit make them unsaleable. Control with insecticides of stages other than the crawlers is made very difficult by the shiny unwettable scale cover. The pest is found throughout the tropics and subtropics

The Florida red scale (*Chrysomphalus aonidum*; Fig. 7.57b) is serious on citrus as well as on many other plants, again throughout the tropics and subtropics. This scale insect appears to inject saliva which causes a particularly damaging plant reaction leading to necrosis. Scales are found along leaf veins and on the surface of the fruit. They are purple, again with a raised central nipple.

On coconut, a complete covering crust of *Aspidiotus destructor* (coconut scale) can be found on the lower leaf surface in a heavy infestation. The leaves yellow and die and flowers and young nuts may also become infested. Other symptoms are much honeydew and attending ants. The life cycle takes little over a month, so there can be 10 generations a year; the pest is found throughout the tropics.

7.6 Order Thysanoptera (thrips or thunderflies)

Thrips are very small (less than 2.5 mm long) and slender fusiform insects, which may be wingless or, if winged, the wings are diagnostically narrow straps fringed with long hairs (Fig. 7.58) looking a bit like little feathers. This adaptation to avoid the wings being wetted is found in several other groups of tiny insects (see Section 2.6.4). The antennae are composed of a few short subequal segments. Colour ranges from white or yellow to brown or black. Some of the predatory thrips (species of *Franklinothrips*) are brighter in colour, with a reddish-orange abdomen. The mouthparts are adapted for sucking. Pest species puncture individual parenchyma cells and suck out the cell contents, leading to pale flecks on the leaves. Also, the leaf surface may be rasped and the exuding sap sucked up. Thrips feeding can result in distorted plant

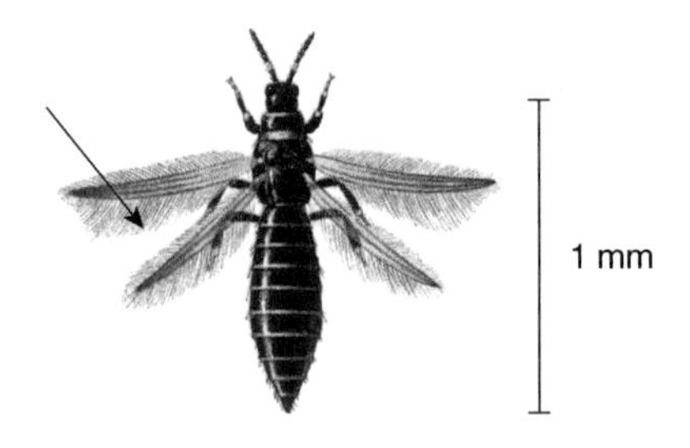

Fig. 7.58 *Limothrips cerealium* (from Zanetti 1977).

parts and scarred leaves, flowers and fruits. The black specks of their faeces around the leaf scars are diagnostic of thrips damage. Many species feed concealed in buds or furled leaves. Other species are harmless and feed on fungal spores and pollen; the Order also contains species that are predatory on other small insects, including other thrips, and on mites. Reproduction is normally sexual, but several species reproduce asexually by parthenogenesis. Males develop from unfertilised eggs, but are usually rare. There are four instars between egg and adult. The first two are active feeding nymphal stages. The late nymph usually moves to the soil to pupate. The final immature stages are called the propupa and pupa because they are non-feeding and display adult characters such as long wing sheaths. Thrips are weak fliers, but like aphids they can easily spread long distances in air currents. They also often spread when infested plants are moved.

7.6.1 Suborder Terebrantia

These have a saw-like ovipositor used to make cuts into plant tissues into which the eggs are then laid. The tip of the abdomen is conical (female) or rounded (male). The wings have obvious veins and small hairs on the membrane (***microtrichia***).

7.6.1.1 Family Thripidae

This Family includes most of the species regarded as crop pests. Whereas the adults are usually black or a dark brown, the nymphs are often much lighter and yellow or red.

Limothrips cerealium (cereal thrips; Fig. 7.58) is commonly found on cereals and grasses and can be a problem on winter wheat in the UK. It is widespread in temperate regions. Florets may become blind, and grains shrivel and are discoloured. On barley, quality for malting may be reduced. The thrips overwinters outside the crop, often under the bark of conifers; there is usually one generation a year. On peas, *Kakothrips pisivorus* causes flowers to abort and the leaves and pods develop silver flecks (Fig. 7.59) where the thrips has fed. These flecks can be dense and unsightly on the pods. Pupation is in the soil; there is again one generation a year.

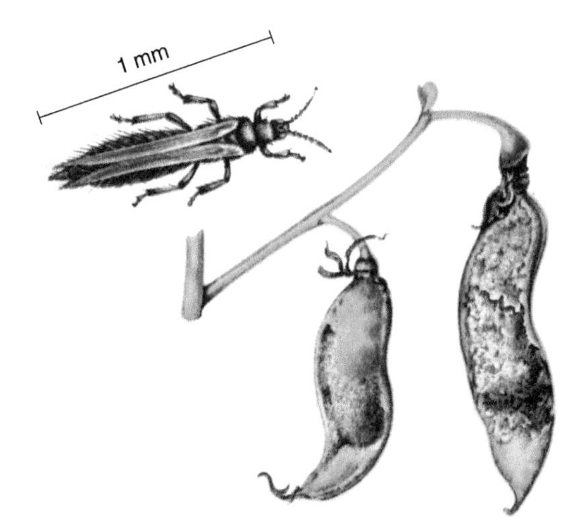

Fig. 7.59 *Kakothrips pisivorus* adult and damage to pea pods (from Bayer 1968, with permission).

One of the best known thrips pests is the onion or tobacco thrips (*Thrips tabaci*), a worldwide pest of vegetable crops. It is only about 1 mm long; there may be dark patches on the body, and the basal segments of the antennae are a much lighter brown than the other segments. Nymphs are much lighter in colour and may be almost white. Most damage is done to onions and tobacco, but it also attacks many other crops including garlic, pepper, cotton, tomato, pineapple, beet, lettuce, peas and cabbages. Leaves and fruits are scarred with silvery flecks, and the tips of shoots die back. Young plants may even be killed. The onion thrips is a vector of *Tomato spotted wilt virus*. On onions, the insects congregate above the bulb concealed by the leaf sheath, and so damage may occur before the infestation is noticed. Also, the feeding scars on onion leaves allow neck rot fungus (*Botrytis allii*) to establish a foothold. Pupation is in the soil and the life cycle only takes about 3 weeks, resulting in many generations each year.

Thrips simplex is the gladiolus thrips, a ubiquitous and often serious problem in gladioli. Damaged leaves are speckled with silver, and there are bleached specks on the flowers. Severe attack causes flowers to fail to open or bleach and wither. The thrips later infest the corms, causing grey-brown patches. Those that move to the soil in cooler areas (including the UK) mostly perish, as the species can only overwinter in the soil outside glasshouses in warmer climates.

On legumes, the bean flower thrips (*Megalurothrips sjostedti*) feeds inside flowers and causes them not to set seed nor form a pod. Damage is greater in legume varieties that are 'determinate' and cannot compensate by continued production of flowers (given nitrogen and water) until pods have successfully formed. Pupation occurs in the soil. The life cycle is completed in less than a fortnight.

Some thrips are really only a problem if growers are aiming for high-quality produce for export rather than for local markets. Thus the orange-coloured citrus thrips (*Scirtothrips aurantii*) breeds on young tissues of leaves and fruits and causes harmless but unsightly brownish rings where the fruit is attached to the stem and similarly cosmetic blemishes elsewhere on the fruit. It is highly polyphagous and has been recorded from plants in 30 Families. It is widespread in Africa and has been introduced to Queensland in Australia. A diagnostic feature of the species, as well as the yellow colour with dark ridges on the abdominal tergites, is that the abdominal sternites are covered in microtrichia. The dark brown banana thrips (*Hercinothrips bicinctus*) is similarly mainly a cosmetic problem, making silvery or brown patches on the skin of the fruit. Only if infestation is severe is real damage done by the skin splitting open allowing rots to set in on the fruit. Another thrips pest of banana is the banana rust thrips (*Chaetanaphothrips signipennis*), found in Central and South America, Asia and Australia. It feeds on the peel between the bananas on the tree, again affecting marketability of the discoloured fruit. However, this thrips also feeds on the banana flower buds and this can cause substantial loss of yield when infestation is heavy.

Yet another thrips attacking banana is *Hercinothrips femoralis*. It has the common name of banded greenhouse thrips, the 'banded' deriving from the two brown bands and a brown spot on the otherwise pale wings. The species attacks a huge variety of crops, including banana, pineapple, tomato, aubergine, sugar cane, beet and celery. It is widely distributed in the world, but in temperate regions it is mainly restricted to greenhouses. It causes typical thrips damage to flowers and foliage, and leaf tips may wither and die. On banana, the scarring may cause economic cosmetic damage, which becomes more extreme when fruits turn a peculiar red (still without affecting their palatability) if the infestation is large or a lesser infestation is combined with red spider attack.

The coffee thrips (*Diarthrothrips coffeae*) sucks the leaves, shoots and berries; irregular silvery patches with small dark spots appear; heavy infestation may defoliate the tree. Adults are dark grey to brown; pupation is in the soil. The life cycle takes about 3 weeks in hot climates, and the pest is found in Central and East Africa.

On rice the thrips problem is *Stenchaetothrips biformis* (rice thrips). This is a pest of India and south-east Asia, and feeding causes yellow streaks which later fuse to discolour the whole leaf. Eventually, the whole plant may desiccate and wilt, but this usually only happens to young seedlings. The life cycle is usually less than 2 weeks, allowing rapid reproduction.

Bean thrips (*Caliothrips fasciatus*) attacks beans, but sometimes also other legumes, in western USA and China, browning and distorting leaves and the shoot of seedlings. The thrips can be recognised by the two white wing bands on the brown forewings.

Frankliniella schultzei (cotton bud thrips) is polyphagous, but mainly damaging to cotton, groundnuts and beans. The leaves are attacked while still in the bud and emerge distorted. Yield is reduced partly because the seedling's growth has been slowed. The thrips transmits *Tomato spotted wilt virus* to groundnuts, which can dramatically reduce yields. In hot dry conditions reproduction and thus the appearance of symptoms can be very rapid with a 2- to 5-week life cycle. Other species of the genus feed on cereals, legumes tobacco, potato, peppers, sweet potato – you name it! Together these species have an almost world-wide distribution, but of special importance is the western flower thrips, *F. occidentalis* (Fig. 7.60). This was accidentally introduced to Europe from the USA and seems to have arrived carrying multiple resistance to insecticides. It has similarly spread on infested plants to other parts of Europe, to South America and Australia. The female is about 1.4 mm long and colour varies from red to yellow and brown. The much paler males are only 1 mm long and are rare, so reproduction is mainly by parthenogenesis. Females lay 40–100 plus eggs into punctures in the plant tissues, often in the flower but also the fruit and leaves, and these punctures can be a major source of cosmetic damage since they develop into 2- to 3-mm diameter white bumps. Plants react badly to the pest's saliva when it feeds, and holes and a silver discolouration result. Nymphs feed extensively on young fruit that is only just developing. Like *F. schultzei, F. occidentalis* transmits *Tomato spotted wilt virus*. Damage to fruits tends to be concentrated where they touch each other, leaves (when the shape of the leaf is etched on the fruit wall by thrips feeding) or other parts of the plant. The insect is an especial problem on tomatoes, peppers, cucurbits, cotton and many ornamentals (especially roses and chrysanthemums); in the UK, it is mainly a problem in glasshouses.

Fig. 7.60 Western flower thrips (*Frankliniella occidentalis*) (courtesy of David Riley).

Another major polyphagous thrips pest in field and glasshouse crops is the melon thrips (*Thrips palmi*). It infests glasshouses in many parts of Asia, from where it has spread to the Americas, the Caribbean, Africa and Australia, but field outbreaks occur only in the tropics. The adults are white or pale yellow, 0.8–1.0 mm in length and appear to have a black line (actually the junction between the wings) along the back of the body. Another diagnostic feature is that the fringe of hairs on the leading edge of the wings is markedly shorter than on the rear edge. Best known as a pest of cucurbits and Solanaceae, the melon thrips also infests most vegetables, pome and citrus fruits, tobacco and ornamentals. Both mated and unmated females lay eggs into a slit cut in the plant tissue. Feeding by nymphs and adults causes yellowing, crinkling and even death of leaves; damage can be so severe that whole crop fields develop a dry bronzed colour. Terminal growth may be stunted and distorted, flowers aborted and fruits scarred.

Even more limited to glasshouses is *Heliothrips haemorrhoidalis*, the greenhouse thrips. Although originally described from Europe, its glasshouse habit has enabled it to live in many parts of the world, though it will 'escape' into the field where the summer or latitude provides warm conditions outside. It is found in shaded parts of plants, primarily ornamentals but also citrus as well as avocados and mangoes. Leaves become discoloured and the vein pattern distorted, and both upper and lower leaf surfaces are covered with small reddish droplets voided by the thrips and thought to deter predators. These droplets slowly turn black; the black spotting of infested areas is diagnostic for the species. On infested immature citrus fruit, blemishes such as russeting or ring-spotting appear, especially where the fruit surface is in contact with a leaf or twig. The adult greenhouse thrips is fairly dark brown or black, but the legs are a pale yellow. Males are seen rarely; the species is parthenogenetic. Eggs are hidden, laid entirely into leaf surfaces, but their location is revealed just before the hatch when a characteristic blister develops, which raises the leaf surface.

The sucking mouthparts of thrips have enabled some species of thrips to evolve the carnivorous habit, taking mites and early instars of phytophagous thrips as prey. *Frankliniella occidentalis* and *Thrips tabaci*, which are normally phytophagous, can also be predatory, but more specialist predators are found in a different Family of Terebrantia, the Aeolothripidae (see Section 7.6.1.2).

7.6.1.2 Family Aeolothripidae

This Family is common in the northern part of the world, but some species occur in the drier parts of the subtropics and the tropics. These largely predatory thrips are often more brightly coloured than the Thripidae. They normally dwell in flowers or on leaves. They pupate in a silken cocoon at ground level.

One major genus is *Aeolothrips*, facultative predators which also take plant food. Indeed, *A. intermedius* appears unable to breed successfully on a diet solely of thrips larvae. Species in the other prominent genus, *Franklinothrips*, such as *F. vespiformis*, are found in the tropics and are ant mimics. Three species are used for biological control of pest thrips in European glasshouses.

7.6.2 Suborder Tubulifera

There is no projecting ovipositor, and the Suborder gets its name from the tubular tenth abdominal segment. Eggs are therefore not inserted into plant tissues, but glued (often in clusters) onto the epidermis. There are virtually no veins on the wing membrane and microtrichia are absent. The Family is unusual in having an additional pupal instar.

7.6.2.1 Family Phlaeothripidae

The wheat thrips (*Haplothrips tritici*) is restricted to the Poaceae; it feeds on the ovaries of wheat flowers leading to sterility and even abortion. Not only is yield affected by this damage, but also the quality of the flour for baking. Flecking on the grain follows, but from the milky ripe stage the insect moves to the pericarp where it causes a brown stain, which can lead to rejection by processors, especially for pasta. The thrips has one generation a year, emerging in spring and ovipositing on the very young ear still developing within the wheat tiller. The nymphs descend to the soil and enter a summer diapause, emerging still as nymphs in early October to overwinter in stubble and dry grass (including thatch). Pupation occurs the following March. The adult female is about 1.5 mm long and black except for light tips to the tibia and the tarsus of the foreleg. The species occurs across Europe, much of Asia and South Africa.

The Family also contains *Aleurodothrips fasciapennis*, which is an important biological control agent of scale insects on citrus in China.

8 Subclass Pterygota, Division Endopterygota, Lesser Orders

8.1 Introduction

We now move on to the Endoterygota, which (Chapter 3) are the 'higher insects' that show a complete metamorphosis with a resting pupal stage between the wingless larva and the adult. As described in Chapter 3, adults and larvae may have quite different mouthparts and may occupy different niches, both in terms of food and habitat.

The Endopterygota includes four very large Orders, the Lepidoptera (butterflies and moths), Diptera (true flies), Hymenoptera (ants, bees and wasps) and the largest of all, the Coleoptera (beetles). Each of these large Orders has its own chapter, and the other and smaller Endopterygote Orders are grouped together in this chapter. Apart from the Strepsiptera (which may be related to the Coleoptera), the others (Mecoptera, Siphonaptera, Trichoptera and Neuroptera) are appropriately grouped together, since they are regarded as the most primitive of the Endopterygota. Rather like the Palaeoptera of the Exopterygota (Chapter 5), their wings tend to have more veins and cross veins than other endopterygote Orders (though the Siphonaptera are apterous).

8.2 Order Mecoptera (scorpion flies) – *c.* 400 described species

These attractive insects have patterned wings with a complex venation, and an obvious long downwardly pointing 'beak' at the tip of which are the biting mouthparts (Fig. 8.1a). The common name of the Order refers to the abdomen of the male, which ends in genitalia forming a bulb very reminiscent of the sting of a scorpion. The female abdomen ends in a simple tapered ovipositor. The scorpion fly *Panorpa communis* is commonly seen in Europe in April, feeding on nectar from the umbels of hedge parsley, though the adults are mainly carnivorous on aphids and other small insects. The larvae

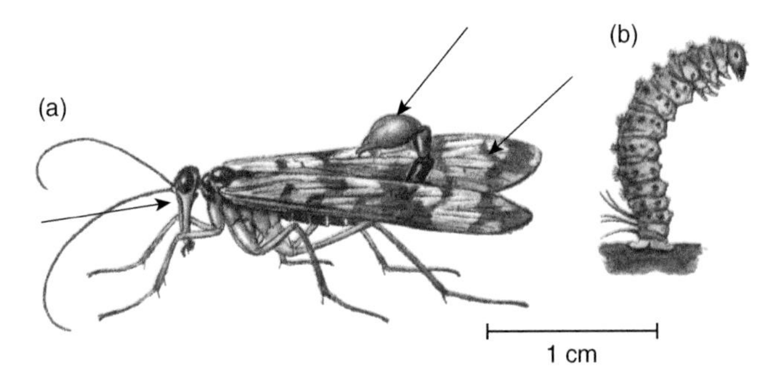

Fig. 8.1 Mecoptera (*Panorpa communis*): (a) adult and (b) larva (from Mandahl-Barth 1974, with permission).

Handbook of Agricultural Entomology, First Edition. H. F. van Emden.

of scorpion flies are also carnivorous. They live in damp soil and moss and the larvae of some species are 'caterpillar-like' (Fig. 8.1b) in that they have abdominal feet rather like the prolegs of Lepidoptera larvae. Pupation also occurs in the soil and the pupae are exarate (i.e. 'free' with the adult appendages dangling and not soldered down to the pupal case (see Section 3.3.2).

8.3 Order Siphonaptera (fleas) – *c.* 2000 described species

These small leathery bilaterally flattened parasitic insects are wingless (Fig. 8.2), and for much of the time that insects have been studied were therefore placed in the Subclass Apterygota. Indeed in the early 1950s, long after their true position in the Endopterygota had been recognised, they were still to be found among the Apterygota in the public display of insects in the British Natural History Museum. Only about 100 years ago were flea larvae identified as such, and they closely resemble some dipteran larvae in appearance. Of course, it had long been known that some Diptera (the keds in the dipteran group Pupipara, Section 10.4.2.3) were wingless leathery flattened blood-sucking parasites like fleas, but the connection had probably not been made as keds are dorsoventrally flattened whereas fleas are flattened laterally; moreover, there was at that time no evidence that fleas had an endopterygote life cycle. So the Order was then placed close to the Diptera, but more recently both genetic and morphological evidence place the Siphonaptera near the Order Mecoptera. Some entomologists actually feel that fleas should not have Order status, but be a Subfamily of the mecopteran Family Boreidae.

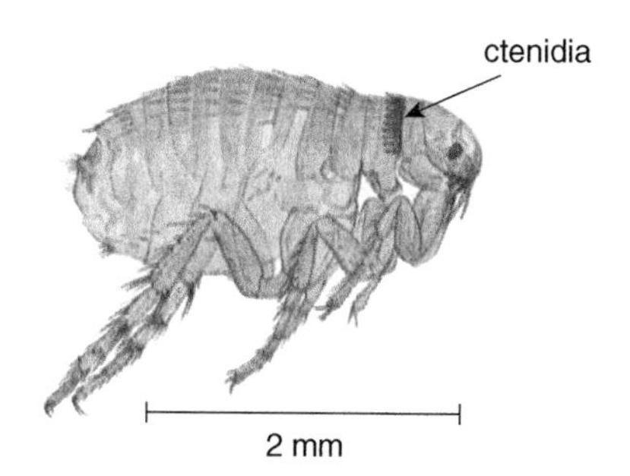

Fig. 8.2 Dog flea (*Ctenocephalides canis*) (from Zanetti 1977).

'Siphon' is a tube, and 'aptera' means 'wingless'. This seems an apt description of this Order, which has lost the wings of other Endopterygota (i.e. like worker ants, fleas are 'secondarily wingless') and have mouthparts adapted for sucking blood.

Adult fleas are blood-sucking external parasites on hosts with fur or feathers (or clothing!), as they would find it hard to attach to or move around on bare skin. Most are less than 5 mm long. The hind legs are adapted for jumping with enlarged coxae, enabling the flea to jump (usually with a somersault) over 50 times its body length. Many have large hard bristles (***ctenidia***) forming a structure that looks like a comb on the head and thorax. The antennae lie in grooves (***antennal fossae***), which divide the head into anterior and posterior regions.

The larvae are segmented, with a few obvious bristles and biting mouthparts (Fig. 8.3). Like dipteran larvae, they are apodous with a much reduced head. They feed on

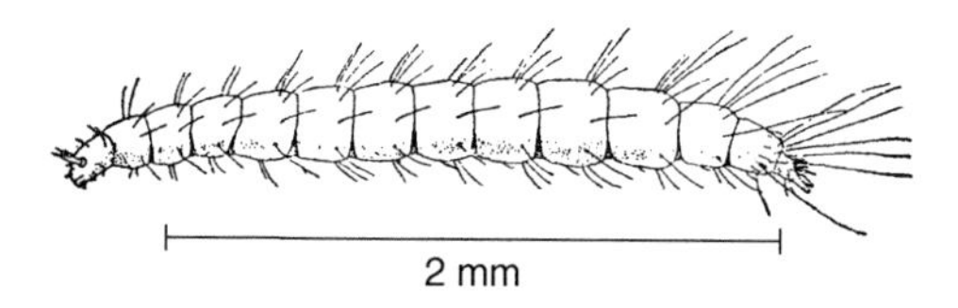

Fig. 8.3 Flea larva (the head is to the left) (from Richards and Davies 1977).

organic matter, including remains of blood in the faeces of the adult, in the nests or bedding of their hosts (where the eggs are also laid), but also under fitted carpets. Therefore they are not parasites of mammals such as ungulates (e.g. deer) which have no regular nest sites, but occur very frequently on rodents (mice, squirrels, rabbits etc.). Pupation (exarate) occurs in a little silken cocoon at the larval feeding site. The life cycle can take less than a month.

The adult females need a blood meal to reproduce. Some then delay the reproductive cycle till it is triggered by hormones from a pregnant female host (e.g. a rabbit in its burrow); this insures a generation of fleas can develop on young hosts before these leave the nest. Then they carry the fleas to new nest sites.

Adult fleas can live away from a host without a blood meal for many months. I can recall returning home after 7 months away, during which time the cats had been kennelled. On entering the house, a myriad of jumping black spots, sensing the entry of a potential host, were bouncing towards me from the edges of the fitted carpets!

Fleas can give an irritating bite, but they can also transmit diseases. The bubonic plague (a bacterium), which killed nearly half the population of Europe in the Middle Ages, was primarily spread by the oriental rat flea (*Xenopsylla cheopsis*). The myxomatosis virus of rabbits is vectored by the rabbit flea (*Spilopsyllus cuniculi*), and cat and dog fleas (respectively *Ctenocephalides felis* and *C. canis*) are intermediate hosts for a tapeworm (*Dipylidium caninum*) that infects both these hosts but also humans. Thus although some species are very host specific, many are not. Cat and dog fleas both attack cats, dogs and humans.

The long lifespan of adult fleas led to the Victorian sideshow known as the 'flea circus'. These circuses were principally staffed by mole fleas (*Hystrichopsylla talpae*) as these at 5.5 mm are a bit larger than other fleas and also do not jump. Although 'jumping' would seem a desirable attribute for a circus performer, it would actually have created a problem in a flea circus. This is because the 'circus act' – such as an upside-down flea juggling a small cotton wool ball with the feet or fleas pulling a paper carriage – depended not on training the fleas, but merely cleverly harnessing them with fine wires. Then, lowering a lamp over the fleas stimulated undirected activity with its warmth, and the harness automatically created the desired performance. As the tradition began to die out (partly because of better hygiene and therefore fewer potential performers available from landladies of seaside guest houses), flea circus proprietors became an increasingly elderly profession. What was noticeable, however, was that they always seemed to be with very young wives. Perhaps this was connected with the maximum flea longevity that can be obtained by feeding them on young female blood?

8.4 Order Neuroptera – *c.* 4700 described species

This Order surprisingly has no popular common name; however, sometimes the translation of Neuroptera, that is 'net-veined insects', is used. They do indeed have many longitudinal and cross veins giving a net-like appearance, especially as veins often double up by forking near the wing margin. Except in the Coniopterygidae, there is a 'ladder' of cross-veins at the front margin (Fig. 8.4a). The wings, which are large and mainly transparent (though coloured patches are common in some groups), are held roof-wise along the sides of the body when at rest. The adults have long simple antennae and biting mouthparts, but they often feed on sources of carbohydrate such as aphid honeydew. The larvae throughout the Order are voracious predators (Fig. 8.4b) with large forwardly pointing mouthparts (= prognathous). The pupae are exarate.

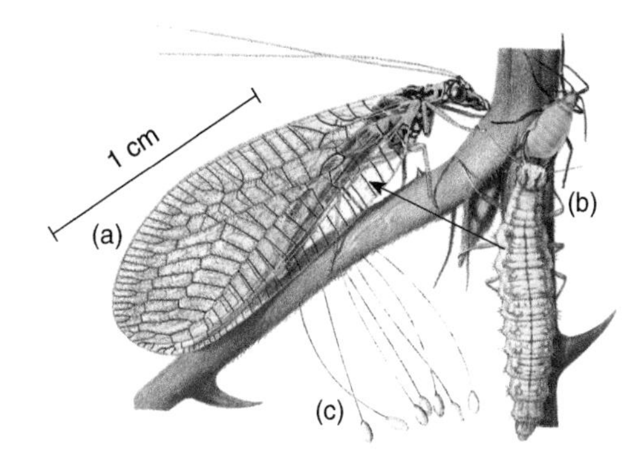

Fig. 8.4 Green lacewing (Chrysopidae) (from Zanetti 1977): (a) adult with ladder of cross-veins at leading edge of forewing arrowed; (b) larva with sickle-shaped jaws; (c) stalked eggs.

8.4.1 Suborder Megaloptera

The wing veins do not bifurcate at the wing margins (Fig. 8.5a) or, if they do, the prothorax is elongated (Fig. 8.6a). The larvae have biting mouthparts (contrast the Suborder Planipennia, Section 8.4.2). The Suborder comprises three Families.

8.4.1.1 Family Corydalidae

The eggs are laid on vegetation or stones near water, in masses of up to 2000–3000 per mass. The larvae move to water and are aquatic with abdominal gills but also possess posterior spiracles, which allow them to live for long periods under stones when the shallow fast-flowing streams they inhabit dry up.

8.4.1.2 Family Sialidae (alder flies)

As with the Corydalidae, the eggs are laid in masses (smaller, up to 500 eggs per mass) on vegetation and stones near water, and the hatching larvae make their way to the water. In the case of the alder flies, this water is slow-flowing streams and lakes with a

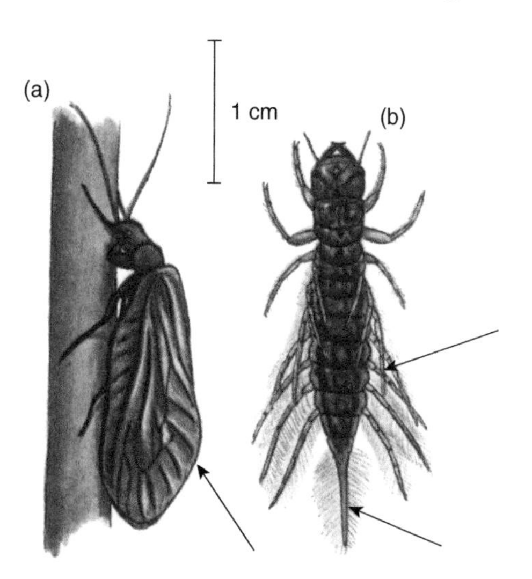

Fig. 8.5 Alder fly, *Sialis lutarea*: (a) adult and (b) larva (from Mandahl-Barth 1973, with permission).

muddy bottom, where the larvae (Fig. 8.5b) prey on other insects and small worms with their sickle-shaped toothed mandibles. The abdominal gills are long and segmented, with a similar central terminal filament at the end of the abdomen. The larvae burrow into the mud to pupate but, when ready for the adult to emerge, the exarate pupae swim to the water surface. Here the adult emerges, often to fall prey to a waiting swallow hawking low over the water. In the UK, alder flies have one generation a year. The adults (Fig. 8.5a) are dark brown with very obvious dark brown wing venation. The wing veins do not fork profusely at the hind margin of the wings.

8.4.1.3 Family Raphidiidae (snake flies)

The adults are easily identified by the elongated prothorax (Fig. 8.6a). The prothoracic legs are right at the back end of the segment, which holds the head rather higher than the rest of the thorax giving the insects a characteristic posture. The larvae (Fig. 8.6b) are predaceous on other insects, and are found under tree bark, particularly in oak and pine woods.

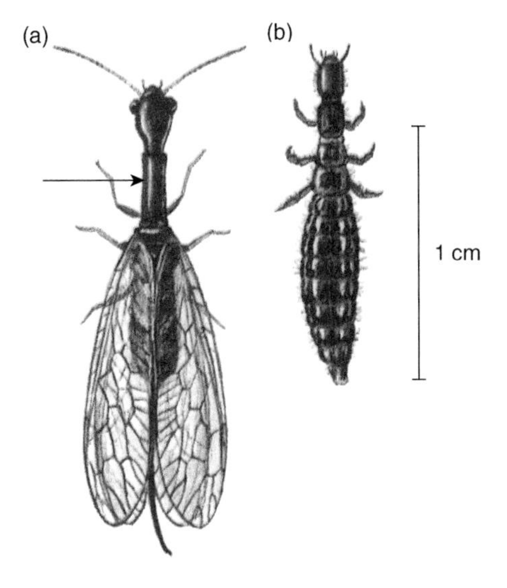

Fig. 8.6 Snake fly (Raphidiidae): (a) adult female and (b) larva (from Mandahl-Barth 1974, with permission).

8.4.2 Suborder Planipennia

The wing veins bifurcate at the wing margins (Fig. 8.4a) except in the Coniopterygidae. The larvae have long suctorial mandibles. The more important Families are described below under their Superfamilies.

8.4.2.1 Superfamily Coniopterygoidea

Family Coniopterygidae

These are very distinctive as adults, being very small and covered with white waxy particles. Wing venation is reduced compared with other Neuroptera, and the veins do not fork at the hind margin of the wings. These insects are usually found in woodland, where the larvae feed on small prey such as aphids, scale insects and mites.

8.4.2.2 Superfamily Osmyloidea – three Families

Family Osmylidae

The adults, found near shaded streams, are quite large and often have beautiful wings with coloured patches. The antennae are rather slender. The larvae are amphibious predators on small invertebrates and are usually to be found among mosses and stones. Their long slender mandibles are peculiar in being curved outwards.

8.4.2.3 Superfamily Mantispoidea – four Families

Family Sisyridae

These small brown insects have a characteristic wing venation – the subcosta and vein R1 are fused together near the tip of the forewing. Eggs are laid in batches, protected with a silk cover, on the underside of leaves that overhang water, and the hatching larvae simply drop into the water beneath. They are aquatic with tracheal gills below the abdomen in the later instars and are predatory on freshwater sponges (*Spongilla* spp.). They have extremely long slender mandibles, which are held more or less parallel to each other until the tips are reached which then (as in the Osmylidae) diverge.

Family Mantispidae (mantis flies)

The mantis flies are so called because they really do look very much like miniature praying mantids with their raptorial forelegs (Fig. 8.7). They catch prey just like the praying mantids, but are active at night. They have a long prothorax like the snake flies, but the forelegs articulate behind the head rather than at the back end of the segment.

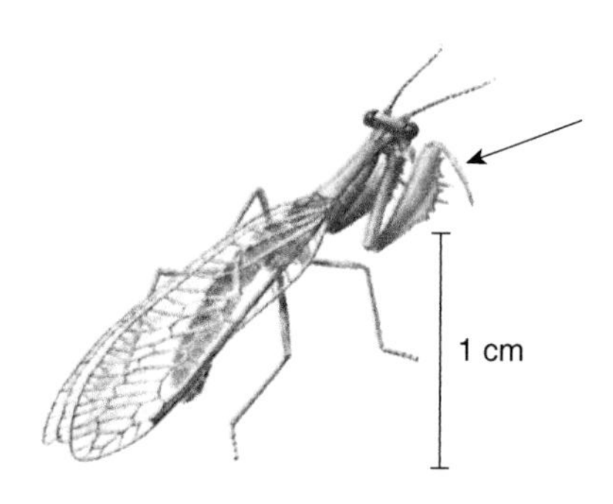

Fig. 8.7 Mantispid adult (from Chinery 1986).

Mantispids occur mainly in the tropics and subtropics, extending to around the Mediterranean. I remember an evening trapping insects at lights on a farm in Queensland and being both delighted and astounded by the vast numbers of mantispids, which I had never seen outside museum collections before!

The eggs are stalked like those of lacewings. The larvae enter the cocoons of wolf spiders and eat the young within, but some species enter the nests of social Hymenoptera and feed on the larvae.

8.4.2.4 Superfamily Hemeroboidea (lacewings) – three Families

Family Hemerobiidae (brown lacewings)

Brown lacewings (Fig. 8.8) are small brownish insects, with often hairy bodies. Some of the cross veins to the costal margin of the wing are forked. Brown lacewings occur on trees and both the adults and larvae are predators on aphids.

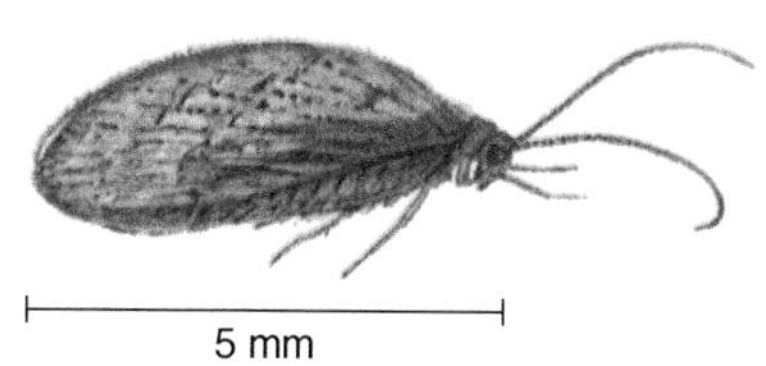

Fig. 8.8 Brown lacewing (Hemerobiidae) (from Mandahl-Barth 1974, with permission).

Family Chrysopidae (green lacewings)

The adults are quite large, beautiful and delicate creatures with bright green bodies and wing veins, and large protruding golden eyes (Fig. 8.4a). The wings have a wide area between the front margin and the first of the few longitudinal veins; in this area there are many parallel cross-veins. The adults are weak fliers and feed on a variety of foods. Many feed on pollen and nectar; others supplement this by also feeding on aphids. The eggs are defended from predation by being laid on the end of 1 cm long stalks of mucus (Fig. 8.4c), which harden on exposure to air and hold the eggs away from the leaf surface. The hatched larvae use these stalks to find their way to the leaf.

The larvae (Figs 8.4b and 8.9) are voracious predators of aphids and small caterpillars. They have stiff hairs on which some species impale their exuvia and sucked-out prey as camouflage. They are considered valuable in the biological control of aphids, especially the widespread and common *Chrysoperla carnea*. Indeed, lacewing larvae can be purchased from producers for release on crops. Plants that are especially rich sources of pollen are sometimes planted at the edges of crops to attract lacewings.

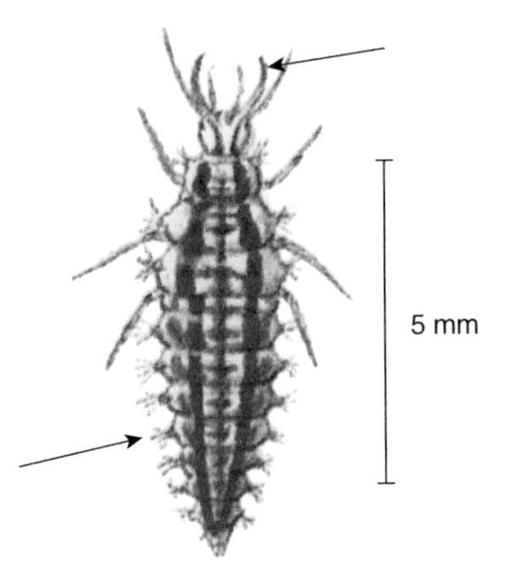

Fig. 8.9 Larva of green lacewing (from Mandahl-Barth 1974, with permission).

8.4.2.5 Superfamily Myrmeleontoidea (ant lions) – three Families

These are interesting predators; the larvae (Fig. 8.10a) of many of which excavate a pit in sand and then bury themselves at the bottom with their jaws protruding. Small insects such as ants slip into the pit and slide down to the jaws (Fig. 8.10b), often helped by the larva using its jaws to throw sand at its victim. Other species merely search on the ground for (or ambush) their insect prey.

The adults are quite large with slender long bodies and large wings (Fig. 8.11). They can thus be mistaken for dragonflies, but differ in having obvious clubbed antennae.

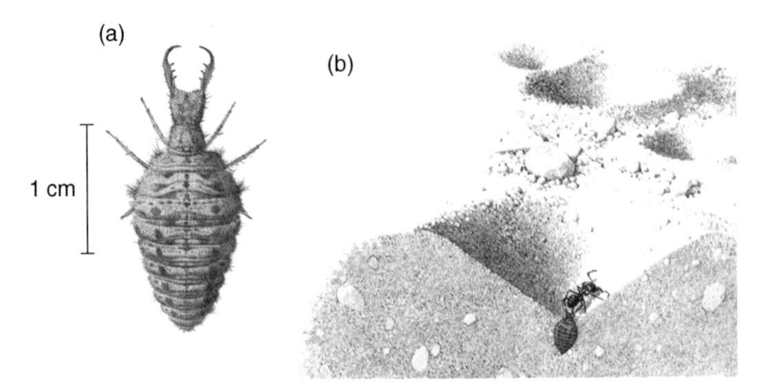

Fig. 8.10 Larva of ant lion (Myrmeleontidae): (a) larva and (b) larva in pit capturing an ant (from Zanetti 1977).

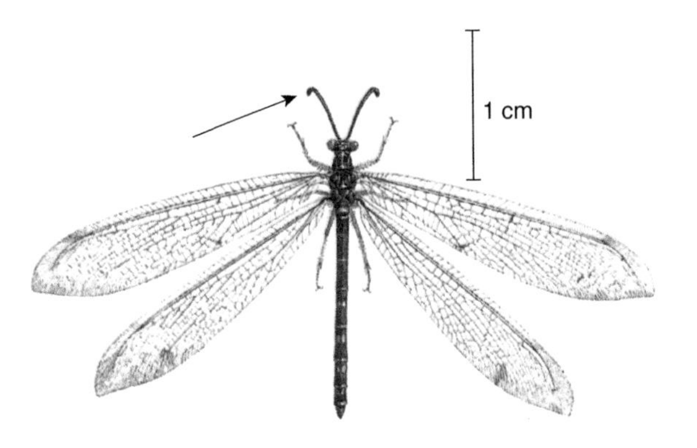

Fig. 8.11 Ant lion adult (Myrmeleontidae) (from Zanetti 1977).

They are often attractive insects with patterned wings, and some (the Family Nemopteridae) are distinguished by their strange hind wings (Fig. 8.12), in the form of a long narrow strap which may have an expanded oval apical tip.

Myrmeliontoidea are mainly tropical insects, though a number of species are found in southern Europe.

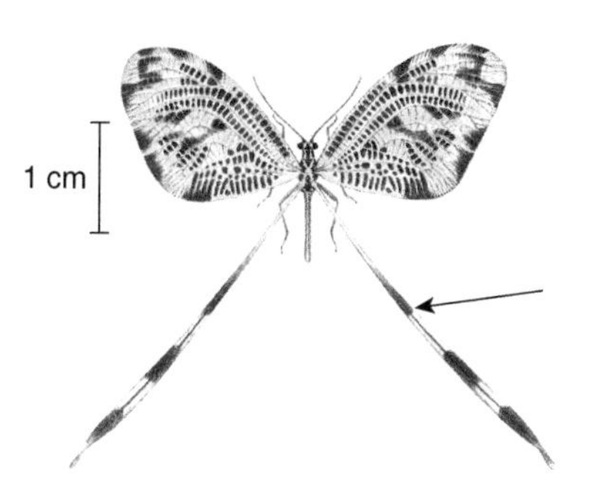

Fig. 8.12 Ant lion adult (Nemopteridae) (from Chinery 1986).

8.5 Order Trichoptera (caddis flies) – *c.* 5000 described species

The adults are weak fliers found near water, the habitat of the larvae. Flight is often nocturnal. The adult mouthparts are adapted for licking fluids and have atrophied mandibles. Adults have rarely been observed feeding. They look a bit like small brown moths (Fig. 8.13), but inspection of the wings will reveal that they carry many hairs, which have not been flattened into scales as in the Lepidoptera. Moreover, the coiled proboscis of the Lepidoptera is clearly absent. Another character is that the forewings are harder in texture than the hind wings. The antennae are long and filiform. The Order comprises 12 Families.

Caddis flies lay their eggs on aquatic vegetation near the water surface, or in mucilage-covered masses on twigs of trees overhanging the water, in which case the emerged first instars drop into the water below. The larvae are therefore aquatic, breathing with long filamentous gills along the body. They are soft bodied except for the heavily sclerotised head, thoracic plates, legs and the hooks at the tip of the abdomen (Fig. 8.14a). They protect their soft bodies from predation by making a case, in which they live, holding on with the hooks and projecting their head and thorax to move around and feed. Caddis cases are beautifully constructed and the main charm of the Order. They are based on a silken cylinder spun from glands in the head and opening on the labium. The cylinder is strengthened and camouflaged by attaching natural materials available in the stream. To some extent the materials used for the case (e.g. stones, leaf portions, small sticks) are indicative of the Family. Thus the Family Limnephilidae usually uses leaf portions and sticks, while the Molannidae use sand grains of mixed sizes (Fig. 8.14b–f). The case is changed by building a new one at every moult. By contrast, the Hydroptilidae live dangerously without cases until the final nymphal instar, when they construct a seed-like silk case with sand or plant particles. Different Families have different habitats from fast-flowing mountain streams to standing water in ponds and lakes. Caddis larvae are largely phytophagous, and so their only recorded pest status is to

Fig. 8.13 Adult caddis fly (Trichoptera) (from Zanetti 1977).

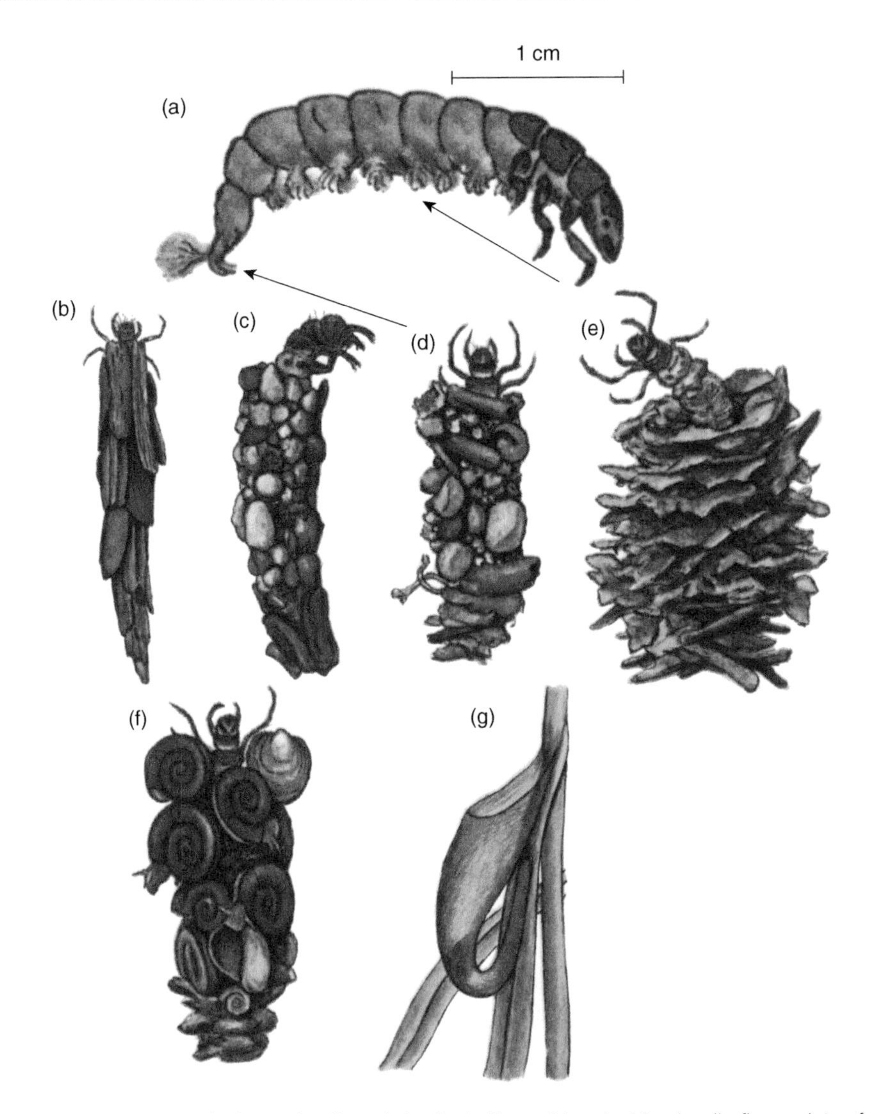

Fig. 8.14 (a) Caddis fly larva showing abdominal gills and terminal hooks; (b–f) a variety of caddis larva cases; (g) net for catching prey constructed by caddis larva (compiled from Mandahl-Barth 1973, with permission).

damage watercress. However, larvae in some Families are carnivorous, and may construct silk nets (Fig. 8.5g) to catch their prey.

The exarate pupae remain in the case until ready to emerge, when they leave the case and swim to the water surface, where the pupa hangs for the adult to emerge and settle on vegetation on the bank.

8.6 Order Strepsiptera (stylops) – *c.* 550 described species

The Order Strepsiptera derives its name from the Greek 'strepsi'=twisted and 'ptera'=winged, after the twisted way the hind wing of the male is held at rest. Indeed the Order used to have the common name 'twisted wing parasites', but now the usual common name is 'stylops' after the main Family, the Stylopidae.

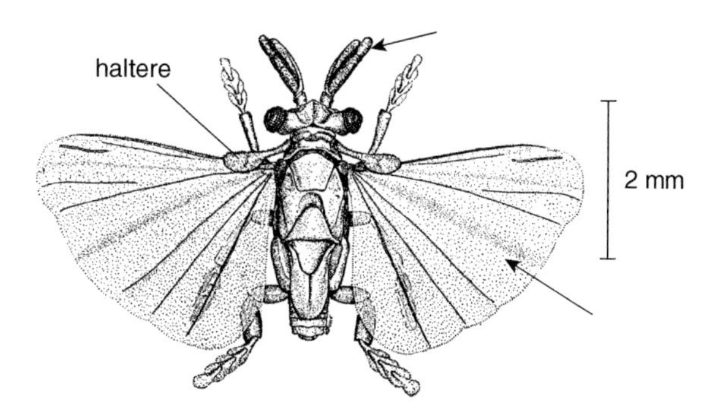

Fig. 8.15 Male stylops (from Richards and Davies 1977).

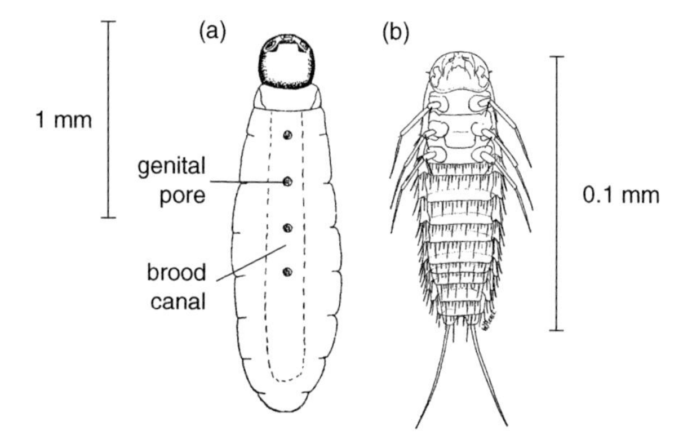

Fig. 8.16 (a) Adult female stylops and (b) ventral view of triungulin larva (from Richards and Davies 1977).

The Strepsiptera are morphologically rather strange obligate parasitoids of other insects. Their taxonomic position in the Endopterygota is still a matter of debate. The obvious connection would appear to be with the Families Meloidae and Ripiphoridae in the Coleoptera (see Section 2.3.11.1, Superfamily Melooidea, Family Meloidae and Family Ripiphoridae). Evidence for this relationship includes the greatly reduced elytra, the branched antennae of the adult male (Fig. 8.15) and the strange phenomenon of 'hypermetamorphosis' during the larval instars, with an active 'triungulin' first instar changing into a very different second instar. However, molecular evidence dos not confirm this, but neither does it suggest affinity with any of the other endopterygote Orders.

Strepsiptera are cosmopolitan internal parasitoids of six other Orders of insects, the Thysanura, Orthoptera, Dictyoptera, Hemiptera, Hymenoptera and Diptera.

The adult female (Fig. 8.16a) is neotenous in form, that is it retains larval characteristics rather than developing adult ones. It is both apterous and apodous, with no eyes or antennae. Females never leave the host in which they developed, and may occupy almost the entire volume of its abdomen. An exception here is females of the Family Mengenillidae, which parasitise Thysanura. In this Family the females are free living and have short legs and antennae. In the other Families the anterior part of the female protrudes from the body of the host. The winged male (see later in this section) breaks into the opening of a canal (the ***brood canal***) and injects his sperm.

One host may harbour several females, which may each lay several thousand eggs. These hatch within the haemocoel of the female and move around there, behaviour which is unique to the Strepsiptera. They are active larvae, and the only life stage (except in the Family Mengenillidae, see paragraph above) other than the male to have legs. These legs are unusual in lacking trochanters. The abdomen has long setae of over a third the length of the body. These first instar larvae (Fig. 8.16b) are known variously as 'triungulins' or 'planidia', and they leave the mother to escape to the outside along her brood canal. They have mandibles, antennae and simple eyes with a single lens. The sternites and tergites have serrated edges, which are regarded as aids for attachment to vegetation and new hosts.

With limited food reserves, these larvae have to find a new host quite quickly. The triungulins of those species that parasitise bees and wasps will climb to a flower to await a host arriving to feed there or collect pollen or nectar. Other species crawl around near the ground. Those larvae lucky to encounter a suitable host jump onto it and burrow into the body, first softening the cuticle at that point with enzymes they secrete. One species in Papua New Guinea is known to enter its orthopteran host through the latter's tarsus! The triungulin then quickly moults into a different form (hypermetamorphosis, as mentioned earlier), a second instar larva with no distinct head, sense organs or legs. In response to the invasion, the host makes a sac-like structure within which the larva feeds and grows, protected from the host's immune defence chemicals by the sac. They undergo four further instars, but do not discard the old cuticles, which form layer on layer around the insect. There is therefore moulting without ecdysis. At the final larval moult, the larva-like females are formed without any pupal stage, but males do undergo a pupal stage from which they emerge as winged adults and escape from the host. Infected insects are said to be 'stylopised'. They are often not killed, but their abdomen changes colour and they may have their sexual organs destroyed, that is they are sterilised by castration. In some cases, there is a reversal of sexual dimorphism, whereby female hosts have some of the secondary sex characters of males.

In the Family Mengenillidae, both male and female larvae pupate outside the host and emerge as free-living adults, though only the males are winged (see earlier in this section).

The males live for only a few hours and do not feed. They are quite remarkable insects in appearance (Fig. 8.15), but are rarely seen in the field. I have only ever seen a single specimen, which I found in one of the water traps I used in my PhD studies over 50 years ago!

The head of the male is small, but carries quite obvious multisegmented antennae, which are strangely branched, a type of antenna termed '***flabellate***'. The compound eyes project from the head and the few large ommatidia give the whole eye an appearance very reminiscent of a tiny raspberry. The forewings are reduced to tiny halteres, while the hind wings are very large and spread out like a fan. As in the triungulin larva, the trochanter is absent, and the number of tarsal segments varies from two to four with Family (the tarsus of the main Family, the Stylopidae, has four segments).

Strepsiptera are in nature too rare to have a significant impact on the populations of their hosts, some of which are important crop pests; these include the brown planthopper *Nilaparvata lugens* of rice and another rice planthopper, *Sogatella furcifera*, as well as a number of important grasshopper pests of oil palm. Yet this Order has received scant attention as a source of potential biological control agents. One attempt was made in Papua New Guinea in the late 1990s with the release against a long-horned grasshopper in oil palm of 2000 stylopised hosts in each of several months. There is, however, no report on the outcome, which suggests it was not successful.

9 Subclass Pterygota, Division Endopterygota, Order Lepidoptera (butterflies and moths) – *c.* 100,000 described species

9.1 Introduction

Largely because of their size and attractive coloured and patterned wings, the butterflies and moths are among the insects best known to the public. The caterpillars are also often obvious as pests in the garden, so again bring themselves to the layman's attention. However, the colourful variation in appearance of adult Lepidoptera hides a low diversity in lifestyle; the majority are rather unspecialised herbivores, though the Order does include some hidden feeders such as root grazers, leaf miners and stem/fruit/pod borers. As pointed out in Chapter 1, coping with the problems of herbivory is unusual in insects, and so this one Order – being dedicated to plant feeding – is of immense importance in terms of damage to crops. Some species specialise on feeding on stored plant products such as flour and grain.

An important characteristic of many species is migration. Many butterflies we sometimes see in the UK have migrated from the continent, and others (including a number of pest moth species) migrate annually between Europe and Africa. Thus they find food in the southern hemisphere when winter grips the north. Perhaps the most famous annual migration of Lepidoptera is that of the monarch butterfly (*Danaus plexippus*) which passes in dense clouds through Mexico on its way between hemispheres in the New World. The key to these migrations is that natural selection favoured those genotypes that took off at either end of the migration path at times when winds and vertical air movements would predictably lead them to the alternate location. Particularly the vertical movements near the equator result in the insects 'changing' at certain points along the route into air streams moving in a different direction.

Lepidoptera means 'scaly winged', and the wings (like also the body and legs) are mostly covered in scales formed from flattened hairs (Fig. 9.1). These scales are hollow and pigmented. They are sculptured with longitudinal ridges, and this leads to iridescence when the ridges are less than 1 μm apart. It is the scales that give the wings of Lepidoptera their colour and patterns; the scales easily brush off to reveal normal transparent insect wings with their venation. The venation pattern is often valuable for identification, but understandably many collectors are reluctant to damage their specimens by removing the scales. Much identification, particularly of butterflies, is therefore based on comparisons with illustrations.

The whole concept of a distinction between 'butterflies' and 'moths' has no real scientific basis, especially as the hobbyist collector then excludes a large number of

Handbook of Agricultural Entomology, First Edition. H. F. van Emden.

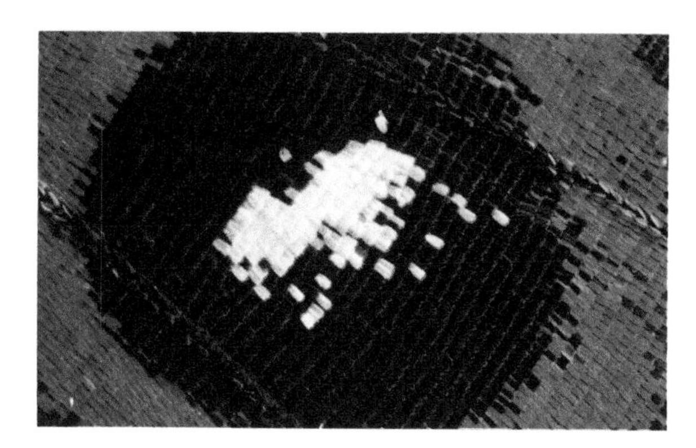

Fig. 9.1 Enlargement of an 'eye spot' on the wing of a butterfly, showing how the colour pattern derives from the different overlapping scales (from Zanetti 1977).

moth Families altogether under the blanket term Microlepidoptera (which I flippantly define as 'too small to be retained in a butterfly collector's net'). Butterflies are the Lepidoptera that are day active, where the males have normal slender slightly clubbed antennae and the wings are held vertically when at rest (revealing on the outside the underside of the hind wings), although they are spread-eagled out sideways when the butterflies are basking in sunshine. By contrast, moths are night active, the males may have elaborate pectinate antennae (to detect the sex pheromone of the female and find her at night) and the wings are held out sideways at rest or reflexed over the abdomen. By these criteria, the Family Hesperiidae (the skippers) are included as butterflies, though by more diagnostic criteria the entomologist would classify them as related to obvious 'moths'. Moreover, the applied entomologist cannot afford to ignore the 'Microplepidoptera', for these include a large number of important pest species.

The caterpillars have biting mouthparts for chewing plant tissue, but the adults of the major Suborder (the Glossata) sip the nectar of flowers through a coiled ***haustellum*** formed by the hooking together of the two maxillary galeae (Fig. 9.2). This haustellum is uncoiled by muscle pressure. The effect resembles the way a coiled squeaker used at parties unrolls into a straight tube when we blow into it and the way that relaxing the pressure immediately re-coils the tube. Some Lepidoptera have a really long haustellum when extended to reach the nectar at the base of flowers with exceptionally long throats. One hawk moth in Australia, which feeds on a particularly long-throated flower growing on granite pavements, has a haustellum some 20 cm long. In some fruit-feeding adult moths (in the Family Noctuidae), the haustellum is hardened for piercing the fruit. Several forms of coupling the wings are found in the Lepidoptera, particularly the jugum and frenulum (see Chapter 2).

Adult Lepidoptera may show considerable sexual dimorphism. This is familiar to butterfly collectors in relation to species such as the orange tip (only the male has an orange tip to the wing) and the Family Lycaenidae (the blues) where only the males are blue and the females are a dark brown. In some moths (e.g. the winter moth – see Section 9.2.3.8, Family Geometridae) the females are completely wingless.

The production of sex pheromones has already been mentioned, as have the pectinate antennae of the males to detect them. Both males and female Lepidoptera produce

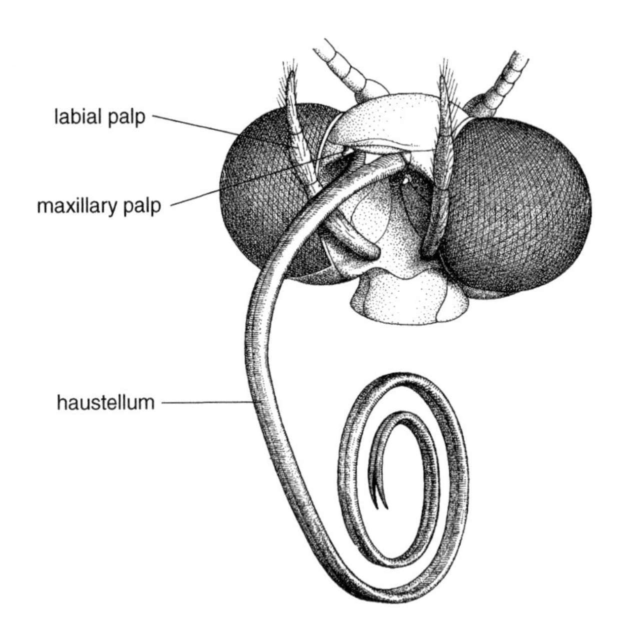

Fig. 9.2 Mouthparts of Lepidoptera (from Zanetti 1977).

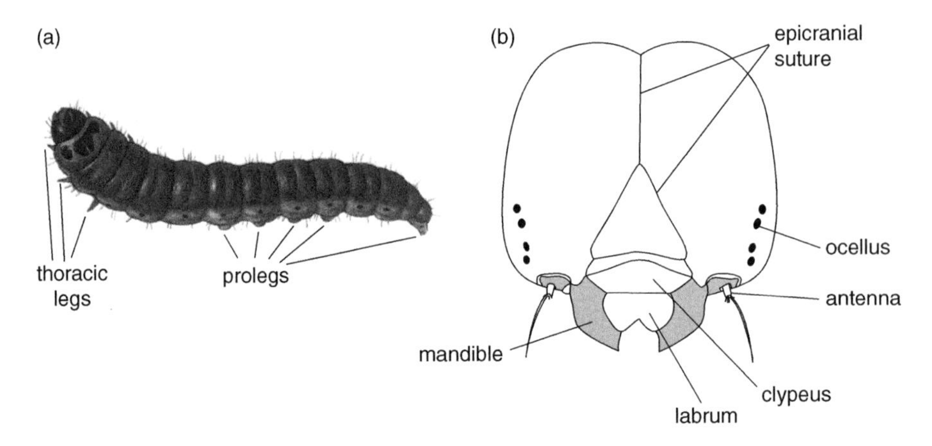

Fig. 9.3 (a) Lepidopteran caterpillar (from Mandahl-Barth 1974, with permission); (b) head of a caterpillar, dorsal view.

scents, the females from extrudable brushes (the ***penicilli***) on the abdomen and the males from specially modified scales (***androconia***). Some moths can be heard to 'squeak' in alarm, and some species possess auditory structures.

Eggs are laid usually in batches on the food. The larva of the Lepidoptera (Fig. 9.3) is popularly known as the 'caterpillar' (from 'cat with hairs' in Old French); as pointed out above, it has normal biting mouthparts and is usually the damaging stage. The ocelli are few and arranged as a well-spaced semicircle, and the epicranial suture is usually clearly visible. The thorax carries the three pairs of 'thoracic legs', so designated because the abdomen also carries 'legs', but of a very different structure. These paired abdominal legs are called ***prolegs*** and are fleshy unsegmented and unsclerotised

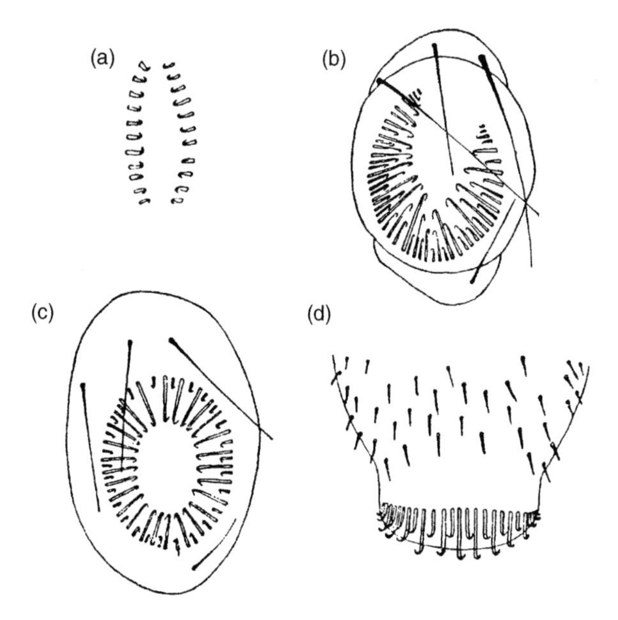

Fig. 9.4 Some crotchet patterns on the end of the prolegs of Lepidoptera (from Richards and Davies 1977): (a) uniordinal uniserial bands; (b) multiordinal uniseries; (c,d) biordinal uniseries.

tubular structures. To provide grip, the ends of the prolegs are equipped with scelorotised hooks called ***crotchets***. Spelled the same way as the musical note of that name, the meaning has to be clear from the context! The prolegs occur on abdominal segments 3–6 and 10, except that the 'looper' caterpillars of the Family Geometridae (see Section 9.2.3.8, Family Geometridae; Fig. 9.21) have prolegs only on segments 6 and 10. Three characteristics of the pattern of the crotchets are valuable aids to classification. Firstly, is the pattern a complete circle or only a curve? Secondly, are the crotchets all of the same size (***uniordinal***) or are there two (***biordinal***) or more (***multiordinal***) distinct sizes? Thirdly, is there one (***uniserial***) or are there more (e.g. ***biserial, multiserial***) rows of crotchets? Examples of combinations of sizes and number of rows of crotchets are shown in Fig. 9.4.

Caterpillars have a variety of defences against predators. Species feeding on plants containing toxins deterrent or poisonous to predators may be especially conspicuous, 'advertising' that they are distasteful with warning colouration (reds, blacks and yellows). Very hairy caterpillars tend to be avoided by vertebrate predators. Other caterpillars can release repellent odours from eversible forked structures (***osmeteria***) between the prolegs or at the back of the head. At the other extreme, palatable caterpillars may use camouflage to escape detection, for example some closely mimic small twigs.

Most caterpillars are able to spin silk from their silk glands. The silk emerges from the mouth of the caterpillar and movements of the head and forelegs are used to create the silken structure required. This may be a 'rope' for lowering the caterpillar to the soil to pupate or a 'lifeline' to enable it to regain the leaf after having dropped off as a defence reaction. Many caterpillars pupate in a silken cocoon for protection; indeed, many species remain as inactive caterpillars within their cocoon for what may be many

Fig. 9.5 Lepidopteran pupa (from Zanetti 1977).

months (e.g. during the winter) and only pupate a relatively short time before they emerge as adults. Some species, especially when the caterpillars are still very small, use their silk to make little 'kites' so that they can float in the wind and disperse the species (see winter moths, Section 9.2.3.8, Family Geometridae).

The pupae (Fig. 9.5) are unusual among insects in being obtect. Obtect pupae (see Section 3.3.2) do not have the adult antennae, wings and legs hanging free (exarate pupae), but have them 'soldered' into the smooth outline of the pupa (often called the 'chrysalis' in the Lepidoptera). Caterpillars that neither pupate inside a cocoon, nor burrow into the ground to pupate, spin a silk pad on a substrate such as a twig and attach themselves to this with hooks (***cremaster***) at the end of the chrysalis.

The Lepidoptera are divided into four very unequal Suborders. The three most primitive Suborders, the Zeugloptera, the Aglossata and the Heterobathmiina have chewing mouthparts like caddis flies and a jugum as the type of wing coupling. They will not be mentioned further. Below, I have selected the more important Superfamilies and Families in the Suborder Glossata as far as the applied entomologist is concerned.

9.2 Suborder Glossata

The Superfamilies can themselves be grouped into higher taxa called 'Infraorders'. The Infraorders Dacnonypha and Lophcoronina need not concern us further. With the Infraorder Exoporia (as in the more primitive Suborders mentioned above) the size, shape and venation of fore and hind wings is similar.

9.2.1 Infraorder Exoporia

9.2.1.1 Superfamily Hepialoidea

Wing coupling is achieved very simply by a simple overlap of the edges of the wings (the jugum mechanism).

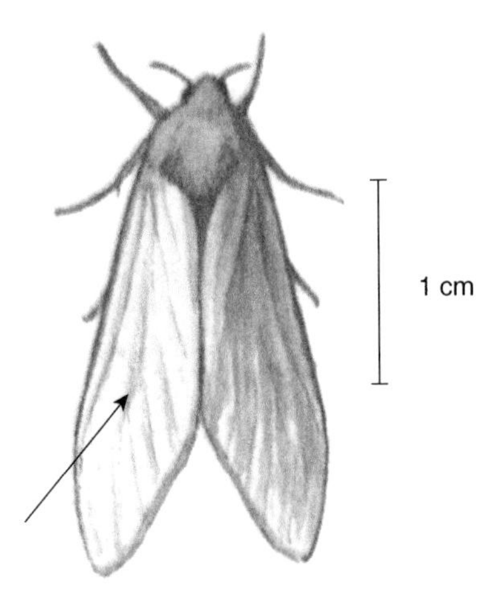

Fig. 9.6 Swift moth (Hepialidae) (from Lyneborg 1968, with permission).

Family Hepialidae (swift moths)

Swift moths (*Hepialus* spp.) are of intermediate size and most species have opaque white scales (Fig. 9.6). The crotchets on the prolegs of the caterpillar form a multiserial circle.

The adults drop their eggs onto the ground while in flight; one female may lay 800 eggs. The hatching caterpillars burrow into the soil and graze on roots. The caterpillars are white and quite large (about 2 cm when fully grown) with a distinctive brown head capsule. They can take 2 years to mature, and continue feeding during the winter. Pupation is in spring in an earthen cell beneath the damaged plants.

Though swift moth caterpillars are notoriously polyphagous, the eggs are frequently laid in grassland and cereal fields. They can thus become seedling pests of wheat and pests of permanent grassland (not only grassland for grazing, but also lawns and golf courses). The larvae are also found tunnelling into potato tubers, and in horticultural crops they eat at the base in the centre of lettuce and strawberry plants. The pest occurs throughout Europe and into the Near East.

9.2.2 Infraorder Heteroneura – Division Monotrysia

In the Heteroneura (99 per cent of the Lepidoptera and the remainder of this chapter) the venation of forewings and hind wings is markedly different (Hetero-neura = different veins). In the first Division of the Heteroneura, the Monotrysia, the abdomen of the female has just one sexual opening, for both mating and oviposition.

9.2.2.1 Superfamily Nepticuloidea

Small moths with a wing span rarely over 6 mm. A spot character is the expansion of the first antennal segment (*eye cap*; Fig. 9.7) over the top of the eyes in front view.

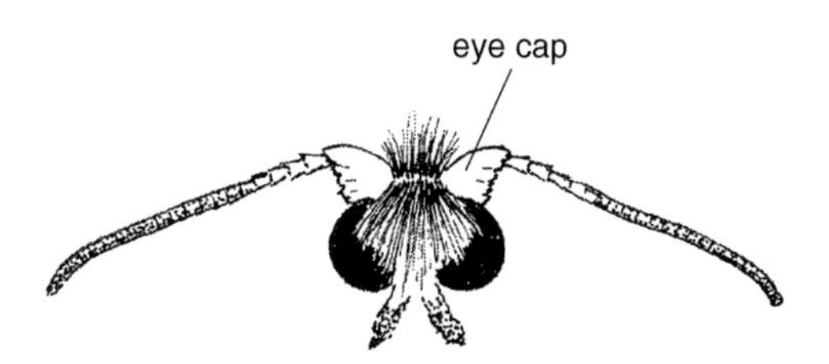

Fig. 9.7 Head of a nepticulid moth (from Alford 1999, with permission).

Family Nepticulidae

These small to minute moths have a metallic sheen. The wings have long hairy fringes and the caterpillars have very small thoracic legs and prolegs; the meso- and metathoracic legs verge on the vestigial. Most species are leaf-miners and a few, for example *Stigmella malella* (apple pygmy moth), may become minor pests.

9.2.2.2 Superfamily Incurvarioidea

There is no eye cap, the wing span is at least 8 mm.

Family Incurvariidae

Another Family of metallic moths, but easily recognised by the often exceptionally long antennae. The young caterpillars are often leaf miners, otherwise the caterpillars move around in cases made of leaf fragments. Again, some species may be occasional pests, for example *Lampronia rubiella* (raspberry moth).

9.2.3 Infraorder Heteroneura – Division Ditrysia

The females have two sexual openings on the abdomen, one for mating and the other for oviposition.

9.2.3.1 Superfamily Tineoidea

The proboscis is not scaly, but the head sometimes has rough scales.

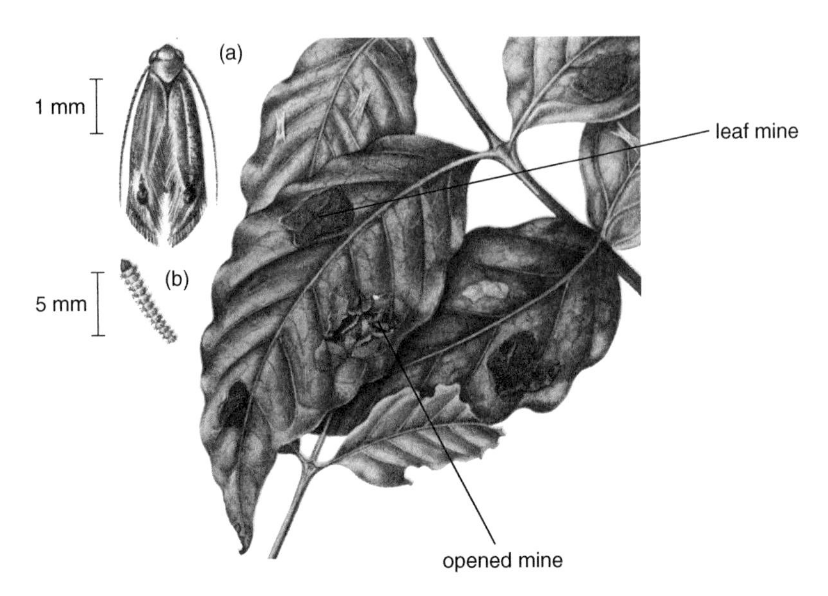

Fig. 9.8 Coffee leaf miner (*Leucoptera* sp.): (a) adult and (b) caterpillar (from Bayer 1968, with permission).

Family Gracillariidae

This is a Family of leaf miners. In the Subfamily Gracillariinae, species of *Leucoptera* are known as coffee leaf miners (Fig. 9.8) but also live on wild shrubs in the same plant Family (Rubiaceae) as coffee. Mined leaves are shed prematurely and the level of defoliation can be significant. Silver eggs are laid on the upper surface of leaves and, as is typical of many leaf miners (both in the Lepidoptera and Diptera) the hatchling bores into the leaf through the base of the egg (thus avoiding any insecticide on the leaf surface!). Further feeding results in brown irregular blotches with the upper and lower leaf surfaces still intact; the mines of several caterpillars may fuse to one larger blotch. The mature caterpillars cut their way out of the mine and descend to the ground on silken threads, where they pupate in a little white cocoon spun on the surface of a fallen dead leaf. The small (about 3 mm long) white moths emerge after 1 or 2 weeks, with the whole life cycle taking about 4–6 weeks; in most places where the moth is found (different species occur in Africa and Central and South America) breeding is continuous.

Phyllonorycter blancardella forms blotch mines on apple leaves, and the caterpillars pupate within the mine. This miner can be damaging in Europe and North America. It has two generations a year; the second generation overwinters as a pupa within fallen leaves.

In contrast with the blotch mines of the Gracillariinae, the Subfamily Phyllocnistinae comprises leaf miners that make serpentine mines. These grow in width as the caterpillar grows, and the faeces are usually visible as a dark line down the centre of the mine. The Family is also unusual for the Lepidoptera in that the caterpillars are apodous. The citrus leaf miner (*Phyllocnistis citrella*) badly twists young leaves, and damaged leaves eventually desiccate. Damage can be serious on mature citrus plants if the infestation is heavy, but young plants are particularly vulnerable. The mature caterpillar leaves the mine to pupate in a turned over pocket at the edge of the leaf. The life cycle takes about 3 weeks. The adult is a small (2–3 mm long) white moth with four black stripes on the forewing and a feathery hind wing; it is distributed throughout India, south-east Asia and Japan.

9.2.3.2 Superfamily Gelechioidea

The proboscis is fairly densely covered with scales at its base.

Family Oecophoridae

The spot character for this Family is the ***pecten***, a tuft of hairs at the base of the antenna. The caterpillars conceal themselves by spinning together leaves and flower heads. Sometimes the attacked plant is a crop species, for example parsnip is attacked by two oecophorids, *Agonopterix nervosa* (carrot and parsley flat-body moth) and *Depressaria pastinacella* (parsnip moth).

Family Gelechiidae

The forewings are characteristically trapezoidal with a narrow fringe (Fig. 9.9). This Family includes several important pests.

Pectinophora gossypiella (pink bollworm) is one of several important lepidopteran pests of cotton (Fig. 9.9). As the name suggests, the caterpillars are pink in colour though closer inspection shows there are pink and white longitudinal bands. By contrast, the adults are a dull brown; their wing span is about 2 cm. Pink bollworm is found in all tropical and subtropical regions of the world. The eggs are laid singly near a bud/boll, and the caterpillars feed on these organs externally, but bore into

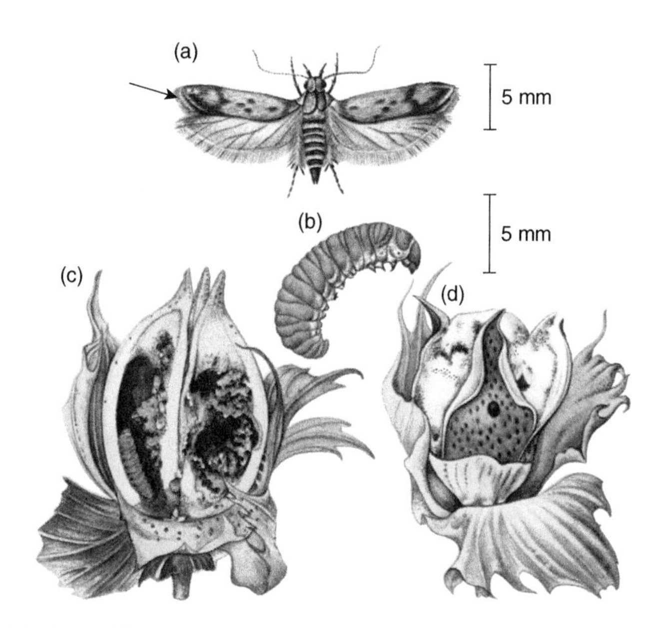

Fig. 9.9 Pink bollworm (*Pectinophora gossypiella*): (a) adult; (b) mature caterpillar; (c) damage to interior of boll; (d) emergence hole of mature caterpillar (all from Bayer 1968, with permission).

bolls that are older, leaving little trace of their entry as they enter close to the base. However, the mature caterpillar (about 1 cm in length) leaves the boll through an obvious circular exit hole. It then falls to the ground and pupates in litter. There is an adult preoviposition period of about 4 days, giving a generation time between 1 month and 7 weeks. Yield loss occurs when bolls are entered; such bolls fail to open properly and become invaded by rots. Pink bollworm is potentially a very serious cotton pest. The life cycle can, however, be broken with a mandatory 'closed season' when no cotton is grown; the problem is that in many countries this measure is not strictly enforced.

The peach twig borer (*Anarsia lineatella*) occurs throughout Europe, the Near and Middle East, North Africa and the USA. It can cause quite serious damage to shoots and fruit of peach and several other stone fruits. Damage commences in the spring, when tiny reddish caterpillars that have overwintered under silk on the bark start to feed by boring into the twigs and buds on new growth. Such feeding stops the growth of the shoot and stimulates bushy lateral growth. The life cycle is only a few weeks, and soon the caterpillars pupate in a silk cocoon on the bark to emerge about 2 weeks later and lay eggs of the second generation. The caterpillars of this generation feed on both twigs and in fruit; with the latter damage, gum is left by the exit hole. The third and fourth generations find the stem tissues have now hardened, and feed almost entirely in fruits.

Phthorimaea operculella (potato tuber moth) caterpillars bore into potato tubers, but also attack tomatoes, aubergines and other Solanaceae; the stems and petioles of tobacco may also be invaded. The forewings are a spotted greyish-brown with a span of about 15 mm, while the hind wings are a greyish white. Eggs are laid singly on the undersides of leaves in the field, and on tubers in storage. The first instar caterpillars hatching on leaves feed as leaf miners. The silver blotch mines are diagnostic. The caterpillars then bore into veins and via these move down the plant, first in the leaf petiole, then in the stem, and they may even reach the tuber of potatoes. Pupation takes place in a cocoon in litter on the soil surface in the field, or in the tuber. A generation takes 3–4 weeks, and in warm climates there can be as many as 10–12 generations a year. Tunnelling in petioles and stems causes the plants to wilt, and the bored tubers may contain caterpillars and/or become rotted by invading pathogens. The moth is also a storage pest as generations continue in store, and then transport of the stored product may result in the movement of the pest to new areas.

Although a pest, the species additionally performs a useful function in crop protection because the caterpillars are easily reared in large numbers and are acceptable as prey when rearing many polyphagous predators in biological control programmes. Many predators will accept the boiled larvae, which can be stored in bulk at low temperatures.

Another gelechiid moth that can be a pest in both field and store is the Angoumois grain moth, *Sitotroga cerealella*. This attacks wheat and maize, but usually only in store, as with its attacks on dried fruit; it is also a field and storage pest of sorghum and other grains. The pink eggs are laid on the grains within which, in the field, the caterpillars first develop at the 'milky' stage. They pupate under a thin 'window' made at the surface of the grain. The emerging moths, in the field or by then in the store, leave by this window, leaving it attached at one point as a flap – a diagnostic identification of the pest. In the store, breeding can then be continuous with a generation time of about 5 weeks.

Family Coleophoridae (casebearer moths)

These are narrow-winged moths; indeed, the hind wings are narrower than their fringes. The moths are recognisable from the way they hold their antennae, sticking out forwards when at rest. The young leaf-mining caterpillars emerge to move around in cases made from silk and leaf fragments and often cigar or pistol shaped. The only large genus is *Coleophora*, which includes *C. anatipennella* (cherry pistol casebearer moth) and *C. spinella* (apple and plum casebearer moth).

9.2.3.3 Superfamily Yponomeutoidea

Family Plutellidae

The economic importance of this Family is dependent on just one insect – one of the world's top ten. This is the diamond-back moth (Fig. 9.10), *Plutella xylostella,* the common name deriving from the way the wavy patterns on the leading edge of the forewings abut when the wings are at rest to form diamond shapes. The wing span is less than 2 cm. The caterpillars are voracious destroyers of crops in the cabbage family. Although the insect occurs in the UK, it rarely warrants control since there is usually effective biological control and the few generations in a year prevent population explosions. However, in warmer climates it can destroy whole brassica fields, and crops can hardly be grown without heavy insecticide use (when the pest quickly shows tolerance to the chemicals). The pest is found on brassicas everywhere, from New Zealand to Norway.

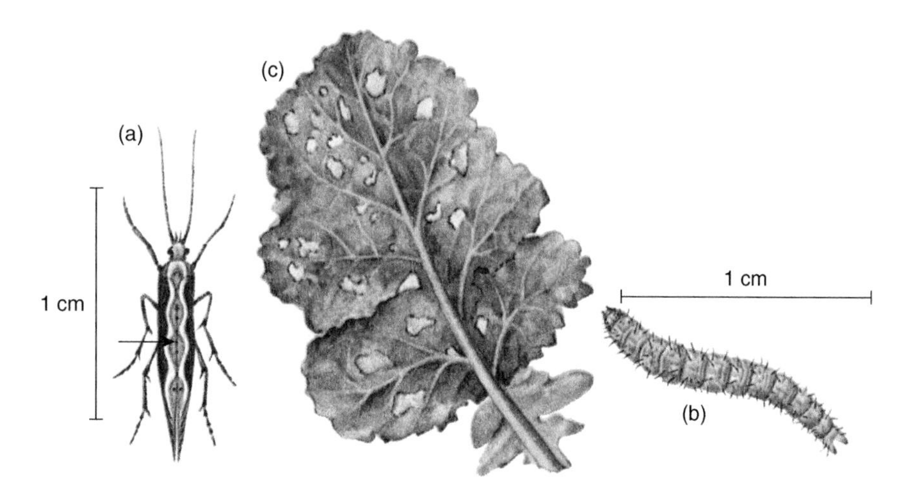

Fig. 9.10 Diamond-back moth (*Plutella xylostella*): (a) adult; (b) fully grown caterpillar; (c) damage (all from Bayer 1968, with permission).

The tiny yellow eggs are laid singly or in small groups on both surfaces of the leaves, and the hatching caterpillars burrow into the leaf on the underside as leaf miners to make little transparent blisters. From the second instar onwards the uniformly green caterpillars are normal defoliators, but on brassicas with thick leaves the caterpillars may not eat through the entire leaf, but leave the upper epidermis uneaten as transparent 'windows'. This makes such attack identifiable as caused by *Plutella* rather than by other cabbage caterpillars. Caterpillars, if disturbed, wriggle and then often drop off, hanging by a silken thread that they then use to regain the leaf. The caterpillars cocoon themselves on the underside of the leaf before they later pupate inside their cocoon. The whole life cycle takes between 3 and 7 weeks, depending on temperature.

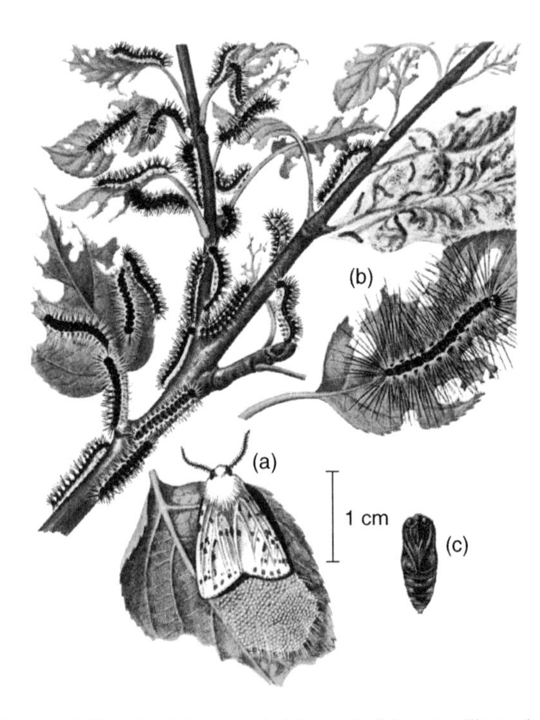

Fig. 9.11 An ermine moth (*Hyphantria cunea*): (a) adult; (b) caterpillars skeletonising leaves and web with young caterpillars; (c) pupa (all from Bayer 1968, with permission).

Family Yponomeutidae

Ermine moths in the genera *Yponomeuta* or *Hyphantria* (e.g. the fall webworm *Hyphantria cunea*; Fig. 9.11) are sporadically serious pests when high populations develop. The grey black-spotted caterpillars are gregarious, and live in colonies under very obvious silken tents, frequently seen in early summer on hawthorn, but also on apple and plum. In years of high abundance of the insects, the tents join up and can form impressive structures covering whole branches or more. Such tented colonies have developed from caterpillars that hatched in August and September the previous year from eggs laid on branches. On apple, the first instars make brown blister mines, but leave these when they next moult. Caterpillars pupate within the tents at the end of June and the moths which have a 2.5 cm wing span of white forewings with small black spots – hence 'ermine' moths – emerge in July and August. Ermine moths are univoltine, and are found in Europe, Asia, Australasia and the USA.

9.2.3.4 Superfamily Pyraloidea

This Superfamily often has the side and hind margins of the wings deeply frayed, and long spurs on the legs at the junction of tibia and tarsus.

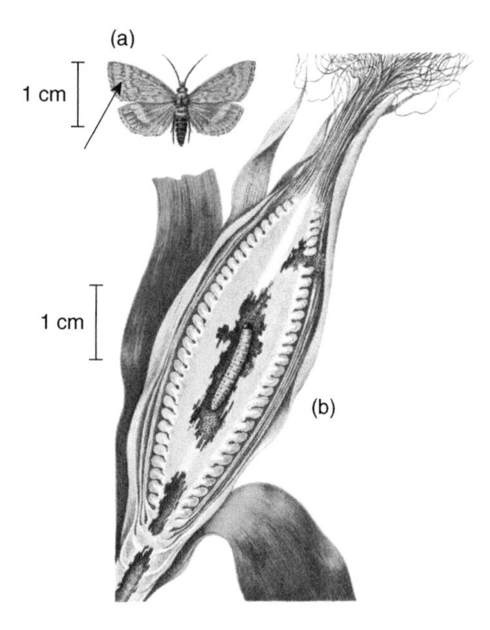

Fig. 9.12 European corn borer (*Ostrinia nubilalis*): (a) adult and (b) damaged maize cob with caterpillar (from Bayer 1968, with permission).

Family Pyralidae

A serious pest of maize in mainland Europe and North America (where it is particularly damaging) is *Ostrinia nubilalis*, the European corn borer (Fig. 9.12). Eggs are laid in clusters on the undersides of leaves where the larvae initially feed, making characteristic elongated holes ('windows') before entering the stem and hollowing out the internodes (sometimes the cobs are also attacked). They overwinter (in diapause) in the stems (if left as stubble) and pupate there or in the soil in spring. This is the normal univoltine pattern, though in warmer climates the species may have two or even three generations. The adults have 1–1.5 cm long buff forewings with dark wavy lines as well as increasing darkening to the leading edge and to the wing tips. In spite of its common name, the moth is actually extremely polyphagous, and has alternative hosts as taxonomically unrelated as sorghum, beet, tomato and soybean.

Many of the important stem-borers of rice are in the Pyralidae, especially *Chilo supressalis* (Fig. 9.13), *C. partellus, C. polychrysus, Tryporyza incertulas* and *T. innotata.* These species may also attack other cereals (especially maize), wild grasses and sugar cane. The eggs are laid in masses on the underside of the basal half of the leaf blades. The young caterpillars move down to the leaf sheaths, where they feed but leave one epidermis to cause 'windows'. The sheaths and attached leaf blades subsequently go yellow and wither. After a few days the caterpillars bore into the stem and feed there, tunnelling until they reach a node. Here they feed more extensively and this leads to stem breakage at nodes. Although some caterpillars eat through the node, others emerge from the stem at that point and enter another internode, sometimes on a different tiller. 'Deadhearts' (the withering of a whole tiller) and 'whiteheads' (the shrivelling of a rice inflorescence) are common symptoms of damage by stem borers. Most

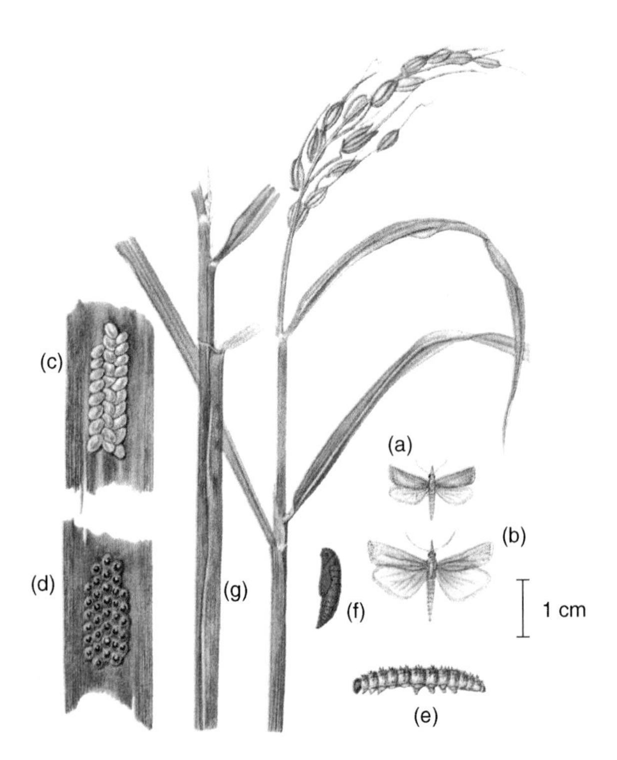

Fig. 9.13 Striped rice stem borer (*Chilo suppressalis*): (a) male; (b) female (dark yellow form); (c) newly laid eggs; (d) eggs ready to hatch; (e) caterpillar; (f) pupa; (g) damage to stem (left) and whitehead symptom (right) (all from Bayer 1968, with permission).

of the species listed pupate within the stems. The length of the life cycle varies with species, but is often in the order of 1–2 months.

Another pyralid rice pest is the rice caseworm, *Nymphula depunctalis* (Fig. 9.14). The name caseworm derives from the case within which the caterpillar lives as it feeds on lower leaf surfaces at or below water level in the paddy; the case is changed as the caterpillar grows. The case is made from leaf tips cut by the insect, and older plants with much of their photosynthetic leaf area well above water level are unlikely to suffer more than slight damage. Seedlings, however, can be so badly damaged that they die. Eggs are laid singly on the leaves, and the caterpillars can live underwater for long periods as they are equipped with filamentous gills on the side of the abdomen. The pupa is formed inside a case on the rice stem, usually just above the water. The life cycle takes about 1–2 months.

The sugar cane borer (*Diatraea saccharalis*) and the sugar cane stalk borer (*Eldana saccharina*) are two pyralids which may also attack cereals, but their pest status mainly derives from their damage to sugar cane. The former is a New World pest, while the latter is a pest in Africa. *Diatraea*, like most of the other pyralid stalk borers above, lays its eggs in the leaf sheaths, whereas *Eldana* oviposits right at the base of the plant and indeed mainly on the soil. Again like many other stalk borers, the young caterpillars feed on the leaf sheath before entering the stem. *Diatraea* bores into internodes near where the eggs were laid; usually only one caterpillar survives per internode.

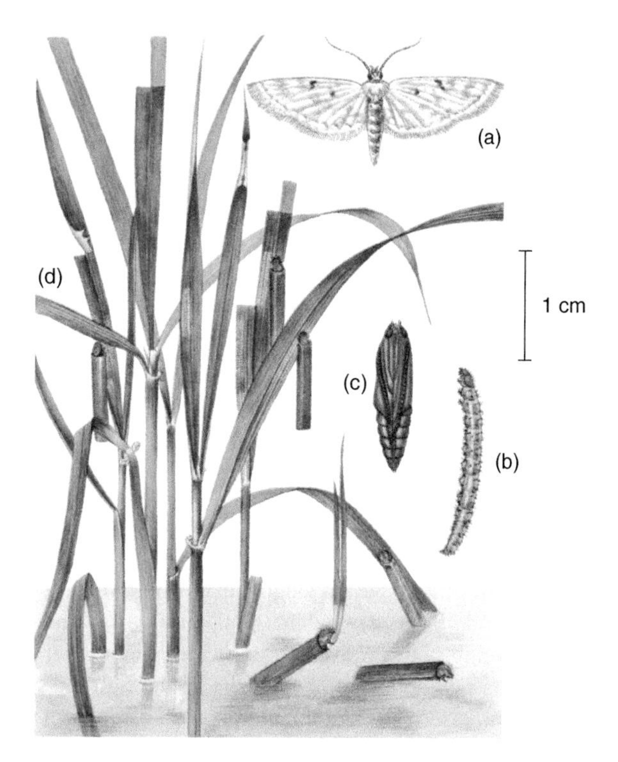

Fig. 9.14 Rice caseworm (*Nymphula depunctalis*): (a) adult; (b) caterpillar; (c) pupa; (d) caterpillars in cases and feeding injury (all from Bayer 1968, with permission).

Eldana caterpillars tunnel mainly low down in the stem, except when infestation is heavy. Young plants may die, and in any case the central shoot is severed, leading to the 'deadheart' symptom. Both species pupate in the plant; the life cycle takes about 6 weeks. The forewings of the two species are rather different. Those of *Diatraea* are yellow-brown with some narrow brown lines, while those of *Eldana* are pale brown with two dark spots near the centre.

Just as the above pyralids are specific to cereals/grasses, so other species are specific to legumes. An important pest of many grain legumes in the tropics and subtropics is the mung moth *Maruca vitrata*. The moth has a wingspan of 2–3 cm; the forewings are brown with two or three irregular white marks, while the hind wings are a light grey with a wide brown border. Eggs are laid singly in flowers and on buds and pods, particularly where parts of the plant are in contact. With cowpeas, the pods of most varieties hang down, and the ends of pods borne on a peduncle are often in contact and provide a specially preferred site for oviposition. The caterpillars feed on leaves and flowers, though most damage is done when the pods are invaded and the seeds eaten. The caterpillars rarely destroy all the seeds within a pod, but appear to damage just a few seeds before they leave the pod at night and bore into a new pod. This behaviour has an important consequence, which also applies to other pod borers such as the Lima bean pod borer (see below in this Section) and pod borers such as *Helicoverpa armigera* in the Family Noctuidae (see Section 9.2.3.12). Since not all the seeds in a pod are damaged with species which move between pods during development, measuring attack by the number of bored pods can be misleading. I have seen worthwhile yields of pigeon pea in India obtained from fields with over 90% bored pods.

The cotton leaf roller (*Haritalodes derogata*), as probably the commonest leaf-feeding caterpillar on cotton, has the potential to be a serious pest in Africa, India and south-east Asia, but is usually kept in check naturally by parasitoids. The cream-coloured adults have brown wavy streaks on both wings, and a span of about 3 cm. Eggs are laid singly or in small groups on the leaves and the larvae, after feeding for a while at the leaf margins, aggregate and create a leaf roll spun together with silk. Although initially each roll may have as many as ten feeding caterpillars, they later leave to make individual rolls at the leaf edges. Pupation is in the roll or on the ground; the life cycle takes 4–5 weeks. The leaf-roll symptom is obvious, but affected leaves also hang down and premature ripening of the bolls occurs.

The Lima bean pod borer, *Etiella zinckenella* (Fig. 9.15), is another important pyralid, attacking many different legumes in the tropics. The adults have a brown forewing only about 30 mm long with a single curved pale vertical band. Eggs are laid in small groups on the young pods, and the young caterpillars enter the pod and then feed on the seeds. Caterpillars also move within the pod to attack fresh seeds, and cannibalism is common. Older ones may even leave the pods where they have fed and penetrate perhaps several other pods before they leave the pods and drop to the ground to pupate in the soil.

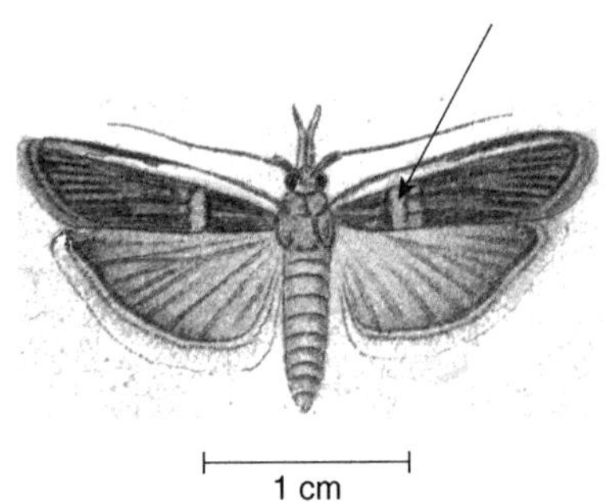

Fig. 9.15 Lima bean pod borer (*Etiella zinckenella*) (from Spuler 1910).

The genus also includes three important pests of stored products. Not only is the stored product eaten, but there is also much fouling caused by cast skins, corpses, faeces and webbing between the seeds/fruit which can cause problems with the handling of the stored material. These three pests are the Mediterranean flour moth (*Ephestia kuehniella*), the warehouse moth (*E. elutella*) which is mainly a pest of stored flour or grain, and the almond moth (*Cadra cautella*). The last named is mainly, though not exclusively, a pests of grains such as wheat and maize in store, while *E. elutella* is more polyphagous and attacks dry stored cocoa beans, legume seeds, nuts, dried fruit and tobacco. The adult forewings span about 25 mm and are grey with a pair of vertical bands. The caterpillars of the two species are very similar, with a dark brown head and a pale body with some dark spots. In wheat, the caterpillars preferentially eat the embryo (germ) of the grains. Pupation occurs in crevices in the store structure or between storage sacks. Both species will breed throughout the year. *Cadra cautella* is cosmopolitan and can be very serious. Moths will even lay eggs through any holes in storage sacks. The forewings of the adult have rather unclear vertical bands and the life cycle is relatively long at 6–8 weeks. By contrast, the bands on the forewing of *E. elutella* tend to be more clearly defined and the life cycle is only about a month. Although cosmopolitan, this species is more a problem in temperate than tropical countries.

Another stored products pest is the Indian meal moth, *Plodia interpunctella*. This attacks stores of flour and other milled cereals, but also dried fruits and nuts. As with

Ephestia and *Cadra*, direct damage is aggravated by webbing and frass. The Indian meal moth is a major pest in warm countries with a life cycle there of less than a month and often eight generations a year. However, it also survives when imported into temperate countries as long as neither humidity or temperature in the store dip too low; there will then be only one or two generations per annum. The forewings have a cream-coloured proximal third separated from a coppery apical part by a dark band; the wing span is about 2 cm.

9.2.3.5 Superfamily Sesioidea

Family Sesiidae (clearwings)

As the common name suggests, scales are largely missing from all wings (Fig. 9.16). The edges of the wings are often lined by a broad dark band, and the abdomen ends in a diagnostic tuft of scales. The antennae are often widened out at the tip. The caterpillars feed concealed inside stems or other plant organs.

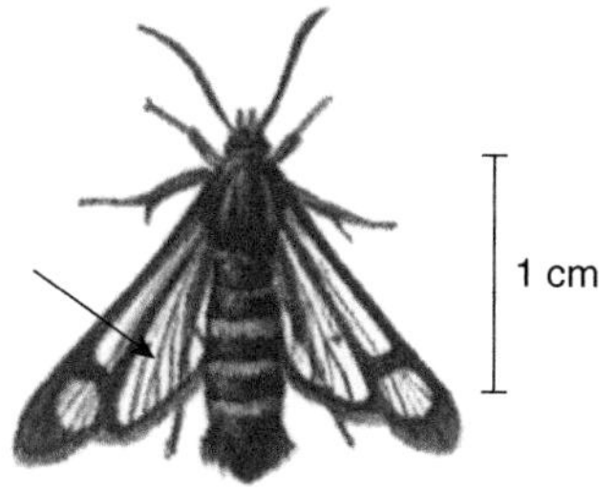

Fig. 9.16 A clearwing moth (Family Sesiidae) (from Mandahl-Barth 1974, with permission).

Synanthedon dasysceles is the sweet potato clearwing of East Africa. The wing span is only about 2 cm, and the moths are a dark brown with a yellow stripe down the midline of the abdomen in dorsal view. Eggs are laid on the stems and petioles, and the caterpillars then enter stems and tunnel downwards till they reach the base of the stem, which becomes swollen. The caterpillars may tunnel further down and get into the sweet potatoes. They can then become pests in storage; the stem attack rarely warrants control. Pupation is in the feeding tunnels.

In Europe, the USA and Australia, the stems of blackcurrant (redcurrant and gooseberry less so) are tunnelled by *Synanthedon tupuliformis*, the currant clearwing. Eggs are laid singly on stems in early summer, and the caterpillars enter the stem and tunnel upwards in the pith of the stem towards younger wood. They continue feeding throughout the winter and pupate in their tunnel around April. The leaves on infested stems wilt beyond the tunnel, and fruit does not develop. The clearwing moths are univoltine.

9.2.3.6 Superfamily Cossoidea

Family Cossidae

The Cossidae are large moths with a wing span of several centimetres; the larvae are also large and are timber borers. With the low nutritive value of wood, development usually has to extend over several years, One of the most familiar cossids in the UK is the goat moth (*Cossus cossus*), the bright red caterpillars of which can reach 10 cm, and which tunnel in branches of willow. This can cause major damage to the trees, with quite large branches breaking off.

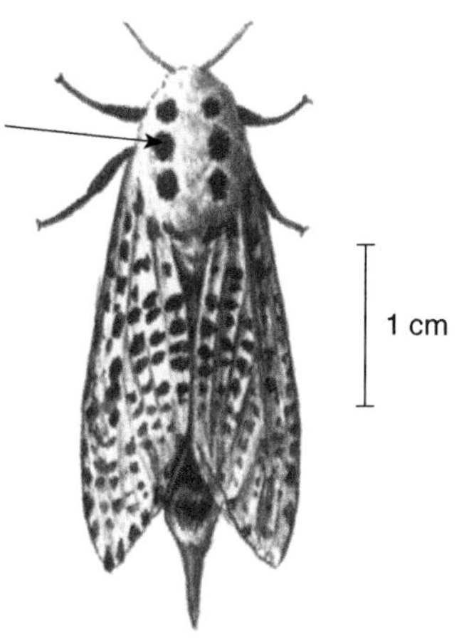

Fig. 9.17 Leopard moth (*Zeuzera pyrina*) (from Mandahl-Barth 1974, with permission).

The branches of many fruit trees, especially apple, plum, cherry and walnut, are tunnelled by the caterpillars of the leopard moth, *Zeuzera pyrina* (Fig. 9.17). The orange eggs are laid singly in bark crevices, and the caterpillars initially bore into the tip of a shoot of the still soft current year's growth. When this dies, the caterpillar leaves and makes a new entry into older growth, and there may well be a final third entry near the base of a branch. The whole development of the caterpillar may take up to 5 years. Attack causes the foliage beyond the tunnel to wilt, turn brown and die; as with the goat moth, branches may well break off entirely. The caterpillars are cream coloured with black spots, and may reach 5 cm in length before they pupate under the bark. Diagnostic of the pest is that it leaves part of its pupal skin projecting from the exit hole in the bark. The adult is white with black markings, including six large black spots on the back of the thorax. The moth is widespread across the temperate regions of the northern hemisphere.

9.2.3.7 Superfamily Tortricoidea

Family Tortricidae

This Family of small usually dull-coloured moths with rather straight-edged distal wing margins (Fig. 9.18) contains many pest species. The most important ones are probably those that feed on fruit or on the seeds within the pods of legumes. In the tropics, *Leguminovora ptychora* is a pest of grain legumes such as cowpea and pigeon pea.

The false codling moth (*Thaumatotibia leucotreta*) has a very wide host range including maize, but is known mainly as a pest of cotton and citrus in central and southern Africa where the caterpillars tunnel in the boll, fruit and seeds. The adult moth has a wing span of about 16 mm, and is brown in colour. Eggs are laid singly or in small groups on the surface of the bolls/fruit, and hatchlings wander over the surface for a time before entering, the entry point being marked by a clump of frass. The caterpillars then descend in the soil to pupate in an earthen and silk cocoon. The life cycle takes about a month.

The oriental fruit moth (*Grapholita molesta*) is a major pest of peach as well as attacking many other temperate stone fruit. It is found in many parts of the world, but in Europe it only occurs in the south. The adults have a wing span of about 14 mm, and they appear around the time the peach trees are in flower. Eggs are laid on leaves and twigs and the caterpillars bore into the shoot tips, causing them to wilt. Later the caterpillars bore into the fruit. When the mature caterpillars leave the fruit, they spin a cocoon in a sheltered spot on the tree or on the ground, and do not pupate till the following spring. There are four or five generations a year.

In temperate climates there is the pea moth, *Cydia nigricana* (Fig. 9.18). There is one generation a year, as eggs have to be laid near the flowers when these are beginning to fade. Indeed, pea varieties that flower particularly early may escape attack. The hatched caterpillars burrow into the young pods and feed on the seeds. Pea moth was never a serious problem when peas were sold in their pods; the few attacked seeds, with the caterpillars and their frass, were merely discarded while shelling. However, peas are now mainly harvested and processed mechanically, and pea moth can therefore be a problem even at low density. A live caterpillar is objected to by the public. However, one that has been boiled, bloated, stained a standard pea green and then tinned is far worse! It has been said that the tolerance threshold for pea moth is related to what frequency of complaints cannot be satisfied by a small bribe of free comestibles. Mature pea moth caterpillars drop to the ground and pupate in the soil. Although male pea moths fly between fields, females tend to stay in the field where they pupated and so populations can gradually increase from year to year.

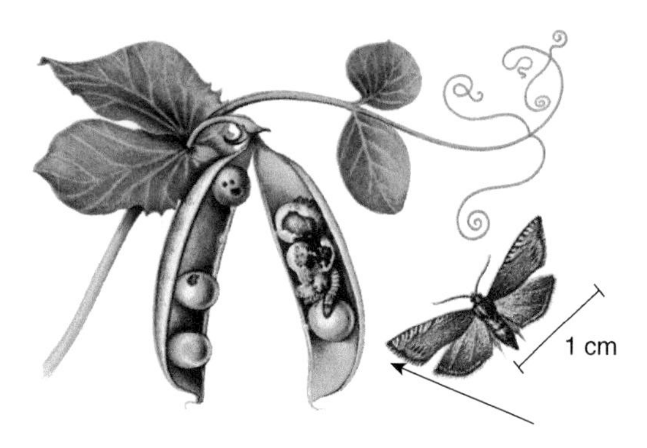

Fig. 9.18 Pea moth (*Cydia nigricana*) and pod opened to show damage with caterpillars (from Bayer 1968, with permission).

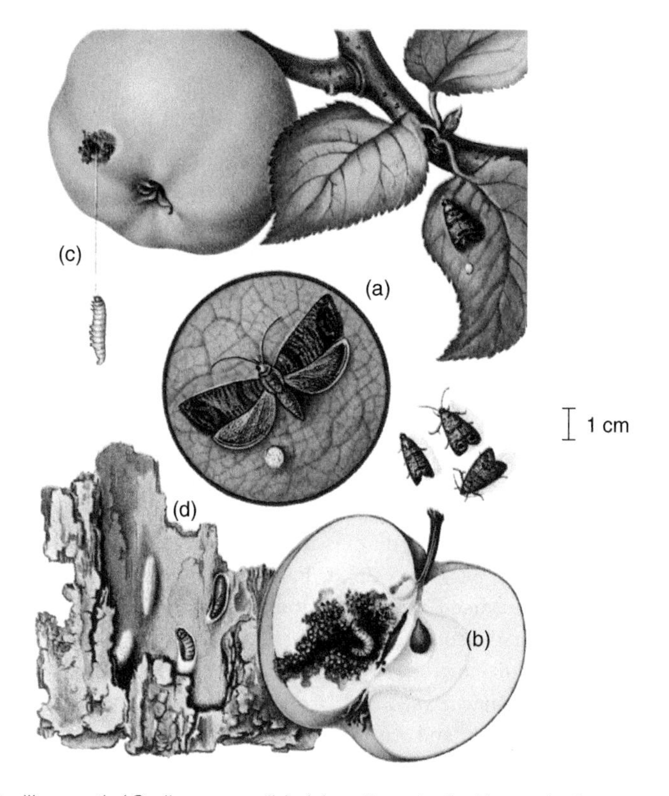

Fig. 9.19 Codling moth (*Cydia pomonella*): (a) moth on leaf with egg (enlarged in circle) and group of three moths; (b) caterpillar in sliced-open apple; (c) mature caterpillar emerged from an attacked apple and lowering itself on a silken thread; (d) bark on trunk removed to reveal cocoons with two torn open to show the pupae (all from Bayer 1968, with permission).

Cydia pomonella (codling moth; Fig. 9.19) is an important tortricid pest of apples wherever these are grown, and is a 'key pest' in the sense that most other pests of apple, some now often more important than codling moth itself, owe their promotion to this insect. This is because the need to control codling with insecticides has damaged the natural biological control of other herbivores on apple trees.

In the UK, adult moths emerge from pupae in bark crevices on the tree trunk in late April, when about 75% of the flower petals have fallen. This emergence is not synchronous, and so eggs are laid on the flower receptacles or adjacent leaves over a period as long as 6 weeks. The young caterpillars bite out a circular hole in the peel and burrow into the young fruit, where they feed on the flesh around the seeds. Many infested fruitlets are abscised and fall from the tree (forming part of the natural 'June drop'). In the apples that stay on the tree, the mature caterpillars bite their way out and lower themselves on a silk rope. When blown against the tree trunk they seek a crevice in the bark, where they overwinter in a silk cocoon to pupate and emerge the following spring.

This univoltine (one generation per year) cycle is typical for UK codling moth, though in exceptionally warm years second-generation eggs may be laid, but the generation will fail with no young apple fruits available. However, in warmer climes more generations may occur and more synchronised egg laying will then happen several times during fruit development. There may then be as many as seven generations a year.

This might suggest that control becomes increasingly difficult as the number of generations per year increases, but the reverse is actually the case. This is because insecticides can only kill caterpillars before they enter the fruit. In warm climates, all the caterpillars from one generation hatch together, so that is possible to have insecticide on the leaves and fruit surface throughout the appropriate time. In the UK, by contrast, a few caterpillars hatch each day over many weeks, and it is almost impossible to protect the foliage and fruit continuously over such an extended period.

A large number of other tortricid species damage leaves, flowers and fruits of fruit crops. Where these have a generation from one summer to the next, they overwinter on the leafless tree as half-grown caterpillars. These attack the new foliage as soon as there are the first signs as the buds begin to burst in the spring. Each 'bite' by the caterpillar then removes a large amount of what would be future leaf area. Later the caterpillars make a great deal of use of silk for spinning leaves together or attaching a leaf tip to a fruit; their feeding activity is then hidden from view. Characteristic of tortrix caterpillars is that they wriggle backwards if disturbed. Like the winter moths (see Section 9.2.3.8, Family Geometridae), the young caterpillars of tortrix can be dispersed aerially for long distances on their silken threads. The caterpillars pupate in a cocoon under their silk webs on the leaves. In the UK, *Adoxophyes orana* (summer fruit tortrix; Fig. 9.20) is probably the most serious such problem. Additionally, *Pammene rhediella* (fruitlet mining tortrix) feeds on fruitlets woven together into a cluster, and *Archips podana* (fruit tree tortrix) feeds on foliage but also attacks maturing apples from late August.

Another tortricid pest of fruit is the eye-spotted bud moth, *Spilonota ocellana*. This has apple and blackberry as major hosts, but is also found on peach, quince, walnut and plums. It occurs more or less wherever apples are grown. Like the fruit tortrix species, the moth is univoltine from summer to summer. Adults emerge in mid-June (northern hemisphere) and lay eggs on leaves, where the larvae feed individually in small silk tubes. They hibernate in silk shelters on the tree bark and emerge in spring to bore into the buds. This is when the real damage is done, especially on young trees where terminal buds may be destroyed. The larvae are dark red with a black head. Later they use silk to sew together leaves and blossom clusters, and feed there till early summer when they pupate under the webs. The adults have a wing span of about

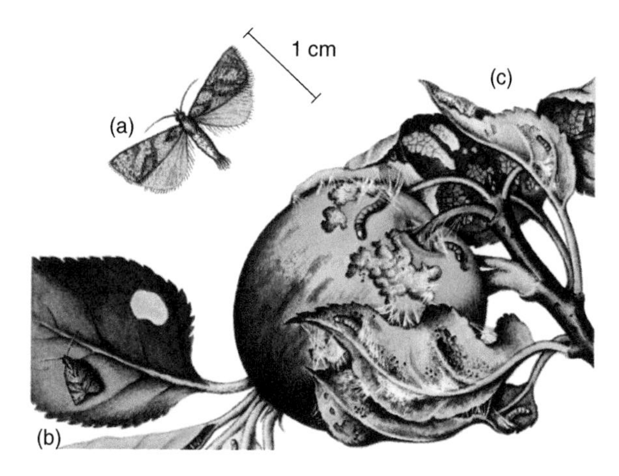

Fig. 9.20 Summer fruit tortrix (*Adoxophyes orana*): (a) adult; (b) moth on leaf with egg mass; (c) feeding damage on plum (all from Bayer 1968, with permission).

16 mm; the forewings have dark bases and tips with a broad pale median band and a small eye spot near the hind edge.

9.2.3.8 Superfamily Geometroidea

Family Geometridae

The caterpillars in this Family are colloquially called 'loopers' from their diagnostic way of walking. Prolegs are only found on abdominal segments 5 and 9, and the caterpillars move by looping the abdomen upwards as they bring the widely separated prolegs together, followed by releasing the walking legs and front prolegs and stretching the body forwards before re-attaching them (Fig. 9.21). The caterpillars of some species are camouflaged to the twigs they move over and mimic a short branching twig by attaching their rear prolegs and remaining motionless. They are then very easy to overlook. Characteristically, the adults have large, rather triangular forewings, with an acute apical angle so that the distal edge slopes back towards the body. This makes the trailing edge of the wing much shorter than the leading edge. They also spread-eagle their wings at rest; with their often cryptically marked wings, the camouflage when the moths are settled on branches or tree bark can be impressive.

Fig. 9.21 A variety of geometrid caterpillars (from Mandahl-Barth 1974, with permission).

Operophthera brumata, the winter moth, is a pest of apples and other fruits (Fig. 9.22). It is very polyphagous and is also found on other trees such as oak. The name 'winter moth' derives from the eggs being laid in November,

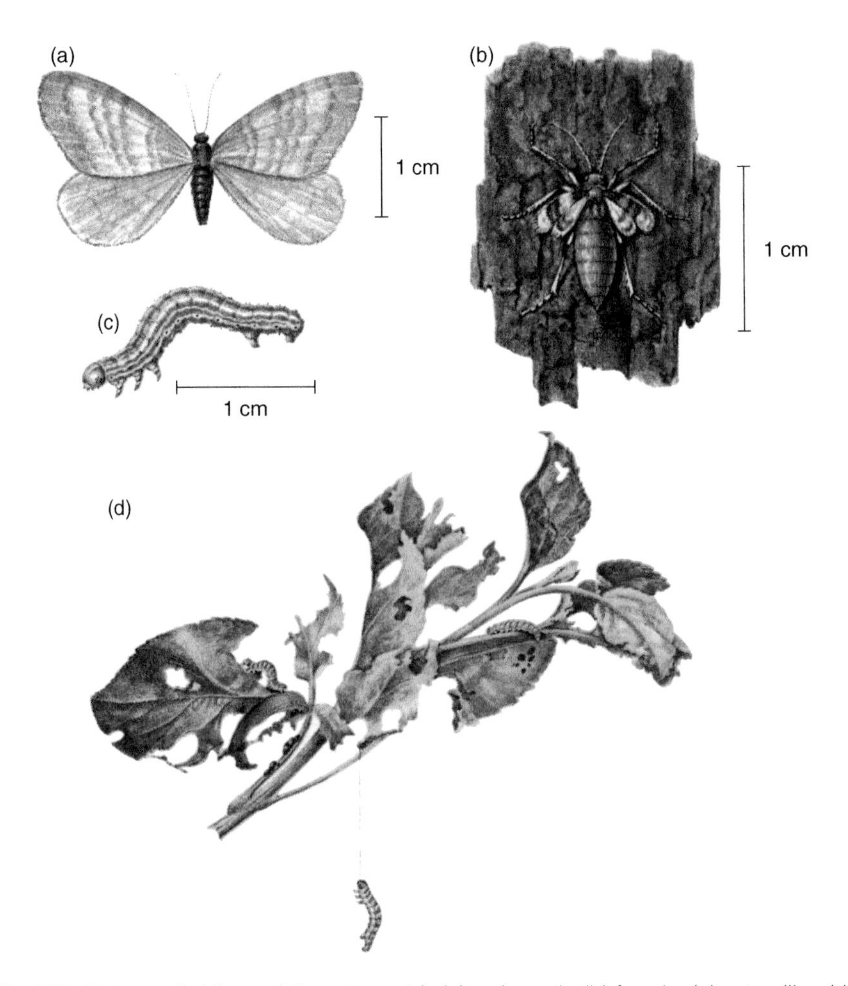

Fig. 9.22 Winter moth (*Operophthera brumata*): (a) male moth; (b) female; (c) caterpillar; (d) damage to apple foliage with caterpillar lowering itself to the ground on a silken thread (all from Bayer 1968, with permission).

a time of winter winds and rain. The females are protected by being wingless, and it is the winged males that seek out the female for mating. The wings of males result in many being battered and stuck down to stems by moisture. The eggs do not hatch till the spring, just as the buds are beginning to burst and show edges of green leaf. A bite of even a tiny caterpillar at this stage can remove much future leaf area (see also the larger tortricids caterpillars, Section 9.2.3.7, Family Tortricidae), and as the leaves expand they can look very ragged. The caterpillars continue eating the foliage as they grow and move on to cause serious damage to the flowers also. They do not attack fruit, but descend to the soil on a silken thread to pupate.

Since the females are wingless, they crawl up the trunk of a tree in the orchard they developed in, making it possible to kill them by trapping them on a sticky band wrapped around the trunk. But how does the species colonise new orchards if the egg-layer is flightless? The solution lies in the 'flying' caterpillars. Young caterpillars launch themselves into the wind, throwing out silken threads to aid buoyancy, and get carried long distances and lifted up on thermals to great heights. The first entomologists to study the aerial plankton were surprised to find winter moth caterpillars attaining altitudes greater than most winged insects!

In fact, 'winter moth' is a general appellation applying not only to *O. brumata*, but also to other geometrids with similar habits and also feeding on fruit trees. Examples are the mottled umber moth (*Erranis defoliaria*) and the March moth (*Alsophila aescularia*).

The caterpillars of winter moths are less than 2 cm long, in contrast with the 5 cm long ones of *Ascotis selenaria*, the 'giant looper' of Central Africa. This is a defoliator of coffee, but also local people value the caterpillars as a quick snack! The wings have a span of a little under 5 cm and can be very variable in the density of their patchy grey colour. The eggs (which can be 2000 per female) are laid on the tree in bark crevices and the caterpillars eat holes in the leaf as well as from the leaf margin when they are older. This latter feeding results in a characteristic 'jagged leaf edge' symptom. Flower buds and young berries may also be fed upon. Normally a minor pest, it is elevated to major status where pesticide use has destroyed its natural enemies. The life cycle takes between 6 and 10 weeks.

9.2.3.9 Superfamily Papilionoidea (butterflies)

See the Introduction to this Chapter (Section 9.1) about the distinction between butterflies and moths.

Family Hesperiidae (skippers)

The skippers are mainly small brown, grey or orange butterflies though some are more black and white or even bluish. They differ from other butterflies in that the antennae thicken gradually rather than abruptly to a club, and the club is hooked backward like a crotchet hook. Skippers are not normally pests but the Essex skipper (*Thymelicus lineola*; Fig. 9.23), which feeds on grasses and is rather a rarity in the UK, reached seriously damaging densities on rangeland in Canada after its introduction from Europe. There is also the rice skipper (*Pelipodas mathias*). The caterpillars initially feed within leaf rolls on rice and sugar cane; they may eat the entire leaf area except for the midribs, but the species is usually a minor pest in Asia and is easily controlled.

Fig. 9.23 Essex skipper (*Thymelicus lineola*) (from Mandahl-Barth 1974, with permission).

Family Papilionidae (swallowtails)

This is a Family of large attractive butterflies with bright colours and a characteristically obvious 'tail' on the hind wing (Fig. 9.24). The larvae are very given to everting the osmeterium (see Section 9.1) behind the head when alarmed. Several species are defoliators of citrus in Africa, Asia and South America and, feeding primarily on the flush growth, can cause enough damage to be a problem on small citrus trees and in nurseries. The orange dog, *Papilio demodocus*, occurs on all types of citrus in Africa. The adult has brown wings with a span of about 10–11 cm and with many yellow patches giving an overall 'orange' impression, but with very short 'tails'. Eggs are laid

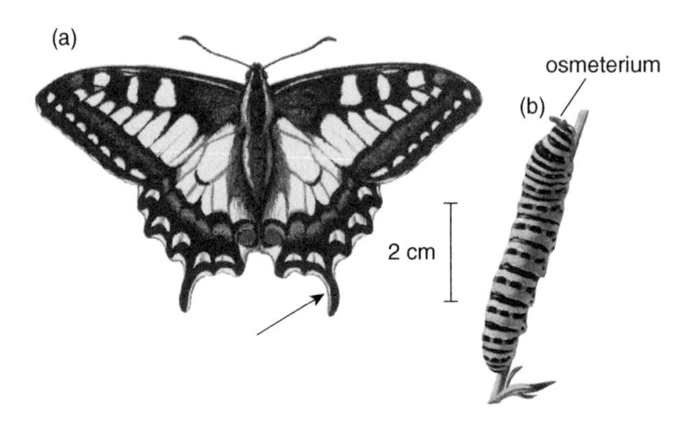

Fig. 9.24 Swallowtail (*Papilio machaon*): (a) adult (from Lyneborg 1968, with permission); (b) caterpillar (from Zanetti 1977).

singly on the flush leaves and the brown and white caterpillars, which resemble bird droppings, feed from the leaf edges until the fourth instar. They then change in colour to green with darker markings until they pupate attached to a branch at the end of the fifth instar. The life cycle takes about 6 weeks.

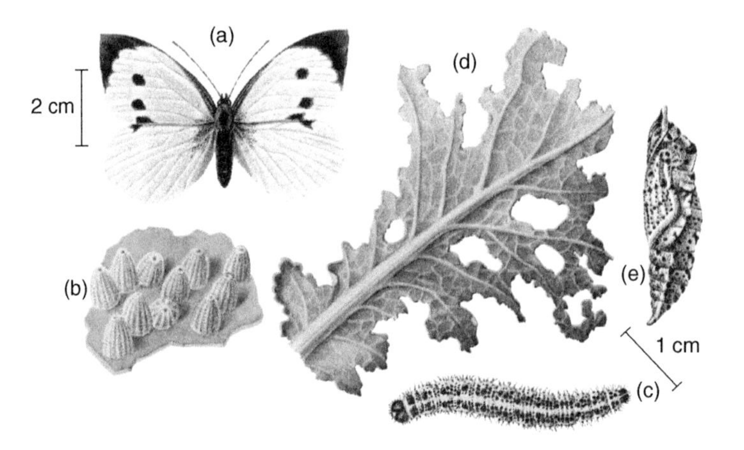

Fig. 9.25 Large cabbage white (*Pieris brassicae*): (a) adult; (b) eggs; (c) caterpillar; (d) damage; (e) pupa (chrysalis) (all from Bayer 1968, with permission).

Family Pieridae (whites)

Pieris brassicae (large cabbage white; Fig. 9.25) and *Pieris rapae* (small cabbage white) are familiar pests of brassica crops in many parts of the world. The former is especially serious as a pest in the Old World; the latter is a much more widespread problem. The caterpillars of both species defoliate the plants, leaving the skeleton of main veins uneaten on mature and old leaves. The caterpillars of *P. brassicae* hatching from the batch of ovoid yellow eggs are uniformly dark and initially feed together. As they grow they become yellow with black markings and disperse over the entire plant. *Pieris rapae* adults and caterpillars are noticeably smaller; the caterpillars are an emerald green from birth. Caterpillars of *P. brassicae* leave the plant to pupate in an upright position on a solid substrate such as a rock, fence or tree trunk, whereas *P. rapae* pupates on the plant. Both species overwinter in the pupal stage and have two generations a year in the UK, with the second generation doing by far the greater damage. In the warmer climates, *P. rapae* can have more generations; for example in Israel the species may have eight generations annually.

Family Lycaenidae (blues)

With these small attractive butterflies, it is generally the males that are blue, while the females are a dark brown. The caterpillars of many species, soon after they hatch on their host plant, are taken by ants into their nests, where the caterpillars – without interference from the ants – become voracious predators on the ant brood.

Blues are normally regarded as subjects for conservation rather than for control, but the caterpillars of *Lampides boeticus* (long-tailed blue; Fig. 9.26) bore into the pods of legumes in the tropics and feed on the seeds. The adults have a wing span of about 3 cm, with 2–3 mm long tails close to two black spots near the rear corner of the hind wing. As with other lycaenids (see Section 9.1), there is marked sexual dimorphism in the colour of the upper wing surfaces. Eggs are laid singly, on or near flowers. The green hatchlings first feed within the flower and then on the seeds within the pod. Pupation is on the leaves and the life cycle takes 8–10 weeks; there can be several overlapping generations per year. The species was once on the British list, and is still occasionally recorded as a presumed migrant from France. On one occasion a 'migrant' was recorded by someone near my home; I suspect it was no coincidence that it was on the pavement outside a greengrocer's shop with a display of green beans from Kenya!

Fig. 9.26 Long-tailed blue (*Lampides boeticus*) (reproduced with permission of CSIRO, Australia).

Family Nymphalidae

This is the Family that includes most of the larger butterflies such as the peacock, red admiral and the colourful tropical *Heliconia* and *Morpho* species. The adults have only the meso- and metathoracic legs as functional and the antennae end in an obvious knob (Fig. 9.27). The Family is not normally considered as of pest importance, but the sweet potato butterfly (*Acraea acerata*) is a common serious problem on sweet potato in East Africa. The caterpillars can totally defoliate different plants several times in a season, since the life cycle takes only 4–5 weeks. The adult has a wing span of 3–4 cm; the wings are orange with a black border. The pale yellow eggs are laid in large batches of over 100 on the leaves.

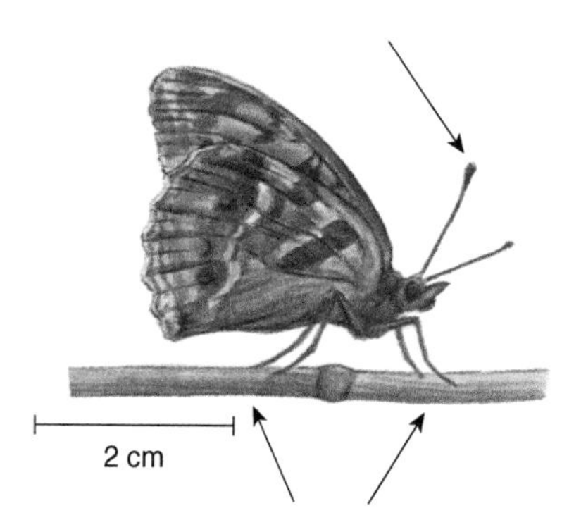

Fig. 9.27 A nymphalid butterfly to illustrate that only two pairs of legs are functional (from Mandahl-Barth 1974, with permission).

The caterpillars are dark green with thick branched spines; they remain together for about 2 weeks and feed on the upper leaf surface under the protection of a silken web. However, for the remaining week before they pupate they feed in isolation on the entire leaf lamina. They then leave the plant and perhaps even the crop area to seek some vertical stem or post, which they climb to perhaps 3 metres before pupating.

9.2.3.10 Superfamily Bombycoidea

These are large hairy moths with broad wings, the proboscis is absent and there is no frenulum wing coupling as in the Sphingoidea (see Section 9.2.3.11).

Family Bombycidae

Bombyx mori is the well-known 'silkworm', which has been 'farmed' by man for centuries, probably beginning in China at least 5000 years ago. The thread is spun from the silk of the cocoon, in making of which the caterpillar may produce 1 km of silk. Although the caterpillars will feed on other plants, the main host is mulberry (*Morus* spp., especially *M. alba* – Moraceae). In the UK, the presence today of mulberry trees often indicates the previous location of a silkworm farm.

Family Lasiocampidae

Lackey moths (especially *Malacosoma neustria* (Fig. 9.28) and *M. americanum*) are a group of several species known in the USA as 'tent caterpillars' (e.g. also the forest tent caterpillar *M. disstria*) because the highly gregarious caterpillars live in large silk tents on forest and fruit trees. The trees can be completely defoliated, and lackey moths can be serious problems in forestry as well as horticulture, though attacks tend to be sporadic. The different species span Europe, Asia and the USA in their combined distribution. The moths are dark to pale brown with stout bodies, with a wing span of 3–4 cm. The forewings characteristically have two parallel stripes running from

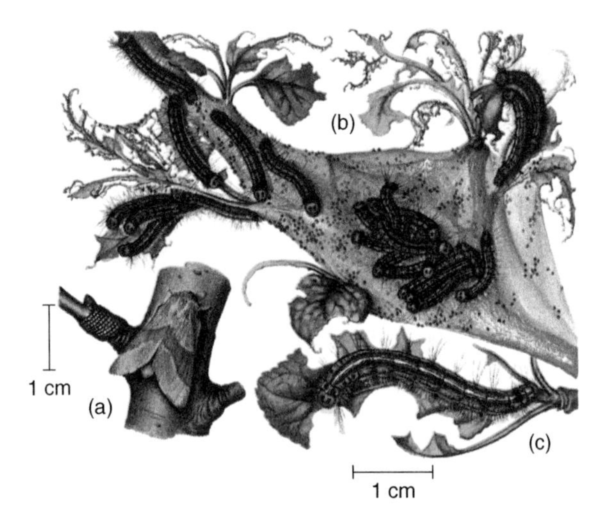

Fig. 9.28 Lackey moth (*Malacosoma neustria*): (a) adult and an egg band around the twig; (b) web with feeding caterpillars; (c) fully grown caterpillar (all from Bayer 1968, with permission).

leading to trailing edge. On dark wings the stripes are lighter, but appear darker on lighter wings. The eggs are laid in autumn in a spiral band around the twig (Fig. 9.28a) and do not hatch till the spring, when the caterpillars begin to feed on the foliage. Initially black, they later become a dark bluish-grey with white and red longitudinal stripes. In mid-summer they disperse, and pupate in a silk cocoon on the leaves or trunk to emerge in late summer or autumn.

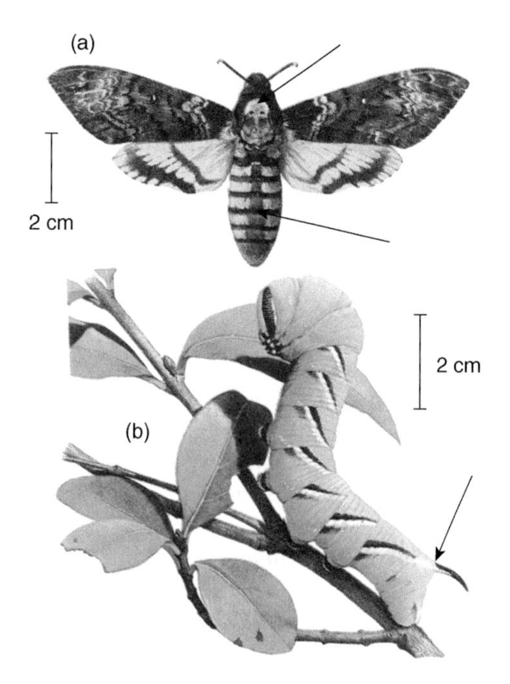

Fig. 9.29 (a) Death's head hawk moth (*Acherontia atropos*) and (b) a hawk moth caterpillar (from Zanetti 1977).

9.2.3.11 Superfamily Sphingoidea

A frenulum is present, and the proboscis is long and coiled.

Family Sphingidae (hawk moths)

These large stout-bodied moths have similarly large and stout-bodied caterpillars, which can be recognised by an obvious curved horn at the end of the abdomen (Fig. 9.29b). Both adults and caterpillars are beautiful showy insects, but one of only a few species feeding on a crop plant is *Acherontia atropos* (Fig. 9.29a). This is the death's head hawk moth, so called because a yellowish patch on the back of the thorax of the adult looks rather like a skull. The adults are unusual for Lepidoptera in being able to produce sounds – high pitched squeaks or chirps. The caterpillar feeds on potato, but is a rare insect in Europe and is only ever a minor pest in Africa and Asia. However, at a research station in China I once saw a display of many death's heads in a 'rogue's gallery' of insect pests of agriculture. When I expressed surprise that it was a pest of potatoes anywhere, my comment was met with equal surprise and 'Does it eat potatoes?' It turned out that the insect was not a crop pest at all, but a pest of beehives, and was *A. styx,* another of the three species of death's head hawk moths. The moths hover outside the hive and rob it of honey by sucking it out through their long haustellum. Indeed, the common name of *A. styx* is the 'bee robber'.

A much more important pest among the hawk moths is the tobacco hornworm, *Manduca sexta* (Fig. 9.30), which is a pest on tobacco, tomato and potato in the New World. The large caterpillars (up to 10 cm long) defoliate crops, and breed rapidly with up to four generations a year. The green eggs are laid singly or in small groups on the leaves, and the yellowish green caterpillars have slanting white stripes on the sides of their abdominal segments with the typical hawk moth terminal horn a red colour. The voracious larvae take only a few weeks to develop and then pupate in the soil for about 2 weeks.

Fig. 9.30 Adult of tobacco hornworm (*Manduca sexta*) (photo by Shawn Hanrahan).

9.2.3.12 Superfamily Noctuoidea

This Superfamily is especially important economically, containing some serious pests of tropical agriculture, including of subsistence crops. The forewings are often a drab grey or brown, but the hind wings (hidden at rest) are a bright red, orange or yellow. This 'flash colouration' startles a predator (e.g. a bird) when the moth takes off to escape.

Family Lymantriidae

These are the tussock moths, a Family of quite large stout-bodied drab white moths; the 'tussock' refers to the anal tuft of scales most females possess, and which become detached to cover the egg batches. The larvae are thick-bodied with bristles and are often brightly coloured, and they defoliate forest trees (both conifers and deciduous) as well as fruit trees.

Lymantria dispar (gypsy moth; Fig. 9.31) is a well-known defoliator of forest trees in northern climes (Europe, Canada and the USA). There are also Asiatic subspecies. The gypsy moth caterpillars pupate, usually in crevices, on the ground or bark. Eggs are laid at these sites when the moths emerge, and so the young caterpillars have to climb the trees to reach the foliage. On fruit trees they can therefore be controlled by intercepting their ascent with a sticky band around the tree trunk. Once on the trees, many young caterpillars will produce silk and launch themselves from the tips of branches to be blown to other trees. The hairy caterpillars are particularly damaging to oak trees. Populations build up slowly over a number of years and can cause the death of mature trees in natural forests with a consequent crash of the moth popula-

Fig. 9.31 Gypsy moth (*Lymantria dispar*): (a) adult male and (b) caterpillar (from Spuler 1910).

tions until the forest regenerates. The pest plays a significant role in causing the cycles of about 50 years that is common in the regeneration of natural forests in central Europe.

There are some other important polyphagous tussock moth pests. The brown tail moth (*Euproctis chrysorrhoea;* Fig. 9.32) has pure white wings and a brown tuft of hairs at the end of the abdomen (hence the common name brown tail moth). These hairs are used to cover the eggs and hide them. The caterpillars are dark with red and white markings and are very hairy; the hairs can cause rashes when they contact human skin. The larvae overwinter communally in a silken tent and pupation occurs in June the next year, in cocoons spun on surfaces such as bark and even walls of houses. The brown-tail moth occurs in Europe, Asia and the USA. *Orgyia antiqua* (the vapourer moth) has a similar distribution, but the genus is unusual in that the females have vestigial or even no wings. The male forewings are brown with a large white spot. There are also many species of the genus *Dasychira* in the Old World, which also defoliate trees and shrubs.

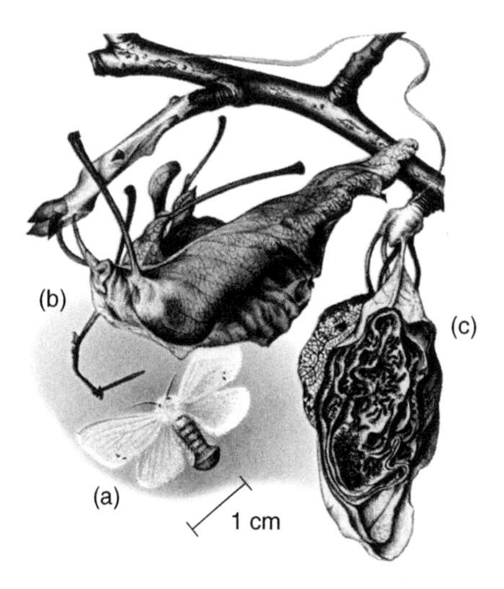

Fig. 9.32 (a) Brown tail moth (*Euproctis chrysorrhoea*) with (b) overwintering nest and (c) section through nest with young caterpillars (from Bayer 1968, with permission).

Family Arctiidae

The colourful garden tiger moth (*Arctia caja*; Fig. 9.33) and its hairy caterpillar ('woolly bear') are familiar to many people.

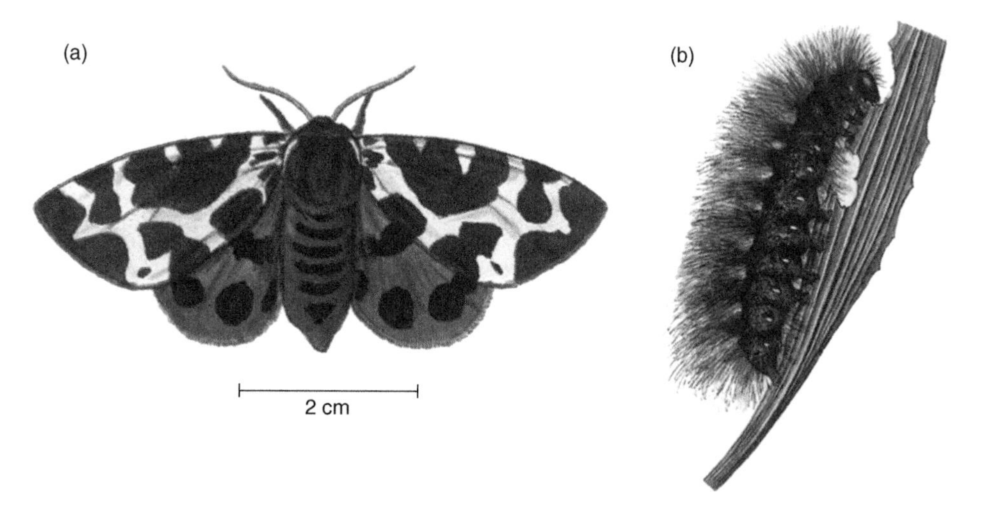

Fig. 9.33 (a) Garden tiger (*Arctia caja*) and (b) its caterpillar ('woolly bear') (from Mandahl-Barth 1974, with permission).

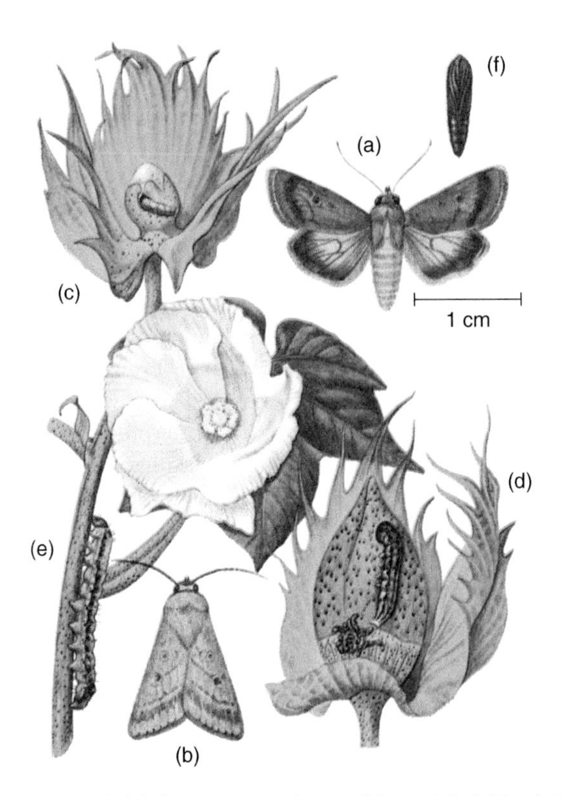

Fig. 9.34 *Helicoverpa zea*: (a,b) the two colour forms of the adult; (c) bud attacked by young bollworm; (d) infested cotton boll; (e) fully grown caterpillar; (f) pupa (all from Bayer 1968, with permission).

Family Noctuidae (owlet moths)

This Family is probably the most important Family of Lepidoptera in terms of the number of heavyweight and mostly polyphagous crop pests that it includes. The crotchets on the prolegs of the caterpillars are a uniordinal uniseries (see Section 9.1).

Firstly, the bollworms of cotton (Fig. 9.34) must rank in the top ten world pests. Once all included in the genus *Heliothis*, the major cotton bollworms are now in the genus *Helicoverpa*. *Helicoverpa armigera* (Fig. 9.35), the major cotton bollworm in the Old World, attacks leaves, flowers and bolls. However, it is very polyphagous and so is a pest of many other crops including maize, sorghum, tomato, legumes, sunflowers and even brassicas. The caterpillars feed out of sight by day, but then wander over the plant to seek new bolls, pods, fruits etc. at night. In contrast to the spiny bollworm (see following paragraph), pupation is in the surface layers of the soil. The moths are attracted by the smell of the cotton volatile gossypol emanating from the

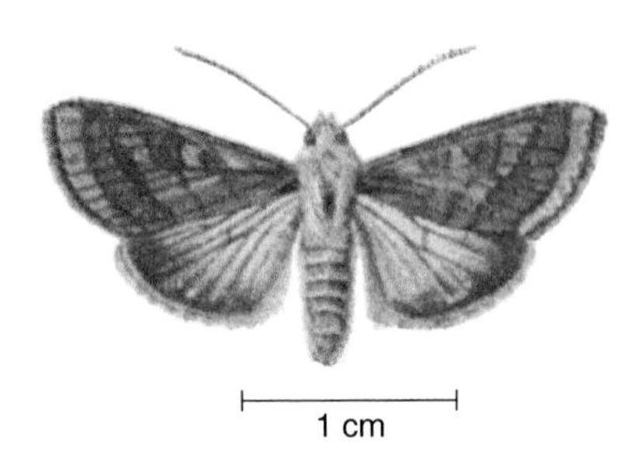

Fig. 9.35 Cotton bollworm (*Helicoverpa armigera*) (from Pearson 1958, with permission).

floral and extrafloral nectaries (on the leaves), and lay up to 200 eggs per female, especially near flowers. *Helicoverpa zea* (Fig. 9.34) is the New World equivalent on cotton and maize. On New World cotton it is joined by *Heliothis virescens* (tobacco budworm) as perhaps the most important cotton bollworm in the USA. The latter species, as its name suggests, is also a problem in tobacco. In Australia, there is *Helicoverpa punctigera* in addition to *H. armigera*. Once, when cotton growers in Queensland were over-applying the potent synthetic pyrethroid insecticide deltamethrin, I asked a grower what he would do when the bollworms showed resistance to the chemical. 'Well', he said, 'I'd probably switch to growing sunflowers'. Actually, if there's one food *H. punctigera* likes better than cotton, it's sunflowers! It also feeds on tomato and, unusually for *Helicoverpa*, broccoli; I was told caterpillars munching broccoli stems could be so numerous that one could actually hear them chewing when standing at the edge of the field.

The Family Noctuidae also includes other important bollworms of cotton, *Earias* spp. (spiny bollworms), especially *E. insulana* (Fig. 9.36). It is the caterpillars that are 'spiny', having pointed projections of the cuticle on the abdomen. The caterpillar attacks the developing bolls within which pupation occurs. The insect will also attack other members of the Malvaceae, including okra; I have known the adults of *E. insulana* to emerge from okra fruit bought in the UK but imported from Pakistan.

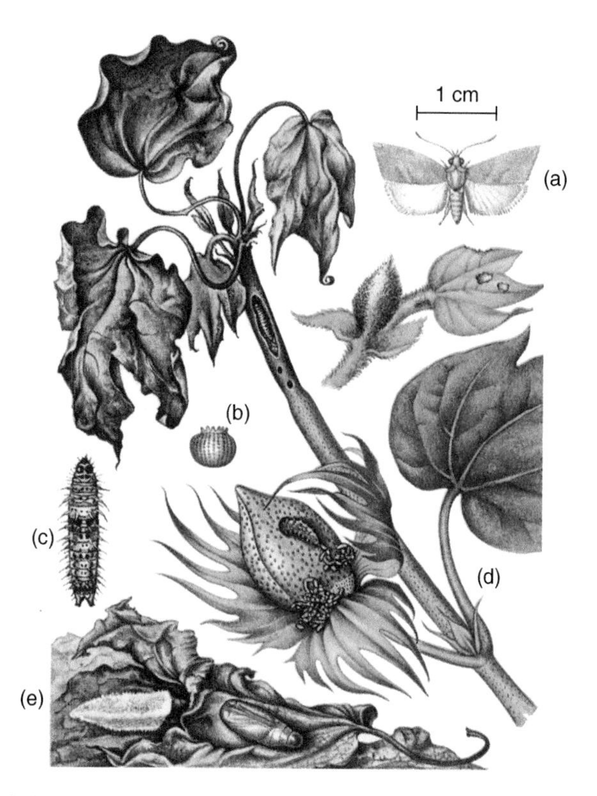

Fig. 9.36 Spiny bollworm (*Earias insulana*): (a) adult; (b) egg; (c) caterpillar; (d) die-back of shoot tip and attacked boll showing accumulation of excrement; (e) pupa and (left) cocoon (all from Bayer 1968, with permission).

Armyworms (species of *Spodoptera*) are 'locusts' by the biblical definition (see Section 6.6.2.1) in that they are plant-feeding insects appearing intermittently in plague proportions and devouring all before them. Like the grasshopper locusts, the permanent population exists in wild grassland but, when the population there increases, the pest disperses to crops leaving little not eaten as they pass through. They attack as many crops as *H. armigera* (above), and then more (importantly including rice). Batches of 100–300 eggs are laid on the underside of leaves, and can be recognised as armyworm eggs because the mother covers them with her scales. The black-headed caterpillars are principally defoliators, especially of cotton, rice, maize and tobacco, but they also bore into pods and fruit such as tomatoes. At first they are gregarious, but later disperse over the plant. Finally, the caterpillars pupate in an earthen cell near the soil surface. The life cycle in the tropics is about 3–4 weeks, and eight generations a year are common.

Spodoptera littoralis (Fig. 9.37) is the cotton leafworm, which identifies the crop where it is the most important of the armyworms. *Spodoptera litura*, the fall armyworm, has a more oriental distribution in the Middle and Far East and Australasia, but is also found in the Mediterranean and Africa. As well as being a regular though minor pest of cotton, it is a familiar pest of cereals such as maize, sorghum and millet.

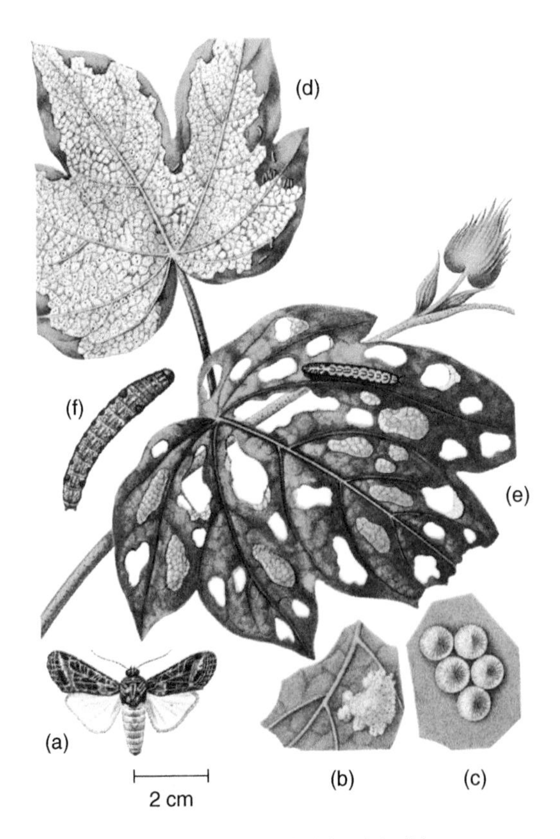

Fig. 9.37 Cotton leafworm (*Spodoptera littoralis*): (a) adult; (b) egg mass covered with woolly felt; (c) eggs; (d) damage by young caterpillars; (e) damage by older caterpillars; (f) fully grown caterpillar (all from Bayer 1968, with permission).

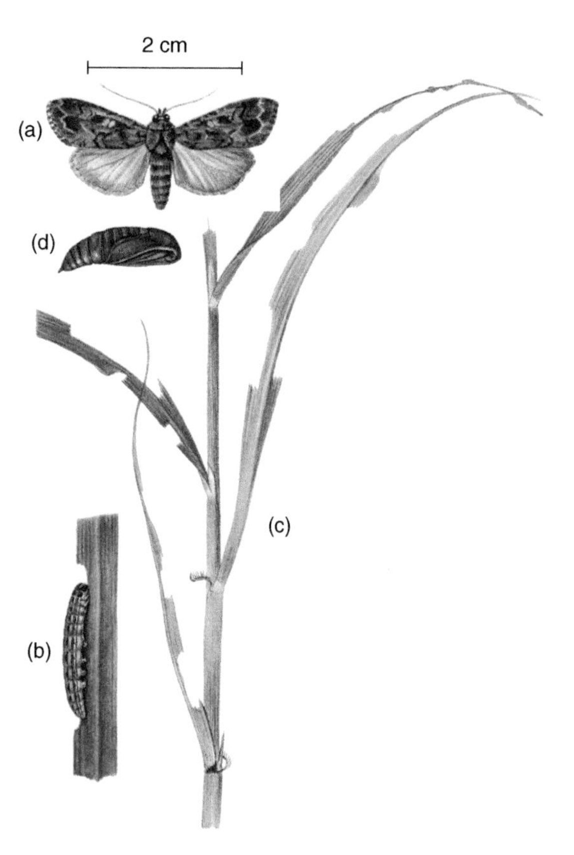

Fig. 9.38 An armyworm (*Spodoptera mauritia*): (a) adult; (b) caterpillar; (c) damage to rice; (d) pupa (all from Bayer 1968, with permission).

It not only defoliates, but may also fell seedlings like a cutworm (see below in this Section). The adults have a 3 to 5-cm wingspan; the forewings are a mottled brown while the hind wings are cream. The adults are virtually impossible to distinguish reliably from those of *S. littoralis*, except by dissection and inspection of the male genitalia. The caterpillars have the same yellow stripes as *S. littoralis*, but are dark green. *Spodoptera litura* has been a defoliator of glasshouse carnations in the UK after accidental introduction.

Spodoptera exigua, the lesser armyworm or beet armyworm, is a rather smaller moth with a 2.5-cm wingspan (note: Fig. 9.38 is of the related *S. mauritia*). The adults do not fly long distances but oviposit close to where they developed themselves. Caterpillars of *S. exigua* appear in swarms the most frequently of all the *Spodoptera* species, and the species is the principal armyworm attacking rice, where it feeds on both the leaves and stems. Young plants may be completely destroyed. As one of its common names suggests, it also attacks beet, but also feeds on alfalfa as well as many other crops as do the other armyworms. The caterpillars are blackish above and paler below. The species has a very wide distribution in the Old World, combining the range of the other two species mentioned above, and adding southern Europe and the southern USA. The Noctuidae also contains two important cereal stem borers, *Busseola fusca*

Fig. 9.39 Maize stalk borer (*Busseola fusca*): (a) adult (courtesy of G. Goergen, Biodiversity Centre, IITA); (b) damage to maize stem (courtesy of Haruna Braimah).

Fig. 9.40 Pink stalk borer (*Sesamia* spp.): (a) adult (courtesy of G. Goergen, Biodiversity Centre, IITA); (b) caterpillar in maize stem (courtesy of Haruna Braimah).

(maize stalk borer; Fig. 9.39) and pink stalk borer (named after the pink colour of the caterpillars of several species in the genus *Sesamia* (Fig. 9.40)). *Sesamia calamistis* is probably the best known species. Most graminaceous crops, including sugar cane, are susceptible. *Busseola* lays the eggs in small batches on the leaves; *Sesamia* lays a linear chain of eggs between the leaf sheath and the stem. The young caterpillars move down to the stalk and bore into the stem at the node there. As they feed and grow they weaken the plant stem so that it breaks, or eat out the living tissues so that the stem above their location wilts and dies. There is an important difference in where the two genera pupate. *Busseola* descends to the soil whereas *Sesamia* pupates on the stem under the leaf sheath. Thus stalk destruction after harvest is effective against *Sesamia*, but is usually too late to affect *Busseola*. If the stalks are not destroyed, as happens where the stalks are used as building material for roofs and fences, a reservoir of *Sesamia* is created to attack the crops the following year.

Cutworms (caterpillars of the large yellow underwing *Noctua pronuba*, as well as of several *Agrotis* spp., especially the dark sword grass *A. ipsilon* (Fig. 9.41) and the turnip moth *A. segetum*) are mainly pests of temperate crops. They literally 'cut' seedlings near soil level at night and drag the felled tops into the cracks in the soil where they live by day at a depth of up to 10 cm, though *N. pronuba* goes even deeper. A variety of crops are attacked, including wheat, sugar beet and brassicas. Damage is often underestimated with the disappearance of seedlings being attributed to other causes such as seedling diseases. As well as felling seedlings of many crops, including cereals and brassicas, cutworms also feed on and hollow out underground organs such as roots and tubers, causing direct damage to the marketed unit in root crops and potatoes. The larvae pupate in a cell in the soil. The moths are mainly dark in colour with a wing span of 4 to 5 cm and a pale white hind wing with a narrow black margin. In temperate regions cutworms are univoltine, but can have four to five generations a year in warm climates. Between them, different species cover most of the world in their distribution.

The tomato moth (*Lacanobia oleracea*) is polyphagous, but is mainly a pest of both glasshouse and field tomatoes. The eggs are laid in batches on the leaves, and the caterpillars (which may be green or brown, but always have a yellow lateral stripe) feed on the leaves and more importantly tunnel into the fruits as they begin to ripen. Pupation is in a silk cocoon in the soil. In the UK there is one generation a year, but the insect is bivoltine further south. The species is common in Europe and western Asia.

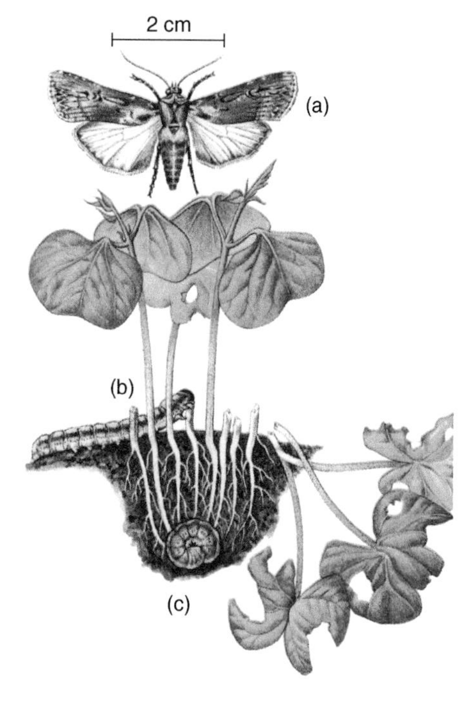

Fig. 9.41 Dark sword grass (*Agrotis ipsilon*): (a) adult; (b) caterpillar (cutworm) and injury to seedlings; (c) caterpillar in soil during day (all from Bayer 1968, with permission).

Another important noctuid pest is *Mamestra brassicae*, the cabbage moth. This is found throughout Europe and right across Asia to Japan. It is a serious problem on crucifers. The young larvae feed gregariously, but later disperse over the plant. They are green when young, but later develop a brown colour with pale green ventrally. They can skeletonise the leaves, but also the large amounts of moist frass in the heart of cabbages (where the larvae prefer to tunnel and feed) is contamination, which is as economically damaging as the removal of tissue. Pupation is in a cocoon in the soil, and there may be one or two generations a year.

Finally, a Subfamily of noctuids, the Plusiinae, has caterpillars with several of the pairs of prolegs before the terminal segment missing; they 'loop' like geometrid caterpillars. They are therefore known as 'semiloopers' (Fig. 9.42). The cabbage looper (*Trichoplusia ni*) is a polyphagous pest with a fairly cosmopolitan distribution. It is especially known as a pest of soybeans and brassicas. With brassicas, the caterpillars not only eat and skeletonise the leaves, but also tunnel into the stem of cabbages and the curds of cauliflowers and broccoli. Once in the stems, the caterpillars cannot easily be controlled with insecticides. The adults have forewings spanning 3–4 cm; there are quite complex patterns of dark and lighter brown and two small white markings, which may resemble the numeral 8. Eggs are laid singly on the undersides of leaves, and the green caterpillars, with four white longitudinal lines, eat out large areas of the leaf. In cabbages they prefer the heart leaves, which they contaminate with large amounts of frass trapped by these leaves. Pupation is in a silk cocoon on the plant, in leaf litter or between clods of soil. The life cycle takes 6 weeks under good conditions with continuous breeding; five or more generations a year therefore occur. In temperate countries there is also continuous, but much slower development with the caterpillars continuing to feed.

The related silver Y moth (*Autographa gamma*; Fig. 9.42) is another important cosmopolitan pest, the caterpillars first skeletonising and later, when larger, totally consuming the leaves of many crops, but particularly sugar beet, lettuce, cabbage, tomato and potato as well as peas and beans. Female moths can lay up to 1000 eggs, in contrast to only 200 for *T. ni*. *Autographa gamma* pupates in a silk cocoon on the plant, and the life cycle takes 7–8 weeks; in Europe there are two generations a year, but up to four or five in warmer climates.

Fig. 9.42 Silver Y (*Autographa gamma*): (a) adult (from Lyneborg 1968, with permission; (b) caterpillar (semilooper) (from Zahradník and Chvála 1989).

10 Subclass Pterygota, Division Endopterygota, Order Diptera (true flies) – *c.* 120,000 described species

10.1 Introduction

The Diptera are the true flies or two-winged (Di-ptera) flies, for their diagnostic feature is that the hind wing is dramatically reduced to a knobbed stalk known as the ***haltere*** or 'balancer' (see Fig. 10.7). This structure vibrates when the forewings are beating. The base of the haltere is equipped with sense organs and has a contractile muscle attached for which there is no antagonist. The sense organs sensitively detect distortions of the cuticle, which occur during turning. Many flies are 'front heavy' with a small abdomen. If the heads of the halteres of such flies are amputated, the flies become unstable and dive; this can be corrected by weighting the abdomen with a small piece of Plasticine.

Powerful flight with emphasis on an increase in the proportion of the fly's volume allocated to the thorax is a main feature of the evolution of adult Diptera. Over a century ago, two gentlemen with stop watches timed a horse fly flying between them at 144 k.p.h., but today most entomologists believe that two horse flies must have been involved!

The adult antennae are variable and important in the major classification of the Order. As the groups become more advanced, and as an adaptation to faster flight, an initially many segmented threadlike antenna becomes reduced to three broader segments and a narrow apical part becomes a bristle (***arista***) whose insertion finally moves back from the tip of the third segment.

The compound eyes of the adults are large and may touch each other (the ***holoptic*** condition), eliminating the area of the head capsule between them. In most flies, males have the eyes set closer together than females. In addition to the compound eyes, there are usually three ocelli arranged in a triangle on the frons.

Adult mouthparts (Fig. 10.1) are typically adapted for licking. The main structure is a hinged labium, which can be lowered onto the food with two flaps (***labellae***, singular ***labellum***) at the tip opening sideways to lie flat on the substrate. Between the labellae are small

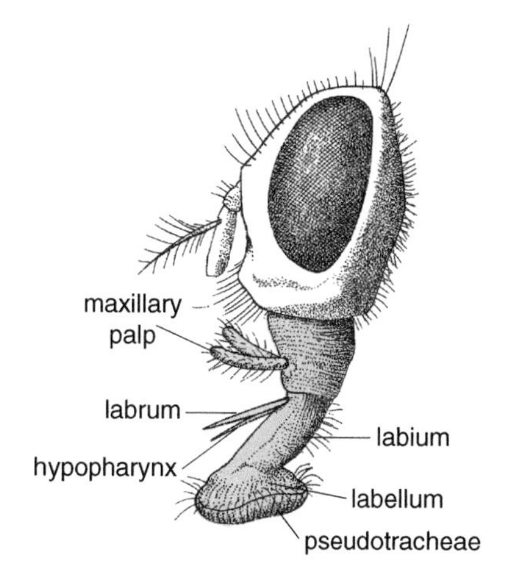

Fig. 10.1 Mouthparts of Diptera (from Zanetti 1977).

Handbook of Agricultural Entomology, First Edition. H. F. van Emden.

sclerotised teeth, which can scrape the surface of the food to release liquid; this is then picked up by capillary channels (***pseudotracheae***) in the labellae and transferred to where the labellae meet and from where the liquid is carried to the mouth. It is well known that many adult flies bite and suck blood. Parts of the mouthparts of such flies become hardened and non-retractable; details will be given later under the relevant Families.

The single pair of wings shows a progression of increasingly reduced venation through the Order. It is easy to confuse small Diptera and Hymenoptera, since the small hind wings of the latter are often hard to distinguish from the forewings at rest or in preserved specimens. Many of these small Hymenoptera have a single darker spot (***stigma***) near the tip of the forewing (Fig. 11.2) – any 'fly' with such a spot merits closer examination!

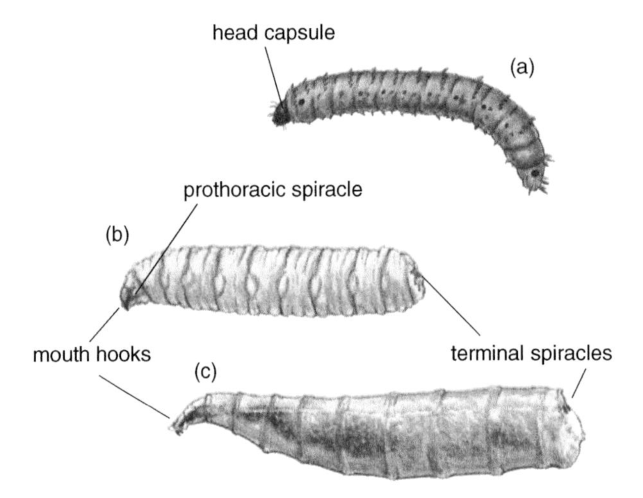

Fig. 10.2 Larvae of Diptera: (a) peripneustic (from Mandahl-Barth 1974, with permission); (b) acephalous amphipneustic; (c) acephalous metapneustic (from Lyneborg 1968, with permission).

Larvae of Diptera (Fig. 10.2) nearly all have no legs (i.e. they are ***apodous***), though larvae in some primitive families do have some very reduced legs. The head capsule and mouthparts become progressively reduced, and most fly larvae have no obvious head (they are ***acephalous***) and the mouthparts are reduced to a pair of sclerotised ***mouth hooks***. There is also a reduction in the number of spiracles. From spiracles on most segments (the ***peripneustic*** condition), they become reduced to a pair at either end of the body (***amphipneustic***) – there are prothoracic spiracles and spiracles on the last abdominal segment. Finally, only the terminal spiracles are retained (***metapneustic***). With all these conditions, the rear spiracles are often enlarged and open backwards rather than laterally.

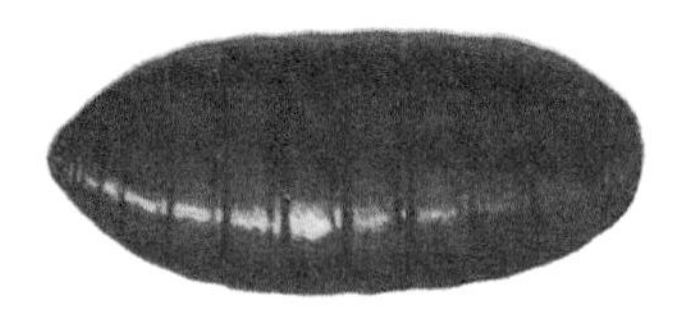

Fig. 10.3 Fly puparium (from Zanetti 1977).

Fly pupae are exarate, but in the higher flies the exarate pupa forms within the smooth case (***puparium***) of the shrunken and sclerotised skin of the last larval instar (Fig. 10.3). The larval mouthparts and spiracles can still be distinguished. Many flies break open the tip of the puparium at emergence by pumping body fluid forwards to expand a soft bladder (***ptilinum***) at the front of the head (Fig. 10.4). If emergence occurs in a dense substrate like soil, the bladder is shrunk and re-expanded to haul the young adult to the surface. Once there, the bladder is withdrawn into the head and a drawbridge (part of the head capsule carrying the antennae) is raised to seal the head capsule. The edge of this drawbridge remains visible as an inverted horseshoe (***ptilinal suture***) around the antennae (Fig. 10.5). We can re-evert the ptilinum by squeezing the head of the fly, and this was cleverly used by scientists some years ago to find out how far cabbage root fly travelled after emergence.

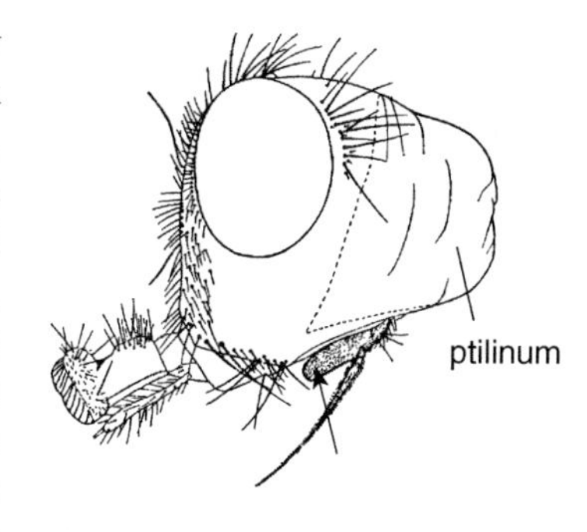

Fig. 10.4 Head of fly with everted ptilinum. The antennae (arrowed) are on the 'drawbridge' behind which the ptilinum everts (from Richards and Davies 1977).

The whole surface of a cabbage field was dusted with a red powder, which was picked up on the surface of the ptilinum and then withdrawn into the head when the fly reached the soil surface. Later, red pigment on the everted ptilinum of flies captured at various distances identified those individuals that had come from the original field.

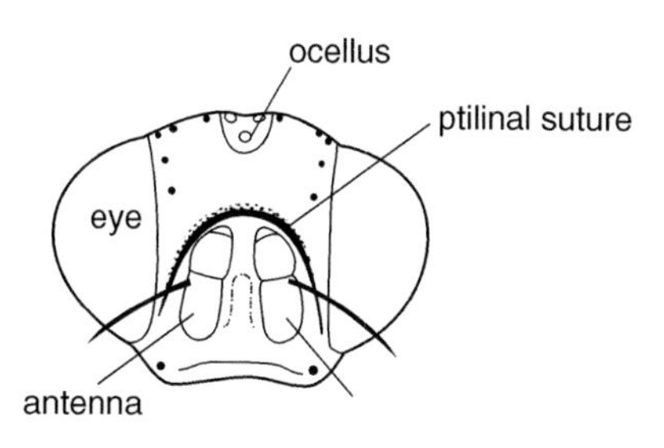

Fig. 10.5 Front view of the head of a fly.

The Order is divided, based on the appearance of the adult antennae, into three Suborders – Nematocera, Brachycera and Cyclorrhapha. The Diptera are one of the largest insect Orders, with very many Families; only a selection relevant to the aims of this book will be presented here. Most plant pests, apart from the Cecidomyiidae (Nematocera), are in the acalyptrate section or in the calyptrate Family Anthomyiidae (Cyclorrhapha).

10.2 Suborder Nematocera

Nemato-cera means 'thread-horns', and flies in this Suborder have antennae of six or more similar segments, usually long and obvious (Fig. 10.6a,b). Larvae are peripneustic, amphipneustic or metapneustic and may have a complete head capsule, an incomplete head capsule (the ***hemicephalous*** condition) or be acephalous. Pupae are free and not enclosed in a puparium.

10.2.1 Family Tipulidae (crane flies)

Also known as 'daddy long-legs', these flies indeed have very long legs and long antennae. The thorax has a distinctive V-shaped suture (Fig. 10.7) dorsally and is vertically deep compared with the abdomen, which ends rather squarely. The wing venation has two obvious long anal veins and is quite complex with many cells, especially towards the wing margin. There is a small discal cell not far from the wing tip (Fig. 10.7); any other Nematocera with such a cell do not have the two long anal veins. The larvae (Fig. 10.7), known as 'leatherjackets', have a distinct head capsule and the rear spiracles are large and point backwards; they are surrounded by fleshy ***papillae***, which protect the spiracles from being blocked by debris or soil. The larvae live in soil, rotting wood or leaf litter, and most are scavengers or carnivores; the larger species, however,

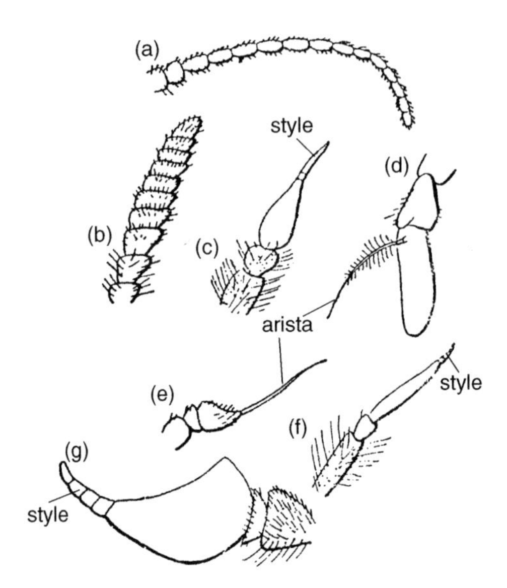

Fig. 10.6 Antennae of (a,b) Nematocera; (c,e,f,g) Brachycera (with a terminal style or arista); (d) Cyclorrhapha (with the arista set back on the third segment) (all from Richards and Davies 1977).

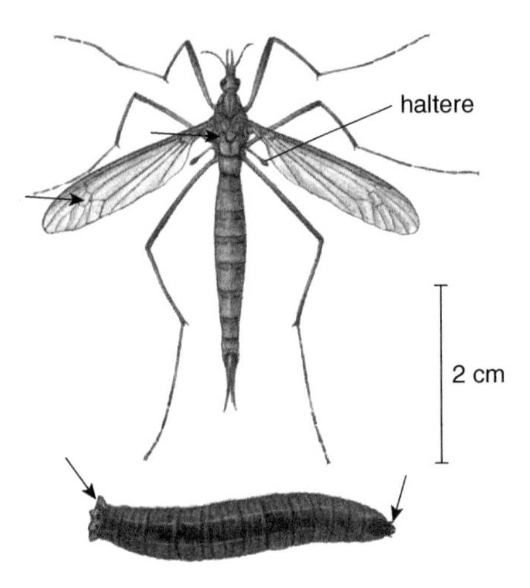

Fig. 10.7 Tipulidae adult (daddy long-legs) with larva (leatherjacket) below (from Lyneborg 1968, with permission).

are herbivores. Several, for example *Tipula paludosa*, extend their ovipositor into the soil to lay eggs in grassland, where the larvae then graze on the roots. In large numbers they can be pests of pasture turf grasses. Densities of 1000 leatherjackets per square metre have been recorded! However, if infested grassland is cultivated and re-sown either with grass or a crop such as wheat, a large number of well-grown 'dispossessed' larvae – which may not have done much damage to the previous grass – now have very few roots available by comparison and can cause real trouble. Leatherjackets pupate in the soil. The pupae are strangely mobile and work their way to the soil surface, where the anterior end projects allowing the adult crane fly to emerge. There is one generation a year. As well as being pests of grassland and seedling cereals, leatherjackets are polyphagous and feed on the roots of seedlings of several vegetables as well as tunnelling into tubers and root crops. Tipulids are pests in Europe, Asia and North America.

10.2.2 Family Psychodidae

These small flies are easily recognised by their pointed hairy wings (Fig. 10.8); the hairy fringe is particularly noticeable. The larvae of *Psychoda* spp. (moth flies or owl-midges) feed on decaying matter with bacteria, usually in water, and are an important ingredient in the mix of larvae of Nematocera, which forms an essential component of the processing of sewage in sewage beds. Huge swarms of adults may be produced from the sewage beds, and can be a nuisance when they enter houses.

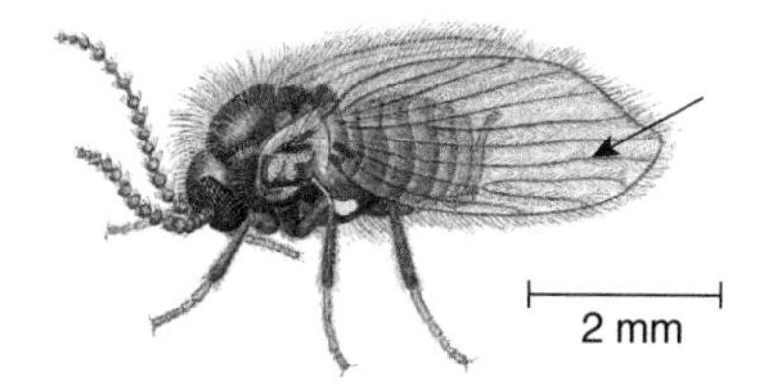

Fig. 10.8 Psychodidae (moth fly) (from Zahradník and Chvála 1989).

The sand flies (*Phlebotomus* spp.), so called because the larvae of many live in sandy soil feeding on dead organic matter, hold their wings vertically above the body. Both sexes feed on sugary secretions of plants and insect honeydew, but the females need to suck blood for egg production and are vectors of the trypanosome parasite *Leishmania* as well as a bacterial and several viral human diseases, as far north as the Mediterranean. Leishmaniasis affects not only humans, but also livestock. The symptoms are skin sores, which can appear weeks or months after the infection has occurred. Usually, these sores (cutaneous leishmaniasis) heal within a few months to a year leaving a scar, but more serious is a potentially fatal progression, which may occur (visceral leishmaniasis) in the spleen and liver, inducing fever and anaemia.

10.2.3 Family Cecidomyiidae (gall midges)

Cecidomyiids (Fig. 10.9) are small delicate flies with long antennae, but the segments are rather spherical like little beads and there is usually a whorl of short hairs at each junction. Males tend to have more feathery antennae. The insects have slender bodies, which are quite frequently reddish or orange, and the slightly hairy wings have no more than four longitudinal veins reaching the wing margin; also there is an absence of cross-veins. The acephalous peripneustic larvae are short and squat, a spot character being a ventral sclerotised patch a bit like a rough cross (***sternal spatula***).

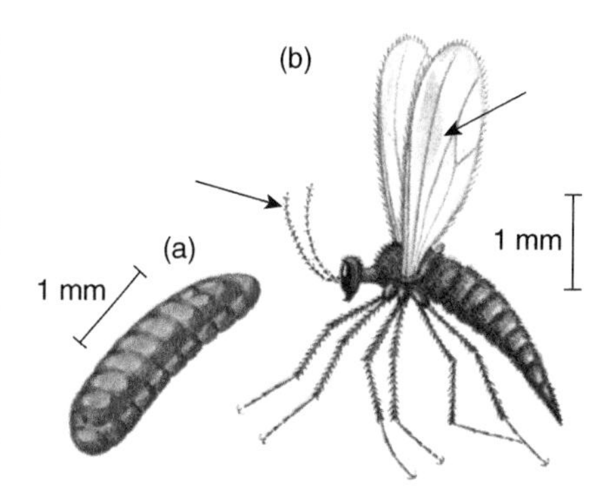

Fig. 10.9 Cecidomyiidae (gall midge – *Sitodiplosis mosellana*): (a) larva and (b) adult fly (from Bayer 1968, with permission).

In spite of the common name of the Family, by no means all cecidomyiids cause plant galls. Several species feed in cereal seeds. Eggs are laid while the cereal spikelet is still flowering; after a few days they hatch and the pink or orange larvae move to the seed's ovary and feed for 7–10 days before pupating there. The larvae of *Sitodiplosis mosellana* (orange wheat blossom midge) feed on the grains of wheat in the inflorescence. Similar attack by the larvae of *Contarinia tritici* (yellow wheat blossom midge) shows as a flattened appearance of the rachis supporting the infested grains. The related sorghum midge (*S. sorghicola*) occurs throughout the tropics and subtropics, and often feeds on the seeds at a density of only one larva per sorghum spikelet. However, in a heavy attack this is enough to cause total loss of grain, as the whole grain head becomes flattened with shrivelled seeds.

Contarinia pisi is the pea midge of Europe. Adults appear in June and oviposit inside pea shoots, young pods or flower buds, and the larvae scrape the plant tissues and imbibe the juices so released. The flowers become deformed (the 'nettlehead' symptom), and the death of shoots reduces yield, though attack also puts a stop to plant growth. This synchronises the ripening of the pods, bizarrely a beneficial effect in relation to the mechanical harvesting of the pods for freezing. As in other *Contarinia* species, the little larvae can curl up and then jump quite long distances, and in this manner they reach the soil to pupate in a silk cocoon. There are two generations a year, but some larvae of the first generation remain in their cocoons and overwinter with those of the second generation to pupate early the next summer.

A European gall midge, perhaps more important on the mainland than in the UK, is the saddle gall midge, *Haplodiplosis marginata*. The larvae are a dark blood red and feed in depressions they make on the stem under the leaf sheath. The stem is weakened and may break so that the head falls to the ground. Mature larvae spin a cocoon in the soil, but do not pupate there till after the winter. The pupal stage lasts just a few days, and the adults emerge in late May to lay their eggs in a short chain on the leaves.

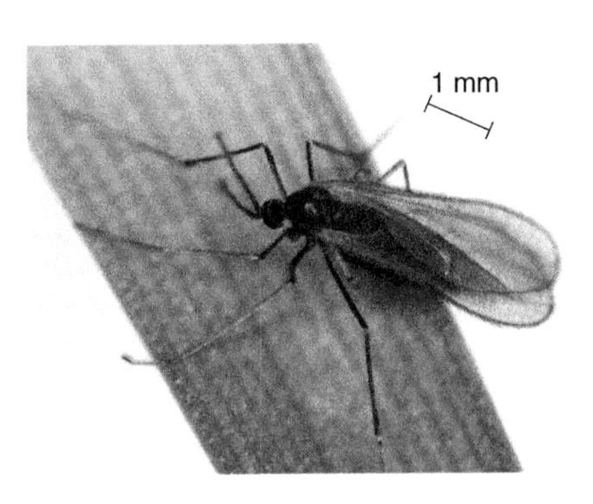

Fig. 10.10 Hessian fly (*Mayetiola destructor*) (University of Nebraska, www.unl.edu).

A major pest of wheat and barley, especially in the USA (but also in Russia and much of Europe) is *Mayetiola destructor* (Fig. 10.10). Eggs are laid on the leaves and the young white larvae crawl under the leaf sheaths to feed on the stem just above a node. The stem may break at this point, but if it does not then the head fails to develop properly and is mainly empty (the 'whitehead' symptom). After about 3 weeks, the larvae form a prepupa. The puparium (last larval skin, see Section 10.1) around it is a dark brown (the 'flax seed') and is adpressed to the shoot of the plant. The second generation stay in the puparium and only emerge the next May and June, and indeed some prepupae may remain in their puparium in the soil for up to 4 years before they emerge as adults. In the USA, it is the second generation that does the most damage to winter wheat, whereas only the first generation does damage in the UK and second generation adults emerge after harvest and oviposit on grasses.

Another important gall midge of cereals is *Orseolia oryzae* (rice gall midge) of Africa and Asia. It has a generation every 3 weeks. Eggs are laid on the leaf sheaths, and the pale larvae (later turning red) move down to feed on the apical bud or the lateral buds at the base of the plant, causing the plant to swell at its base to a silvery onion shape (the gall). The pink larva changes to a red pupa, which projects from the gall. Where rice is not grown continuously in the year, the pest maintains itself in grasses and wild rice in swampy areas, and spreads to rice crops with the rains.

On oilseed rape, brassica pod midge (*Dasineura brassicae*) lays eggs inside the pods, apparently requiring previous damage (e.g. by weevils) for successful oviposition. The white larvae feed on the seeds, and infested pods become yellow and sometimes swollen and distorted ('bladder pods'). These pods split open prematurely, and the larvae fall and pupate in a cocoon in the soil. Heavy infestations causing crop loss can occur, thought the fact that damage is heaviest at the edges of the crop can give a misleading impression. There are at least two generations a year.

A group of cecidomyiids once placed in a separate Family (Itonitidae) is unusual in that the larvae are carnivorous. *Aphidoletes aphidimyza* lays its eggs in aphid colonies and the larvae then feed voraciously on the aphids, biting into their prey (often at a leg joint) and sucking out the juices. They are so effective that they are bred commercially for release in glasshouses.

10.2.4 Family Culicidae (mosquitoes)

Mosquitoes are slender and long legged and the wings have few veins, which like the hind margin are covered in scales (Fig. 10.11). The labrum and labium are all drawn out to form a proboscis (Fig. 10.12), which houses the hardened and needle-like mandibles and maxillae. The female uses these for piercing the skin. The labrum and another mouth structure, the ***hypopharyux***, also enter the wound and form a food canal for sucking up blood. At the same time, saliva is injected. This inhibits blood clotting. By contrast, males feed on nectar and other sugar sources. The antennae are composed of several short segments with hairs at the junctions for detecting odours; females hunt their hosts by detecting CO_2 and volatiles in sweat. The males have very plumose antennae and hairy palps that are longer than the proboscis (Fig. 10.13) and the vibration of the hairs on the antennae translates into sound; thus the male mosquito hears the flight whine of the female through its antennae. The antennae carry 16,000 receptors, about the same as the human ear! The body of adult mosquitoes is slender with a patchy dark/light grey pattern, but the body of the female swells up noticeably with a blood meal.

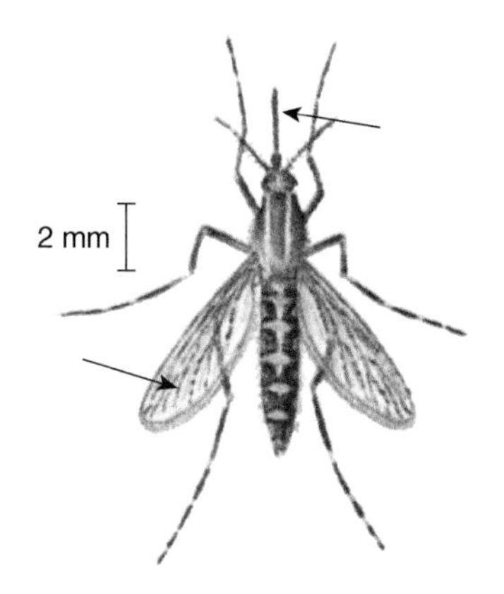

Fig. 10.11 Culicine mosquito (from Lyneborg 1968, with permission).

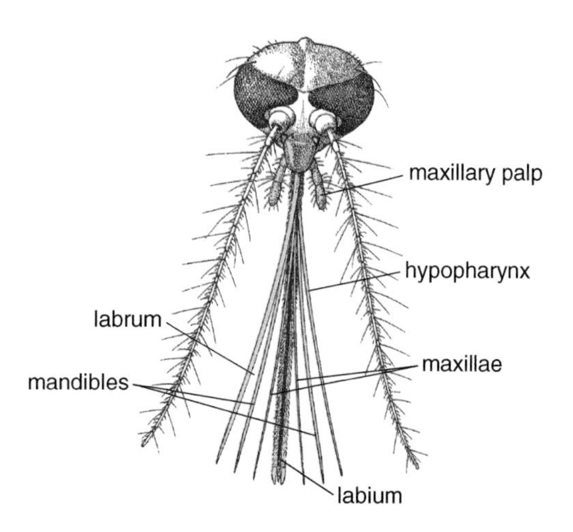

Fig. 10.12 Mouthparts of mosquito (from Zanetti 1977).

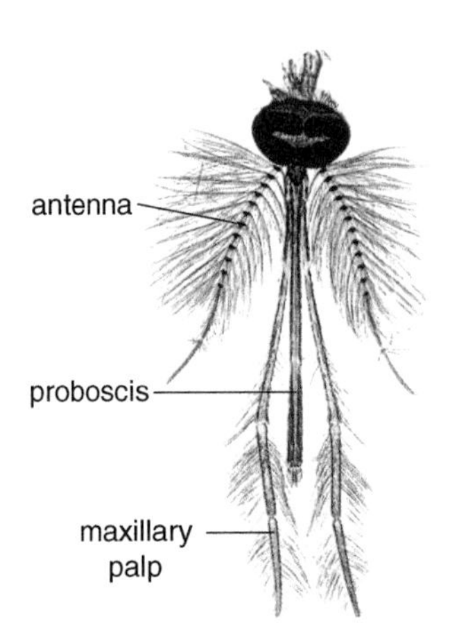

Fig. 10.13 Head of male mosquito (from Zanetti 1977).

In most species, the females must have a blood meal to obtain protein and iron needed for the production of fertile eggs, though they also feed and can survive on nectar. Eggs are laid on the water surface, often glued together as a raft. The larvae live in water and have gills, but at rest suspend themselves at the surface, where they breathe air through a terminal siphon (Fig. 10.14a) or abdominal spiracles (see later in this Section). They are filter feeders, with mouthparts like brushes sweeping in the water from which the tiny particles of food such as algae and bacteria are removed. Mosquito larvae only dive when disturbed. Like the larvae, the comma-shaped pupae can swim but also mostly hang at the water surface, breathing air through prothoracic spiracles shaped like a trumpet (Fig. 10.14b). Both the larval siphon and larval and pupal spiracles are equipped with hydrophobic hairs, which prevent wetting and breaking of the surface tension; however, if you float oil or detergent on the water, the insects sink. The water on which the adult female deposits her eggs cannot be fast-flowing, so mosquitoes breed in the still water of lakes, puddles, tree-holes, discarded tin cans etc. Adults emerge from the pupae hanging at the water surface; the whole life cycle can be as little as 5 days, but more usually takes 3–6 weeks, depending on temperature and success in obtaining a blood meal. In most species, the males swarm in a cohesive cloud at dusk, and the females fly into the cloud of males to mate. Adult mosquitoes can live for up to 8 weeks, but normal life expectancy in the field would be less than 2 weeks.

The most notorious characteristic of mosquitoes is that they transmit serious diseases of humans and animals. Different species, and even subspecies within species, transmit different diseases with different efficacy.

In relation to these diseases, the important distinction is between anopheline and culicine mosquitoes, and the easy ways to differentiate females and larvae are shown

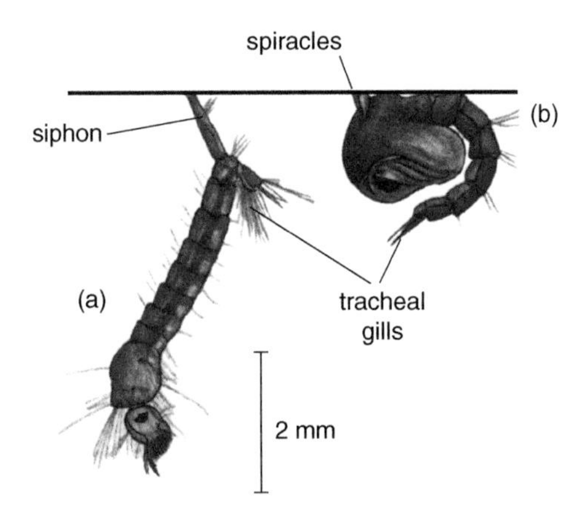

Fig. 10.14 (a) larva and (b) pupa of *Culex pipiens* (from Mandahl-Barth 1973, with permission).

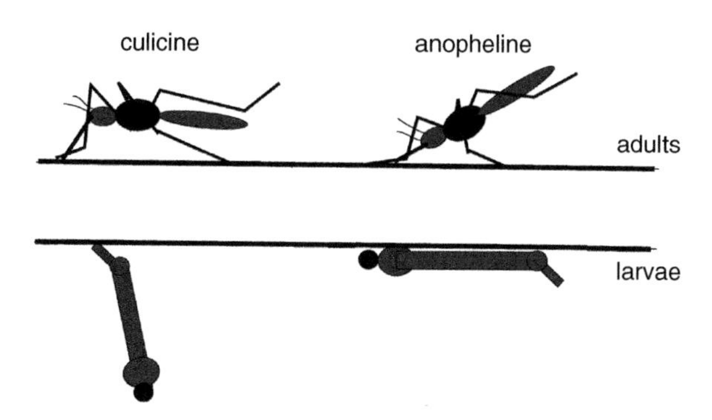

Fig. 10.15 Cartoon showing how to distinguish culicine and anopheline mosquitoes as females and larvae.

in Fig. 10.15. The angle at which the female hold the body at rest is steep with anophelines, and the larvae breathe through abdominal spiracles. Therefore they lie parallel with the water surface rather than hanging down as do the siphon-breathing culicines. The three most important mosquito-vectored diseases are malaria, yellow fever and dengue fever. Malaria is vectored by the genus *Anopheles* (an anophiline), whereas the culicine *Aedes aegypti* vectors yellow and dengue fevers. Both genera have species that have adapted to temperate climates by hibernating, but most mosquitoes are tropical and subtropical where they remain active throughout the year. In the UK, there are about 33 species of mosquito, and there the culicine *Culex pipiens* is probably the usual culprit of mosquito bites on humans.

The causal organism of malaria, vectored by species of *Anopheles*, is the protozoan *Plasmodium*, especially *P. falciparum* for severe cases of the disease. *Anopheles gambiae* is probably the most efficient vector. There are still more than 225 million cases of malaria annually, resulting in over 750,000 fatalities, most of children in sub-Saharan Africa, the region where 90% of all malaria-related deaths occur. Otherwise the disease is widespread throughout the tropics and subtropics, including much of Africa, Asia and the Americas. There are a few *Anopheles* species in Europe with five in the UK, where malaria was known in the Middle Ages and later; it was the 'ague' of which Falstaff complained in Shakespeare's plays. It seems feasible that climate change may cause 'ague' to reappear in the UK in the future.

The symptoms, fever, headache, sudden sweating, and coma in severe cases, result from asexual multiplication of the protozoan parasite in the red blood cells, usually about 6–14 days after the bite by an infected mosquito. The parasites, injected into the victim's blood from the salivary glands of the mosquito, first migrate to the liver where they multiply asexually before escaping into the blood again to enter the red blood cells. When a mosquito takes a blood meal from an infected human or other vertebrate, the parasites lodge in the salivary glands. Here they reproduce into the progenitors of gametes, which transfer to the mosquito's haemolymph and form the gametes for sexual reproduction which fuse and lodge in the gut wall as an oocyst. The asexual progeny emerging from this structure migrate to the salivary glands and are the stage of the parasite which may infect a new human.

The ideal control measure would be a vaccine, and the search for this has been energetic for many years, with the researchers getting ever closer. At time of writing

(November 2011), it appears the researchers are on the cusp of success using traditional methods. However, the best prospects may lie in the new science of synthetic biology, where the DNA sequence required to produce a novel protein can be created on a DNA synthesiser. The synthetic DNA so created can be introduced as a gene into a self-replicating organism such as a bacterium by standard genetic modification procedures. Prophylactic drugs have long been available, but *Plasmodium* rapidly develops resistance to them. There are times when no really effective prophylactic has been available. On one of my trips to Nigeria, I was expecting to take one paludrin tablet daily, only to be told on arrival that paludrin was no longer effective and that I should switch to the quinine-based drug chloroquin. I then went to the north of the country where I found that fellow Europeans still using paludrin had upped the dose to 14 tablets a day! What is still probably the most reliable safeguard is to sleep under an insecticide-impregnated bed net.

Yellow fever is a virus disease transmitted by female culicines of the genus *Aedes*, particularly *A. aegypti* but also other species including the 'tiger mosquito' *A. albopictus*. There are some 200,000 cases of the disease in humans annually, with 30,000 deaths in non-vaccinated populations; unlike for malaria, there has long been a safe and efficient vaccine. The disease has caused epidemics in Europe, but the normal distribution is in the tropical and subtropical areas of Africa and South America; 90% of infections occur in Africa.

After transmission to a human from a mosquito in its saliva, the virus replicates in the lymph nodes and later reaches the liver. In most patients, initial fever, nausea and pain die down after a few days. In others, unfortunately, the disease enters a more toxic phase with damage to the liver resulting in jaundice (hence 'yellow' fever) accompanied by internal bleeding; death often results. It also seems possible that the virus can be passed down from the mosquito to her eggs and hence to the next generation.

Dengue fever is another virus disease transmitted by *A. aegypti* and occasionally also by *A. albopictus*. Symptoms are fever, headache, a skin rash and muscular pain; in a small proportion of sufferers the disease can advance to be life-threatening, with fever resulting from internal bleeding and what is called 'dengue shock syndrome' (organ failure resulting from exceptionally low blood pressure). The disease occurs in more than 100 mainly tropical countries, and incidence has increased greatly since the middle of the 20th century with 50,000–100,000 cases annually. There is at present no vaccine against the virus, and there are considerable efforts to reduce vector populations by reducing suitable habitats.

About 120 million people in the tropics suffer from the mosquito-vectored disease elephantiasis, the great enlargement of limbs or genitals. The disease is caused by nematode worms (filariae) blocking the body's lymphatic system. The threadlike worms can reach 10 cm in length and may live for 3–8 years, though exceptionally longevities of 20 and once of 40 years have been recorded. Although three species of filaria cause elephantiasis, 90% of cases are caused by one species, *Wuchereria bancrofti*. The filariae are transmitted by mosquitoes in the genera *Aedes*, *Anopheles* and *Culex*.

10.2.5 Family Chironomidae (non-biting midges)

This Family is very large, with over 5000 described species, and a global distribution. The adults (Fig. 10.16) look rather like mosquitoes, but do not have elongated sucking mouthparts or scales on the wings. They are very small flies; some have a wing length of less than 1 mm, and 7.5 mm is about the maximum. The small species are quite

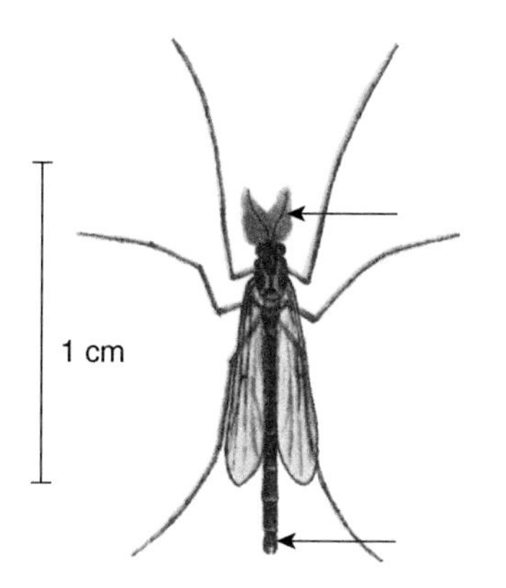

Fig. 10.16 Chironomidae (non-biting midge) (from Mandahl-Barth 1973, with permission).

hard to distinguish from the biting midges (see Section 10.2.6), but they normally rest with the wings held along the sides of the body; the biting midges fold theirs flat over the abdomen. The thorax is much humped and often the head is not visible at all in dorsal view. The front legs are longer than the other pairs, and the male antennae are conspicuously feathery. The wings are narrow and at rest the abdomen often projects well beyond their tips. The mouthparts are much reduced and the adults of many species do not need to feed at all, since they live for only a few days.

The larvae (Fig. 10.17) are elongated (usually more than 12 mm long) and often somewhat curved, often with one anterior and one posterior pair of fleshy appendages (***prolegs***). They can be found in virtually any 'liquid' habitat, not only in streams and lakes, but also in tree holes, in leaf axils that accumulate water (e.g. bromeliads) and even rotting vegetation and wet soil. They have adapted to almost totally anaerobic conditions as in the mud at the bottom of lakes; here the larvae of many species actually contain haemoglobin as an oxygen absorber and are therefore blood-red (bloodworms). Others have tubular gills at the end of the abdomen; these species live in tubes and wiggle their bodies to create a current of water with fresh oxygen through the tube. Lakes with good oxygen content can produce huge populations of chironomid larvae; these generally form more than half the total number of macroinvertebrates present in most aquatic habitats. They are the dominant group in the Arctic, where the low temperature causes the larval stage to last at least 2 years, and sometimes seven. Two species in Antarctica are the most southern free-living endopterygote insects. Elsewhere the larval stage may vary from 2 weeks to more than a year in very deep and therefore cold water. In temperate situations, the number of generations per year varies from one to four.

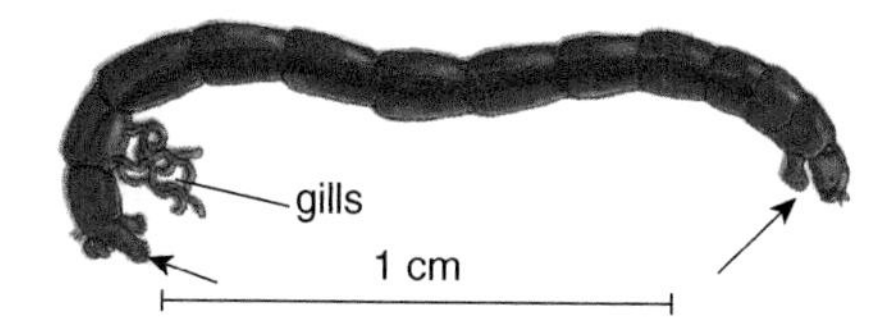

Fig. 10.17 Chironomid larva (bloodworm); the prolegs are arrowed (from Mandahl-Barth 1973, with permission).

The pupal stage lasts just a few days and 'swims' up to the water surface where the adults emerge, frequently synchronised and swarming in huge numbers. As they emerge, many are picked off by insectivorous birds such as swallows strafing over the water.

10.2.6 Family Ceratopogonidae (biting midges)

These flies are less than 0.5 mm long, and resemble the non-biting midges in having a humped thorax (though less so) largely obscuring the head in dorsal view (Fig. 10.18). As pointed out above, however,

Fig. 10.18 Biting midge (*Culicoides* sp.) with wings not held as in life (from Colyer and Hammond 1951, with permission).

they can be distinguished by the fact they fold the wings flat over the body when at rest, and the wings then extend over most of the abdomen. The common name of the Family is only too accurate; the 'bite' (so called 'bites' by blood-sucking flies are of course by piercing and not by chewing) can be very painful and the resulting swelling can last for more than a week. This is certainly the case with midges of the common genus *Culicoides*. The pain and itching is a reaction to proteins in the female's saliva. The approach of the insect is undetected by the vertebrate host and the pain comes as a sudden sharp shock; hence the North American colloquial term 'no-see-ums'. An allergic reaction of horses to the bite of *Culicoides* is known as 'sweet itch'.

The females of most species, including the genus *Culicoides* and some species of *Forcipomyia,* suck blood of birds and/or mammals, while other *Forcipomyia* species and the genus *Atrichopogon* are ectoparasitic on larger insects. Males are not blood-feeders. Other genera are predatory on small insects, but the genus *Dasyhelea* feeds only on nectar. Ceratopogonids are economically beneficial as important pollinators of some tropical crops, for example cacao.

Ceratopogonids are found in almost any aquatic or semiaquatic habitats in the world, and the blood-feeders are often pestilential nuisances on beaches or in mountain areas. Indeed the larvae live mainly in water, particularly where the organic content is high. Thus they are found in swamps and sites such as rotten wood, tree holes, compost, mud and in the water-holding axils of plants.

The blood-sucking habit means that female biting midges are capable of transmitting a range of protozoan, virus and filaria worm diseases. However, the most important disease they vector is undoubtedly the blue tongue virus which affects sheep, cattle, goats as well as a range of related wild hosts (including deer and dromedaries!). Major symptoms in the animals are fever, excessive salivation, nasal discharge and swelling of the face, together with swelling and blue colouration of the tongue; infected animals may also show foot lesions. Only some infected animals display symptoms, but death of the sickest ones may occur within a week of developing symptoms; these appear within a month of the date of infection.

Blue tongue is vectored particularly by *Culicoides imicola*, but also by other *Culicoides* species. The disease has always been known from the tropics and subtropics and has regularly occurred in the Mediterranean region; though here it subsides each year as *C. imicola* cannot survive the winter there. As infected animals recover during these winters, it appears that the virus depends on the midge over winter.

Blue tongue has recently spread north very rapidly, and the first suspected case in the UK was reported in September 2007. Since then the disease has spread to even colder climates; it reached Denmark in 2008 and Norway in 2009. The recent rapid spread is probably an effect of warming due to climate change, but also the disease has jumped to two new vectors, *C. obsoletus* and *C. pulicaris,* which are found over much of Europe.

Another disease believed to be vectored by *Culicoides* is the Schmallenberg virus, a new disease first reported from a town of that name near Dortmund in Germany in November 2011. The virus affects sheep, goats and cattle and the affected young are still born and/or have limbs fused together. The first cases in the UK were reported in spring 2012 in Kentish lambs, probably because ewes had been bitten the previous autumn by midges that had been blown across the Channel.

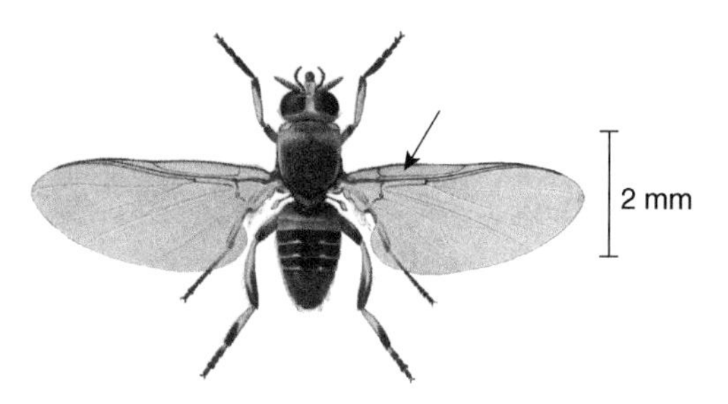

Fig. 10.19 Black fly (*Simulium* sp.) (from Edwards *et al.* 1939).

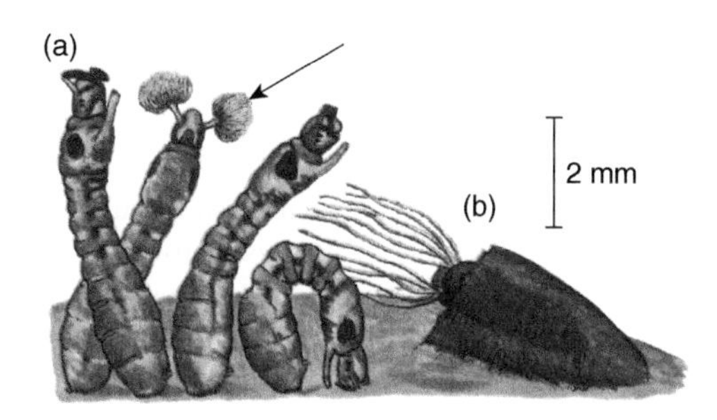

Fig. 10.20 (a) Simuliid larvae with (b) pupa (from Mandahl-Barth 1973, with permission).

10.2.7 Family Simuliidae (black flies)

This is a Family of small rather stout-bodied and black or grey flies (Fig. 10.19), with antennae not noticeably longer than the head and with very deep wings (i.e. a large distance between the leading and trailing edges). The wings, like those of the next Family, the Sciaridae, have thickened veins near the leading edge with the other veins much fainter. Most of the over 1800 known species belong to just one genus, *Simulium*. The females feed on blood, while the males feed on nectar. Black flies are cosmopolitan in distribution.

Eggs are laid in fast-flowing unpolluted water, and the larvae (Fig. 10.20) construct a little silk mat on a stone and anchor themselves to this with small hooks at the tip of the abdomen. Fan-like structures around the mouth filter small organic particles from the water as it rushes past, and every few seconds the fan's catch is scraped into the mouth. The larva pupates under water, and the emerging adult floats to the water surface in an air bubble. The adults may range many miles from their water origin in search of a blood meal. There may be one or two generations a year.

Simuliids are day-active, and their 'bite' is painful. Huge populations can be found in the Arctic and parts of Canada in the summer, to the extent that human activity out-of-doors becomes impossible. *Simulium equinum* can distress horses, as well as causing weight loss and sometimes death in cattle. There is a peculiar recurrent human nuisance of black flies in the UK, specifically around the Dorset Town of Blandford Forum, though it also found at a few other locations. The species responsible, *S. posticatum*, is actually known as the Blandford fly.

However, the biggest threat presented by black flies is the transmission of several serious human diseases, pre-eminent of which is onchocerciasis (river blindness) in Africa, caused by the parasitic nematode *Onchocerus volvulus*, for which simuliids act as the host for the filariae. *Simulium* spp. vector the disease; the most important has the no doubt heartfelt specific name of *S. damnosum*! A successful 14-year black fly control programme in West Africa has saved many thousands of people suffering from the disease.

10.2.8 Family Sciaridae (mushroom flies, dark-winged fungus gnats)

These small dark flies mostly have an obvious fork in the venation, located pretty well in the middle of the wing (Fig. 10.21b), the whole of which is often dusky. Diagnostic is that the eyes join up as a bridge over the antennae (Fig. 10.21a). Eggs are laid in the soil, and the rather elongate larvae generally feed on decaying plant material. Breeding in glasshouses can be continuous, and a 4 to 5-week life cycle means that populations can grow rapidly. Unfortunately, they are not totally dedicated to dead material, and the name 'mushroom flies' is indicative of their pest status in mushroom beds, where the larvae eat the mycelium and even burrow into the fruiting bodies. They will also graze living roots, and have become a regular problem on ornamentals in nurseries and glasshouses. As well as the adults flying in, the larvae can be brought in on inadequately sterilised compost or its invasion after sterilisation.

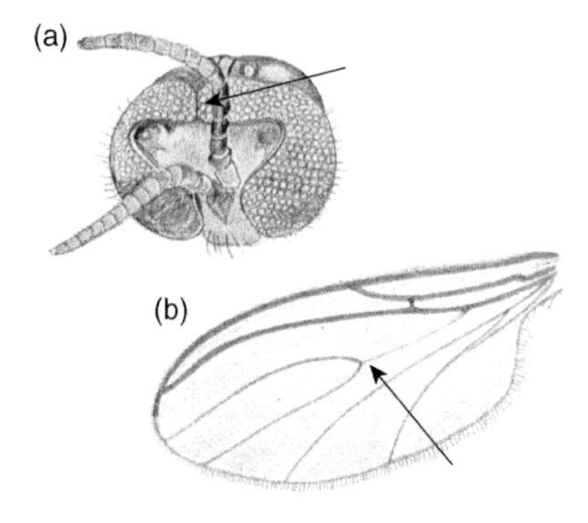

Fig. 10.21 (a) Head and (b) wing of *Sciara* sp. (from Colyer and Hammond 1951, with permission).

10.2.9 Family Mycetophilidae (fungus gnats)

Closely related to the Sciaridae, the fungus gnats are small delicate flies without the eye bridge over the antennal insertions. These antennae are long and slender as are the legs, particularly the hind ones (Fig. 10.22). Like the Chironomidae and Ceratopogonidae, the thorax is humped to conceal the head in dorsal view. The larvae feed in fungi or are predatory on other insects there, and are often abundant though not a problem in commercial mushroom production. The Family is only included here because fungus gnats are often encountered and it is important to distinguish them from sciarids, which are of economic importance.

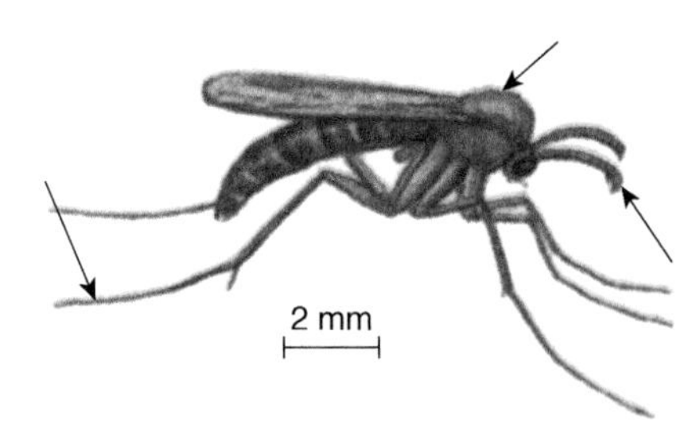

Fig. 10.22 Fungus gnat (Mycetophilidae) (from Mandahl-Barth 1974, with permission).

10.2.10 Family Bibionidae

This Family is only included here because the members are distinctive and commonly encountered. The St Mark's fly (*Bibio marci*) is a common sight resting on flowers around St Mark's day, the 25th April. Bibionids are stout quite large black flies which are obviously hairy, including the compound eyes, and have a spur on the front tibia or circlets of small thorns at the tip of the tibiae (Fig. 10.23). The males have eyes that meet up to occupy the entire

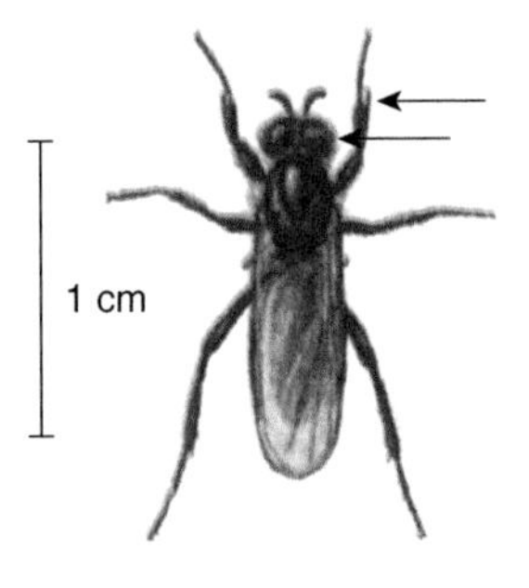

Fig. 10.23 Bibionidae (*Bibio* sp.) (from Lyneborg 1968, with permission).

head in front view. The larvae have a partial head-capsule and live in decaying matter or the soil; they graze plant roots and so can be minor pests.

10.3 Suborder Brachycera

'Brachy' means 'short', and this Suborder is distinguished by the antennae being reduced to three segments. The third segment is larger than the others and continues at its tip into either an annulated projection (***style***) or a terminal bristle (***arista***) (Fig. 10.6c,e–g). The larvae are hemicephalous or acephalous and typically amphipneustic, though peri- and metapneustic ones may also be found. The pupae are exarate, but in just one Family (Stratiomyidae) it remains enclosed in the last larval skin.

10.3.1 Family Tabanidae (horse flies)

These are flies whose bite is painful, and which attack humans and cattle. They do not suck blood, but have a normal labium and labellae to lick up the blood welling from the cut in the skin slashed by the elongate and knife-like mandibles (Fig. 10.24). The cut is then opened up by the stirrer-like maxillae while an anticoagulant is stirred in to keep the blood from clotting. The larger the horse-fly, the easier it is to hear it coming. Cessation of the buzz means it may have settled on you and you may be able to brush it off in time. That the pain of the bite is actually inversely proportional to size, and proportional to stealth of approach, means the smaller tabanids are more unpleasant than the larger ones. Thus a bite by the large *Tabanus sudeticus* can often be avoided, the more painful bite of the common cleg *Haematopota* spp. is more likely, and you get no notice of the searing pain inflicted by the small *Chrysops* spp., easily recognisable by the iridescent green eyes with black spots.

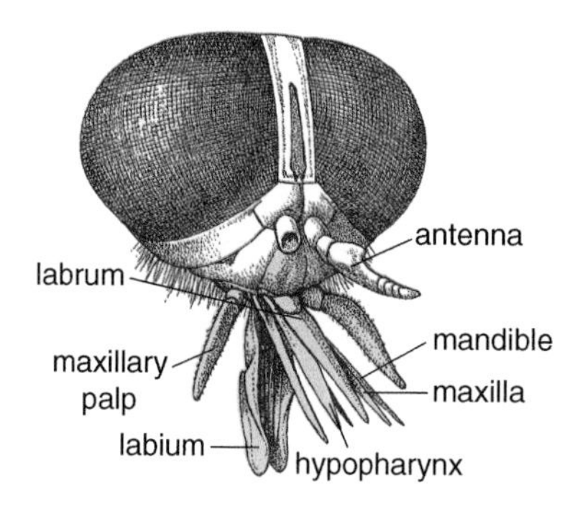

Fig. 10.24 Mouthparts of horse fly (from Zanetti 1977).

Tabanid antennae have that annulated style at the tip of a wider third segment (Fig. 10.6g) mentioned above. They are stout-bodied flies with large compound eyes and a rather blunt tip to the abdomen (Fig. 10.25). The wings have complex venation with many cross veins. The larvae are aquatic and filter feeders in fast-flowing water, even waterfalls. They attach themselves to stones with an anal sucker and breathe the oxygen in the water through filamentous thoracic gills.

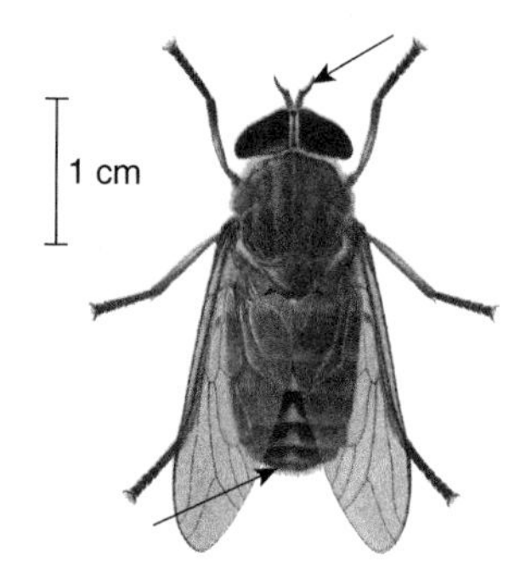

Fig. 10.25 Horse fly (*Tabanus sudeticus*) (from Edwards *et al.* 1939).

10.3.2 Family Stratiomyidae (soldier flies)

As mentioned above in Section 10.3, this is the Family where, as in the Cyclorrhapha (see Section 10.4), the exarate pupa remains enclosed in the last larval skin (forming the puparium). They are only mentioned here because of this peculiarity in their life history, and because their often bright metallic colours make them noticeable.

Stratiomyids are bristleless flies, which are often brightly coloured with a metallic sheen (Fig. 10.26). The third antennal segment is narrow and entirely annulated to its tip, so no separate style is readily discernible. Most species spend much of their time resting on vegetation rather than flying, and are often found near water, where they lay their eggs on waterside plants or on the water surface. Larvae are carnivorous or scavengers in water or soil. The aquatic larvae hang from the water surface with a ring of hydrophobic hairs over the rear spiracles. These hairs trap an air bubble when the larva submerges (see plastron respiration in Section 7.5.1.1, water boatmen).

Fig. 10.26 Soldier fly (Stratiomyidae) (from Colyer and Hammond 1951, with permission).

10.3.3 Family Asilidae (robber flies)

These are again robust largish flies (Fig. 10.27) with an antenna of three segments and an apical style. The tapering abdomen is longer and more slender than that of horse flies. The hardened mouthparts point slightly forwards; this and the stout legs set forward on the thorax are adaptations to hawking for prey on the wing. The colour of the flies may be a dark grey or brightly coloured with red or yellow and black. The larvae are scavengers on decaying vegetable matter in a variety of relatively dry habitats including leaf litter, under the bark and in the wood of fallen trees.

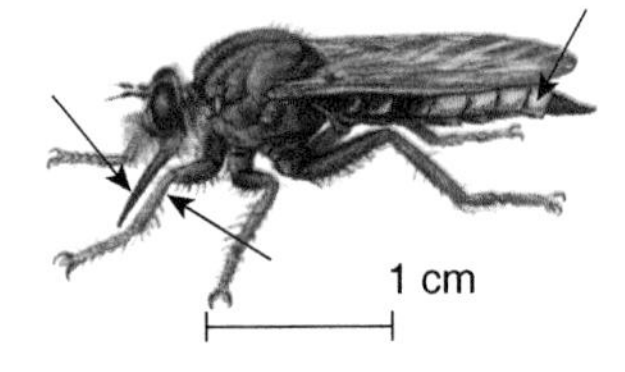

Fig. 10.27 Robber fly (Asilidae) (from Lyneborg 1968, with permission).

Adult robber flies are certainly carnivores, but are not regarded as potential biological control agents of crop pests.

10.3.4 Family Empididae (assassin flies)

These are much smaller than robber flies and one of the several brachyceran families where the antenna ends in either a terminal arista or a style. The adult flies have long hardened predatory mouthparts, which are considerably longer than the head is deep and project down vertically to well below the depth of the thorax (Fig. 10.28). The larvae are also predatory and are found in soil, rotting wood etc., though little is known about the immature stages of many species. Swarms of adult empids have been known to arrive on crops and cause a noticeable reduction in the aphid population there.

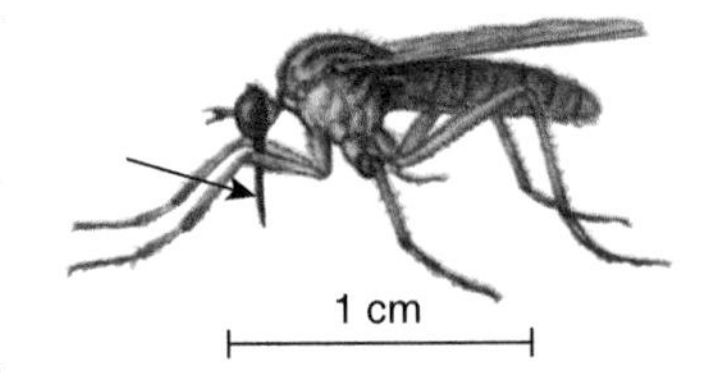

Fig. 10.28 Assassin fly (Empididae) (from Lyneborg 1968, with permission).

10.4 Suborder Cyclorrhapha

The Cyclorrhapha are those flies that pupate within the last larval skin, that is they emerge as adults from a puparium. The word Cyclorrhapha means 'circular break', referring to the circular hole left at the end of the puparium by the emerging fly. The adults are defined by antennae of three segments with the third enlarged and bearing an arista (often feathery) **set back** from the tip (Fig. 10.6d). In most Families, the third antennal segment takes the distinctive form of a sausage-shaped lozenge hanging down vertically against the front of the head capsule (Fig. 10.5). Larvae are acephalous with mouth hooks, and the majority are amphipneustic (Fig. 10.2b). The Suborder is divided into two taxa, the Aschiza and the Schizophora. I shall refer to these taxa as 'Series'.

10.4.1 Series Aschiza

These Cyclorrhapha manage to crack open the puparium at emergence without the aid of a ptilinum; the Series is therefore defined as not possessing a ptilinal suture. Whether or not there is such an inverted horseshoe-shaped suture would seem an easy character to judge, but in reality it can prove rather difficult for two reasons. Firstly, some Cyclorrhapha are very small with uniformly black heads. Secondly, even in the Aschiza, the antennae are recessed in a depression in the front of the head capsule, and the edge of this depression often looks for all the world like a ptilinal suture. I therefore offer you the following pieces of advice:

1) The best place to look for a ptilinal suture is to start on the third antennal segment at the point of insertion of the arista and follow an imaginary horizontal line to the edge of the compound eye (Fig. 10.29). Have you crossed a sculptured line (i.e. the line projects from the surrounding cuticle) and not just an edge formed by a change in angle of the cuticle?

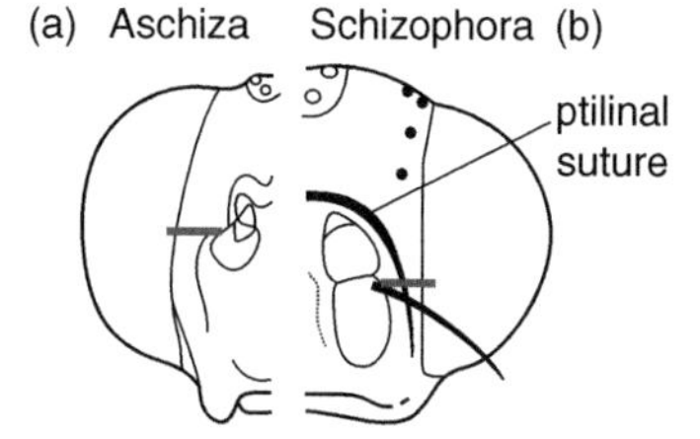

Fig. 10.29 Half-heads of (a) Aschiza and (b) Schizophora to show the imaginary line (in red) that identifies the presence or absence of a ptilinal suture.

2) Any very small fly (i.e. much smaller than a housefly) is unlikely to be one of the Aschiza.

3) Syrphidae (hover flies) are by far the most common Aschiza. There is an easy spot character for this Family on the wing (see Section 10.4.1.1), and hover flies are large enough to make this character quite easy to see. If your fly is a cyclorrhaphan and doesn't have this wing character, you are normally safe to put it in the other series (Schizophora). Perhaps it is also worth looking at a picture of a bee fly (Family Bombyliidae). You might just meet one of these Aschiza, but they are very different with their long mouthparts pointing forwards.

10.4.1.1 Family Syrphidae (hover flies)

The adults (Fig. 10.30) are mostly stout and colourful flies with orange, black and yellow patches or stripes. They feed as adults on the pollen and nectar of flowers and on aphid honeydew. Some closely resemble bees and are often mistaken for these, but syrphids can and do hover and remain stationary in mid-air for several seconds in a way bees cannot.

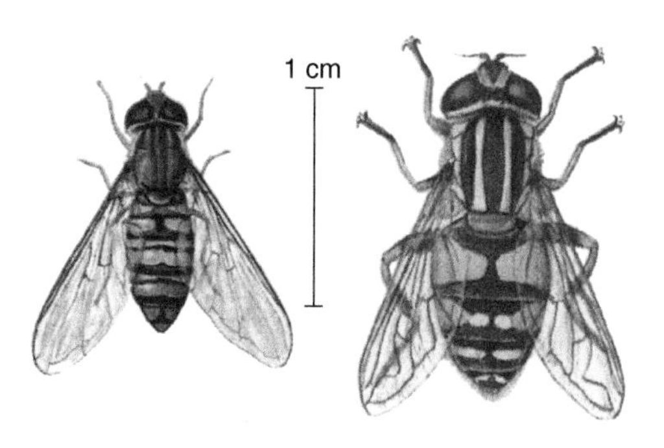

Fig. 10.30 Hover flies (Syrphidae) (from Lyneborg 1968, with permission).

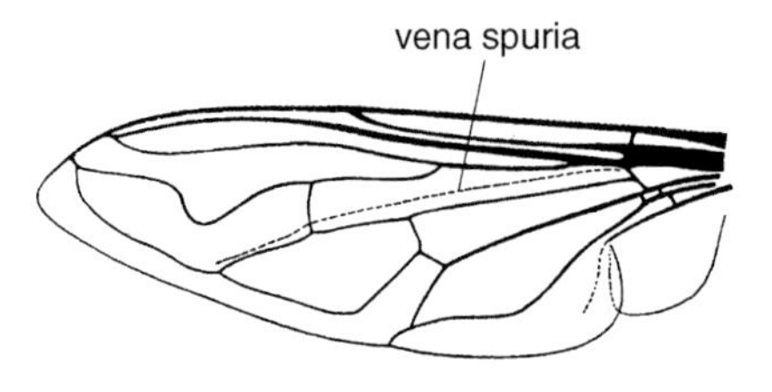

Fig. 10.31 Wing of hover fly to show the vena spuria.

The spot character for a syrphid is near the middle of the wing. Here there is a fold, which looks very much like an extra vein (Figs 10.31 and 10.33) through the upper of the two main longitudinal cells. However, it is a false (spurious) vein known as the ***vena spuria***. But how can you tell this vein is 'spurious'? Easy. True veins never cross cross-veins. Veins and cross-veins are like the bonding of bricks, a vertical join never continues across the horizontal mortar layer. But the vena spuria goes straight across the cross vein at the end of the long cell, like the intersection on a St George's cross.

Syrphid larvae show a variety of feeding habits. Many are aphidophagous (e.g. *Episyrphus balteatus* (Fig. 10.30, left), *Eupeodes corollae*, *Sphaerophoria scripta*, *Syrphus ribesii* and *Platycheirus* spp.) are among the most important natural controlling agents of these pests. Apart from *Platycheirus* spp., the adults lay their white elongate eggs in or close to aphid colonies and the often green, slug-like larvae (Fig. 10.32a) suck dry several hundred aphids (150–500 depending on the size of the syrphid species) during development. The puparium (Fig. 10.32b) is flask-shaped.

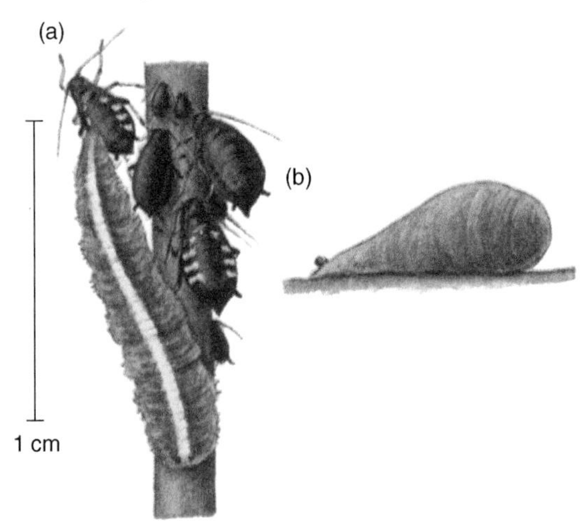

Fig. 10.32 (a) Larva and (b) puparium of an aphidophagous syrphid (from Lyneborg 1968, with permission).

Other syrphids (the large narcissus bulb fly *Merodon equestris* (Fig. 10.33)) and two species of the genus *Eumerus* (small narcissus bulb fly) are pests in the flower bulb industry. Oviposition is from April to June in the UK. A single egg is laid into the soil very close to a bulb; indeed, the female may crawl down the gap left in the soil around the bulb by the withering of the leaves. The hatched larva enters the bulb through the base plate, and gradually eats out the centre of the bulb, leaving the following spring for a 5 to 6-week pupation in the soil. The mature larva of *M. equestris* is nearly 2 cm long. Small bulbs are killed; larger bulbs grow out, but the foliage is distorted and the flower is not formed.

Larvae of several genera of rather stout hover flies, for example *Helophilus* (Fig. 10.30, right) and *Eristalis*, feed on decaying organic matter at the bottom of pools of

Fig. 10.33 Large narcissus bulb fly (*Merodon equestris*). The vena spuria diagnostic for the Family Syrphidae is arrowed (from Colyer and Hammond 1951, with permission).

stagnant water such as in tree-holes. The larvae adjust to the wide fluctuation in water level by breathing air with their rear spiracles down a long telescopic siphon to the surface – hence the colloquial name 'rat-tailed maggot' (Fig. 10.34). *Volucella* spp. scavenge in the waste that accumulates at the bottom of wasp and bumble bee nests. Perhaps the strangest hover flies are *Microdon* spp., whose larvae are taken into ant nests (like some lycaenid caterpillars; see Section 9.2.3.9, Family Lycaenidae) where they hide away by day but move around at night feeding on the ant brood unmolested.

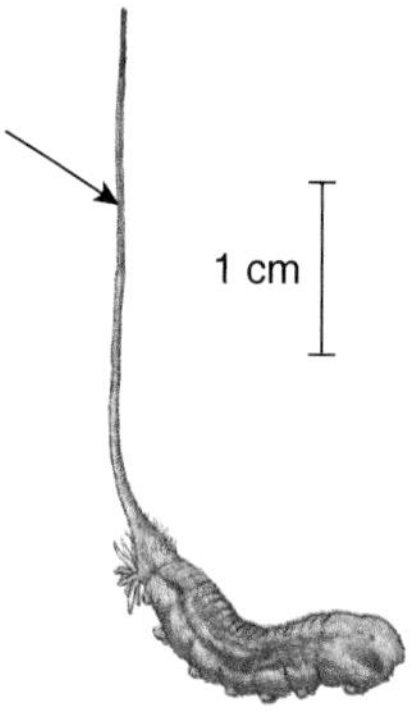

1 cm

Fig. 10.34 Larva of *Eristalis* sp. (rat-tailed maggot) (from Mandahl-Barth 1973, with permission).

10.4.2 Series Schizophora

In this Series a line of weakness on the hardened last larval skin (puparium) is cracked off as a smooth 'nose cone' by a ptilinum to allow the adult already emerged from the exarate pupa to finally emerge from the puparium. These are the flies that possess a ptilinum and therefore a ptilinal suture (see Section 10.4.1 for recognition of this character). As pointed out previously, most crop pest Diptera are to be found in this Series. Damage is by the larvae and nearly always takes the form of feeding on roots, tunnelling in stems (stem borers), mining leaves or feeding on seeds and fruits. The Series is divided into three taxa, Acalyptrata, Calyptrata and Pupipara. I shall call these taxa 'Divisions'. Below I shall therefore describe only how to separate adult calyptrates and acalyptrates, ignoring the third Division (Pupipara) which are atypical calyptrates that nurture all larval stages within their body and finally 'lay' larvae ready to pupate.

Although the terms 'calyptrate' and 'acalyptrate' refer to a morphological character on the wing, the taxonomic criterion that decides the taxon of your specimen is a character on the second antennal segment. This segment may be very small indeed, yet you need to decide whether it has (calyptrate) or has not (acalyptrate) a vertical cleft in it (Fig. 10.35). This character is often hard to distinguish in small flies, but your best chance is to look for the start of the cleft just above the junction of that second antennal segment with the much larger third. Often the cleft widens out here like a river mouth on a map; it is therefore wider, deeper and more easily spotted. However, as with the Aschiza/Schizophora determination (see Section 10.4.1), there

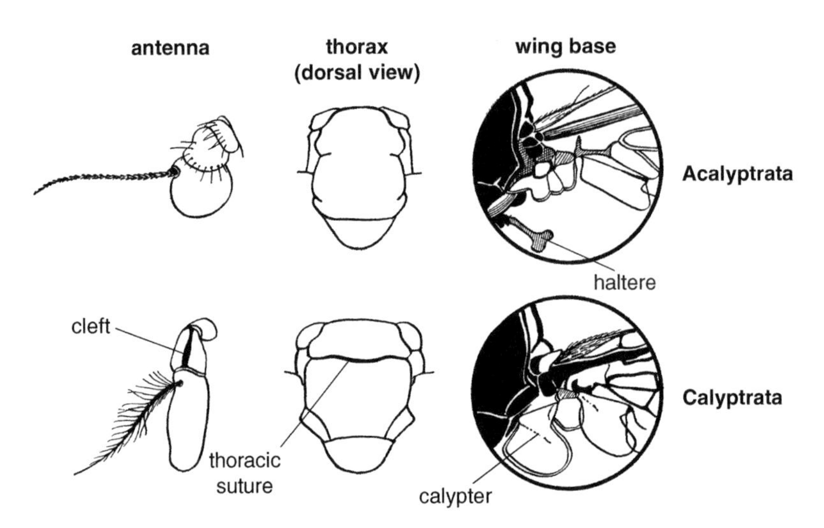

Fig. 10.35 Schizophora: characters on antenna and thorax to distinguish the Acalyptrata and Calyptrata (assembled from Colyer and Hammond 1951, with permission).

are some additional guidelines I can offer you. The second and third guidelines work very well for all but the calyptrate Family Scathophagidae (see Fig. 10.44) and the bot flies, which have the acalyptrate incomplete thoracic suture (see Section 10.4.2.2, Family Oestridae and Fig. 10.56). These flies are, however, placed in the calyptrates because they show that second antennal segment cleft, yet my guidelines would cause you to mistake them for acalyptrates. With the Scathophagidae, the problem is compounded since also calypters (see Section 10.4.2.2) are missing; the bot flies have very small ones. Nevertheless, here are my supplementary guidelines for the majority of Families:

1) Very small flies (i.e. noticeably smaller than a housefly) will be acalyptrates.

2) Look at the thorax directly from above. You may have to tilt it back and forth to catch the light on the relevant area. In the Calyptrata there is a complete groove (***thoracic suture***) right across the dorsal surface of the thorax in front of the wing insertions. In the Acalyptrata, this groove is absent or partial at the sides – it never crosses the midline of the thorax from one side to the other.

3) If the wing is stretched out sideways, the haltere is still visible from above in the Acalyptrata. In the Calyptrata it is hidden from above by an extra opaque lobe (the ***calypter***) which pulls out with the wing at the junction of its trailing edge with the thorax.

10.4.2.1 Division Acalyptrata

Family Psilidae

Psila rosae (carrot fly) is a small shiny black fly with few hairs on the pointed abdomen (Fig. 10.36). The fly roosts in vegetation adjoining carrot fields and on warm damp days flies at low level into the crop to oviposit on the soil around carrot plants. Recently thinned crops have looser soil around the plants which aids the movement of the young larvae into the soil. The white elongate eggs have characteristic vertical ribs. The larvae first feed externally on the developing carrots, leaving unsightly open

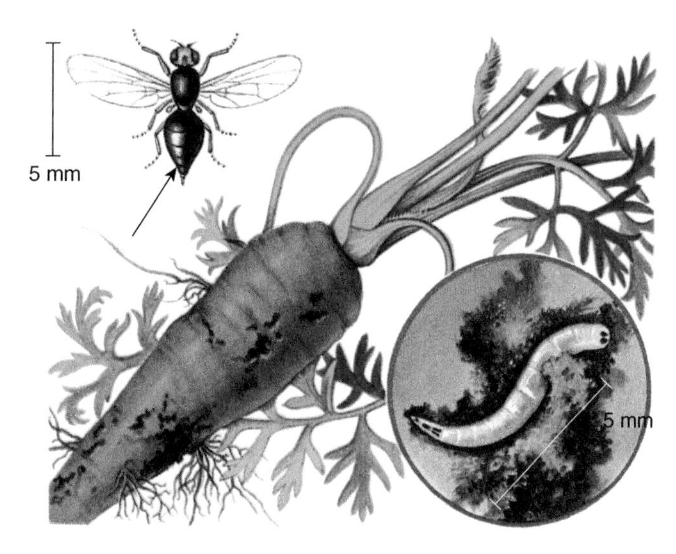

Fig. 10.36 Carrot fly (*Psila rosae*) adult fly with damage to carrot and feeding larva (inset) (from Bayer 1968, with permission).

tunnels which get invaded by fungal diseases normally excluded by the integrity of the epidermis of the carrot. After the first moult the larvae tunnel more deeply into the carrot. Pupation is in the soil. Carrots near the edge of the field adjacent to the roosts receive the most severe damage. As farmers tend to inspect edge plants to avoid trampling into the field, damage can be overestimated and control undertaken needlessly. The insect is found in Europe and North America.

Family Sepsidae (semaphore flies)

This Family is mentioned only because semaphore flies are both common and very easily identified. They are shiny black with a large dark spot near the tip of each wing. Also, when walking, they wave their wings about at different angles as if they were signalling with semaphore flags.

Family Ephydridae (shore flies)

These are small black flies usually found in damp situations. There are no really diagnostic single characters for this Family, and a permutation of characters is required. The costa is distinctly broken twice, which only occurs in a few acalyptrate Families.

Species of the genus *Hydrellia* are found boring grass stems in many parts of the world, including the UK, and a number share the common name of 'rice whorl maggot'. The rice whorl maggot *H. griseola* is an important pest of paddy rice in the USA, Europe, North Africa, Malaysia and Japan. The adult has a pale shining frontal ***lunule*** (the small area between the insertions of the antennae) on an otherwise black face. The curved eggs with a reticulate sculpturing are laid singly on leaves near the water surface, and the hatching larvae bore into the leaf to begin a linear mine, but later the larvae eat away at the edges of their mines to enlarge them into blotch mines. Attacked leaves shrivel and characteristically lie flat on the water; heavily attacked plants may die. Pupation is within the leaf, and the dark pupae are clearly visible in the mine. The duration of the life cycle varies with climate from 2–3 weeks; similarly the number of generations in a year can vary between eight and 11. Another species of rice whorl maggot, *H. philippina*, occurs solely in the Philippines and is most unusual for a dipteran in that the larva does not mine but feeds openly on the leaf surface, scraping at the tissues in the unexpanded central whorl of leaves of the rice plant. The result is stunted plants, which do not tiller, with the expanded leaves disfigured with white areas.

Family Chloropidae

This is a Family of flies with stem-boring larvae. The adults of the genus *Oscinella* are black; some species have reddish patches on the legs, while adult *Chlorops* are black and yellowish-green. All share the Family character of a triangle of ocelli set on a large shiny black ***ocellar triangle***, the base of which at the back of the head more or less bridges the entire distance between the compound eyes (Fig. 10.37a).

Fig. 10.37 (a) Chloropid fly (photo by James Lindsey licensed under the Creative Commons Attribution-Share Alike 2.5 Generic license); (b) *Oscinella frit* (photo by Sarefo, licensed under the Creative Commons Attribution-Share Alike 3.0 Unported, 2.5 Generic, 2.0 Generic and 1.0 Generic license).

Chlorops pumilionis (gout fly) is a European pest mainly of barley, but also of wheat and rye. Eggs are laid close to the central shoot with usually only one egg per shoot, and the larva works its way down the ear, leaving a 'feeding groove'. Attacked shoots are stunted, but swell ('gouty'). If ears are produced, they are deformed and the grain on the side of the feeding groove is missing. When mature, the larva works its way back up the groove to pupate near the top. There is a second generation on winter barley from September to March, as well as on wild grasses. *Chlorops oryzae* (rice stem maggot) bores the central shoot; infested plants can be recognised by the jagged edges of the leaves that eventually emerge from the shoot.

Up to the 1970s, *Oscinella frit* (frit fly; Fig. 10.37b) was important as a pest of oats in southern England as well as elsewhere in Europe. The insect bored the stems of grasses and laid eggs on the oat crop sown in the spring. The larvae tunnelled in and killed the central shoot causing the 'deadheart' symptom, but more seriously the second generation laid large numbers of eggs on the inflorescence and the larvae fed on individual grains. Then there were three major changes in agriculture. Firstly, oats were no longer grown in the south and, secondly, a new technique of direct drilling for grass leys and cereals was introduced. The technique replaced the ploughing up of grassland by using the new bipyridyl herbicides as a chemical flame gun to kill the grass. Thirdly, once the grass was dying, a new design of seed drill then sowed new grass or wheat without any additional soil disturbance. With new varieties, this enabled wheat to be sown in late summer or early autumn ('winter wheat') rather than waiting for the following spring for ploughing to occur and a seed bed to be cultivated after the winter frosts (the up till then traditional 'spring wheat'). The mature frit fly larvae in the herbicide-killed grass migrated into the soil to find far fewer grass or wheat plants as new seedlings. The new grass seedlings were invaded and new leys often failed. Although the frit fly had not previously attacked wheat, the species clearly contained a genotype that could do so, and the frit fly switched to become a pest of autumn-sown wheat from then on.

Family Opomyzidae

Opomyzids are brown flies (Fig. 10.38) with brown shadowing at the cross veins (as also found in a related Family, the Geomyzidae). They are grass stem borers, and have a late summer generation, which laid eggs in grassland until the switch to winter wheat (see previous Section) was made. *Opomyza florum* (grass and cereal fly) then became a problem in wheat and earned a common name for the first time.

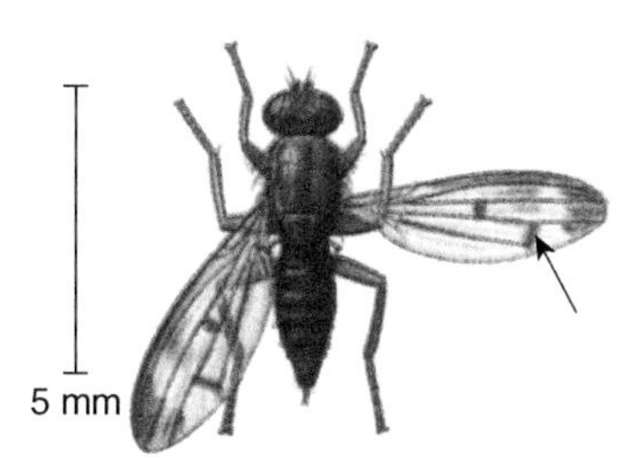

Fig. 10.38 Opomyzidae (from Chinery 1986).

Family Agromyzidae

These are small black flies, but not uncommonly with touches of yellow or green. There is a distinct single break in the costal vein just before the subcosta reaches the leading edge of the wing (Fig. 10.40).

A major pest of beans of various species in the tropics and subtropics of the Old World is the bean fly *Ophiomyia phaseoli* (Fig. 10.39). The adult is only about 2 mm long. After hatching from eggs inserted singly into the upper leaf surface near the petiole, the larvae bore into the stem (often first entering the petiole) and then tunnel to its base (Fig. 10.39b), where they pupate. Damaged plants produce few pods and seeds; the leaves turn yellow and the stems are thickened with vertical cracks. On older plants, attack is limited to tunnelling in the petioles. The life history takes 2–3 weeks.

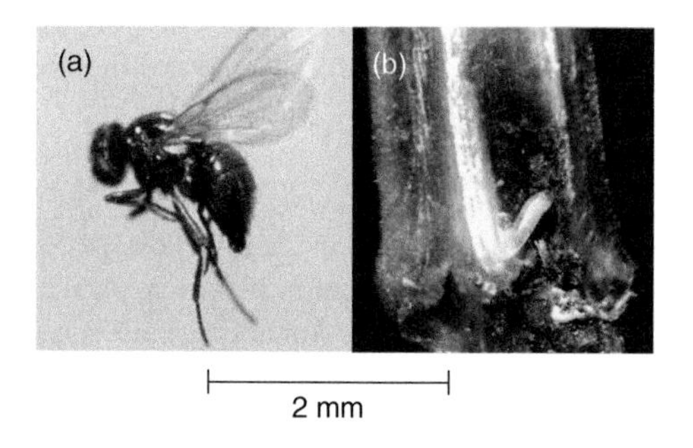

Fig. 10.39 Bean fly (*Ophiomyia phaseoli*): (a) adult (Asian Vegetable Research And Development Center); (b) damaged base of plant with larva (courtesy of AVRCD – The World Vegetable Center).

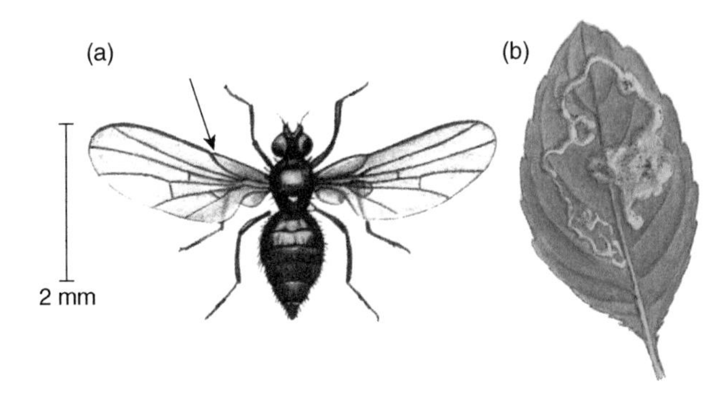

Fig. 10.40 Agromyzid leaf miner: (a) adult (from Bayer 1968, with permission), with (b) typical serpentine mine (from Mandahl-Barth 1974, with permission).

Many agromyzids are leaf miners (Fig. 10.40). The female fly makes punctures in the plant tissue with her ovipositor. She will feed from these punctures, but she also inserts her eggs in some of them. Inserting her eggs in this way counters the plant's attempts to eject the egg by applying turgor pressure to the new cells produced at the insertion as a reaction to wounding. This ejection of eggs is rather like squeezing a piece of wet soap in the hand. The larva then mines the leaf between the upper and lower epidermis, the transparent serpentine mine getting wider as the larva grows, with a trail of frass left behind it, until pupation finally occurs at the end of the mine. The chrysanthemum leaf miners, *Phytomyza syngenesiae* and *Liriomyza trifolii,* can cause financial loss by the cosmetic damage they do to pot chrysanthemums and those for the cut flower trade. As the damage is only cosmetic, I was surprised to be asked, on just about the first day I took up an appointment in the Horticulture Department at Reading University, how to control the insect in chrysanthemums that were not commercial, but were in experiments on the initiation of flowering. It turned out that the transparency of the mines caused the photographic leaf area meter to underestimate the true leaf area – a new one on me, but I can include the insects here as 'pests of leaf area meters'!

Family Diopsidae (stalk-eyed flies)

This Family deserves mention because the flies are so bizarre in appearance. The front femora are thickened, and in most species the eyes are at the end of long slender stalks. In *Diopsis thoracica* (Fig. 10.41) the stalks hold the eyes about 15 mm apart, which is half as far again as the whole length of the fly! The species is a pest of rice and sorghum in Africa, particularly in Swaziland, Cameroon and Sierra Leone. The females cement their eggs singly onto young rice leaves to resist rain-wash, and the larvae move down the leaf sheath to feed on the central shoot above the growing point, killing the central shoot so that it emerges as a 'dead-

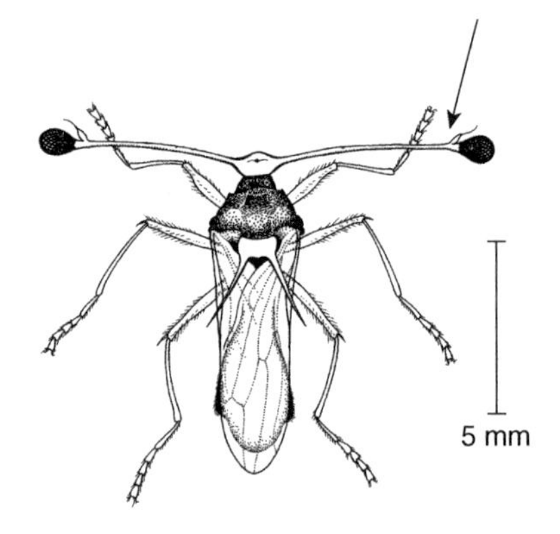

Fig. 10.41 *Diopsis thoracica* (Diopsidae) (from Hill 1983, with permission).

heart'. Later, the larvae feed on the flower head before it emerges from the leaf sheath. Pupae form within the rice stem, and are often nearly triangular in section due to pressure of the surrounding stem tissues. The life cycle takes 5–7 weeks.

Family Tephritidae (or Trypetidae) (fruit flies)

To many, the term 'fruit fly' equates with *Drosophila melanogaster*, famously used in genetics research. This laboratory insect is in the acalyptrate Family Drosophilidae, whereas the Tephritidae are the 'real' fruit flies, which damage fruit in orchard and plantation crops throughout the tropics and subtropics. The females have an easily recognised hardened ovipositor, which makes the end of the abdomen (seen from above) appear to taper to a point; the other spot character, which applies to both sexes, is a characteristic black patterning on the wings ('picture-wing' flies is another common name for the Family; Fig. 10.42).

The female uses her strong ovipositor to make a slot in the fruit, and here she inserts a packet of two to ten eggs, with a total fecundity for one female of 800–1000. Often bacterial and fungal rots then cause secondary damage. The fleshy fruits of several of the attacked plant species produce a ball of resin at the point of oviposition; this can indicate that attack has occurred, though it is a defence response and sometimes prevents successful oviposition. The larvae, which can number as many as 100 in one fruit, feed on the pulp (Fig. 10.43b) for up to 6 weeks before they are ready to pupate. At this point they may drop to the ground, but more often the infested fruit will already have fallen early. Either way, the larvae of most species then pupate in the soil. Because of their rapid reproduction, high infestation rate of the fruit and that the attack is on the marketable unit of the crop, fruit flies are such potentially serious problems that they are targets of major quarantine exercises at the borders both of whole nations and of States within the USA and Australia. Major pest species and the crops in which they are most important are as follows:

Anastrepha ludens (Mexican fruit fly; Fig. 10.42). All citrus, but also peach, pear and apple, are attacked. The adult is yellowish-brown, and the female's ovipositor is particularly long in relation to body size. The life cycle is 3–5 weeks. The pest is endemic to Mexico and Central America, but is a constant serious threat of spread to the USA, a threat which is continually resisted by quarantine and eradication measures.

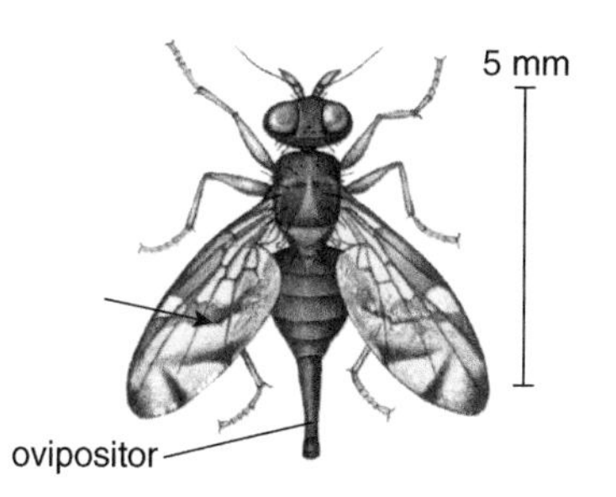

Fig. 10.42 Mexican fruit fly (*Anastrepha ludens*) (from Bayer 1968, with permission).

Ceratitis capitata (Mediterranean fruit fly, Medfly for short [USA]). This is a pest of citrus and peach, but also of many other fruit. The adult fly has a black thorax with yellow patches, a black scutellum and a yellow abdomen with grey markings (Fig. 10.43). The male has a peculiar feature – the end of the antennal arista carries a triangular expansion, so that the antenna rather resembles a fly's haltere (see Section 10.1). The life cycle takes about 4 weeks, with eight to ten generations a year, and the pest occurs in the Mediterranean region, Africa, and Central and South America.

Trirhithrum coffeae (coffee fruit fly). This pest of coffee feeds on the mucilage around the bean but this causes no real direct damage, and the insect is a pest because feeding

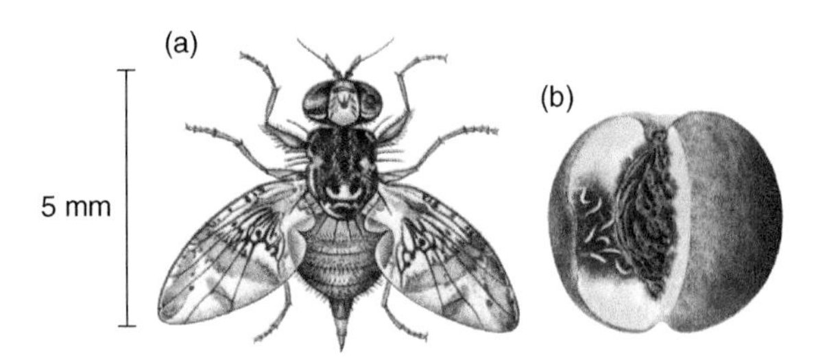

Fig. 10.43 (a) Mediterranean fruit fly (*Ceratitis capitata*) with (b) larvae feeding in the pulp of a peach (from Bayer 1968, with permission).

induces early shedding of the attacked berries. The fly is a small dark fruit fly; a generation takes 5 weeks and the pest is restricted to East Africa.

Bactrocera dorsalis (oriental fruit fly). This attacks all fleshy fruits, particularly citrus, avocado and banana. Larvae often drop to the soil to pupate before the attacked fruits fall. The fly is dark brown with yellow marks on the thorax and a yellow scutellum. The entire wing membrane between the leading veins is slightly dusky. In the tropics the life cycle is only a little over 3 weeks, and the insect is found in Asia from India to China, Indonesia and the north of Australia.

Bactrocera cucurbitae (melon fly).This is a very important pest of melon and other cucurbits in East Africa, India and Hawaii, though it also occurs elsewhere in Asia to Japan, Indonesia and Australia. Masses of larvae can be found in the pulp of one fruit. The adult is large for a fruit fly, brown with three yellow stripes on the thorax and again a yellow scutellum. The life cycle takes 3–4 weeks.

Bactrocera oleae (olive fruit fly). This is the most serious pest of olives in the Mediterranean region including Egypt, though it also occurs in South Africa. Attacked olives show a mottled exterior, fall prematurely and the oil is tainted with an unpleasant taste. The pulp of an attacked olive may contain several larvae – although a female will only lay a single egg on a fruit, other females may lay eggs on the same fruit. In the summer the larvae pupate within the olive, but the last generation pupates in leaf litter under the tree. The adult is a small dark brown fly with a small dark spot at the wing apex. The life cycle takes about a month, and there are three to four generations a year.

Bactrocera tryoni (Queensland fruit fly). This species is a pest of citrus and avocado. The fly does not breed continuously and overwinters as an adult after four or five overlapping generations in the summer and autumn. It occurs in Queensland, New South Wales, Victoria and South Australia. Outbreaks in Western Australia were eliminated by a combination of measures, including the release of sterile males. The adult is brown, marked with yellow, and the front corners of the thorax are also yellow.

Rhagoletis pomonella (apple fruit fly or apple maggot [USA]). This is a pest in the USA and Canada on apple, but also on pear, plum and cherry. The winding tunnels that the larvae make in the flesh of the fruit go the brown colour that apple flesh develops when exposed to the air, and infested fruit fall prematurely. Adults have a pale scutellum and narrow pale transverse bands on the abdomen. Larvae of the related cherry fruit fly (*R. cerasi*) feed near the cherry stone; infested fruit are deformed and unmarketable. Adults are rather black, but with yellow legs and head, and there is a single broad transverse band at the front of the abdomen. The mature larvae of both *Rhagoletis* species leave the fruit to pupate in the soil, and there is just one generation a year.

10.4.2.2 Division Calyptrata (see start of Section 10.4.2 for definition)

Family Scathophagidae (or Cordiluridae)

This is the calyptrate Family mentioned earlier as having the incomplete thoracic suture more typical of the Acalyptrata; also the calypters usually found in the Calyptrata are missing. The most frequently encountered member of this Family is the quite large yellow hairy dung fly (Fig. 10.44), which is easily recognisable.

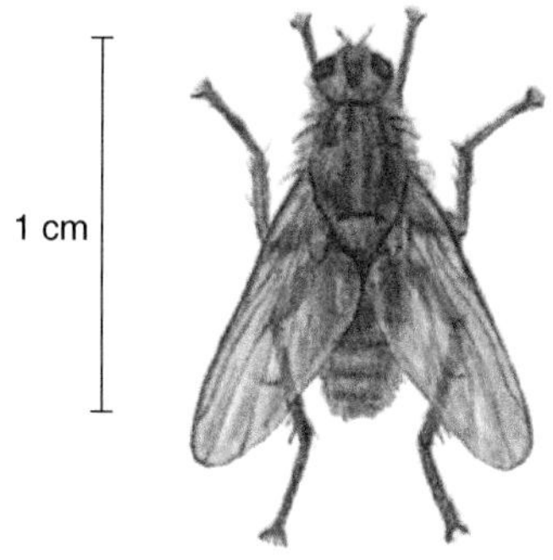

Fig. 10.44 Scathophagidae (from Lyneborg 1968, with permission).

Family Anthomyiidae

These insects have the appearance and grey colour of small house flies and many of them are important crop pests.

Delia radicum is the cabbage root fly (Fig. 10.45). The larvae feed on the roots of cabbage, cauliflower, Brussels sprout, broccoli etc. to the extent that whole plants may wilt and die. The adults feed on nectar of weed flowers and roost outside the crop in the shelter of hedgerows; they visit the crop on dull warm days to oviposit. Several eggs are laid on the soil near the stem of a plant or even on the stem soon after the

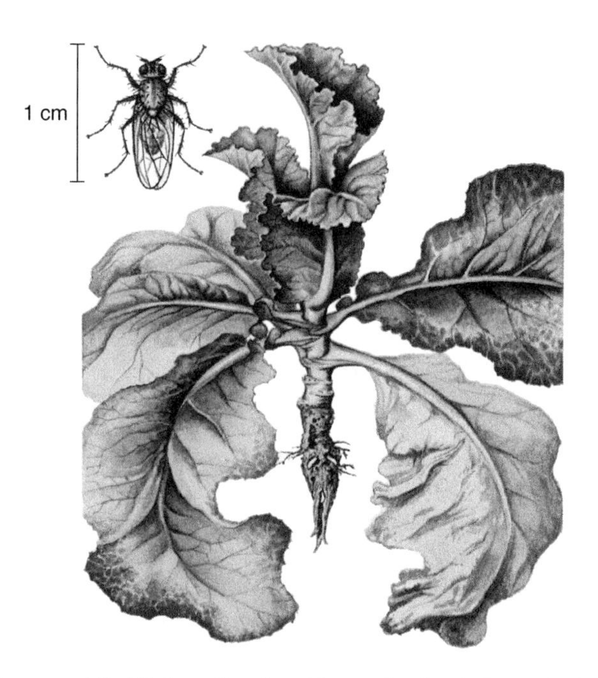

Fig. 10.45 Cabbage root fly (*Delia radicum*) and larvae feeding in the roots of a cabbage plant (from Bayer 1968, with permission).

time the seedlings are usually planted out in the field. Many of the eggs may be eaten by ground beetles (Carabidae) and their larvae, and so egg lay and subsequent damage are not always correlated. The larvae burrow down and graze the tap root; that there are often many larvae attacking one root means that the damage can be very severe. First signs of attack are that the outer leaves of the plant begin to wilt. Digging up the plant will reveal the white maggots at the roots. The larvae feed for about 3 weeks before they pupate in the soil. A second generation occurs in July, but by then the plants have a root system that is already so well developed that significant damage to the roots rarely occurs. More serious is that the flies also lay on the developing sprout buttons, which the larvae then enter and feed there undetected until the consumer boils the sprout! Most pupae of the second generation diapause overwinter, but in the UK some adults emerge as a partial third generation.

In Europe, the USA and parts of Asia, onions are attacked by *Delia antiqua* (onion fly; Fig. 10.46), the larvae of which feed in and rot the bulb. Eggs are laid in the soil close to the young plants or in the leaf sheath. The hatched larvae burrow down and enter the plant from underneath. There may be 30 larvae in one bulb, and they reach about 1 cm in length and maturity in about 3 weeks before they pupate in the soil. There can be a second generation, and even a partial third, yet some first-generation pupae remain in the soil till the following spring.

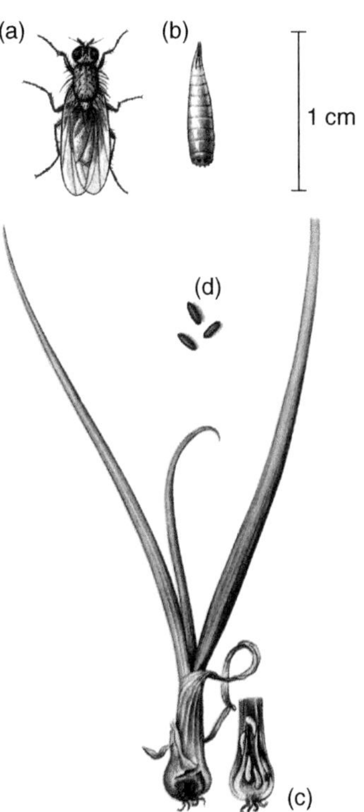

Fig. 10.46 Onion fly (*Delia antiqua*): (a) adult; (b) larva; (c) larvae feeding in onion bulb; (d) puparia (all from Bayer 1968, with permission).

Delia coarctata (wheat bulb fly) lays it eggs in July and August on the soil, but paradoxically only in crops that are not its host plant (i.e. not in wheat!). Thus the female will lay in brassica, sugar beet and potato fields. It seems to me that this must be a human-driven evolution, whereby the practice of crop rotation, practised for so long that wheat did not follow wheat, selected a strange genotype that laid its eggs in the 'wrong place'. As with cabbage root fly, wheat bulb fly eggs are a target for natural biological control by ground beetles, especially since the eggs remain unhatched till the start of the next year. Surviving larvae then bore into the base (the 'bulb') of the wheat (or die within a week if there are no wheat seedlings and there are no alternative weed grasses) and feed there causing the central shoot to wilt and die (deadheart). In heavy infestations, plants are killed and patches devoid of plants develop in the field. Improved disease control means that wheat now regularly follows wheat and as yet the wheat bulb fly problem is not serious in the UK. The insect is found across Europe and into central Asia.

Delia platura is pretty cosmopolitan, and has two common names. In the USA it is known as the corn seed maggot. In the UK, the same insect is the bean seed fly since the larvae feed on the sown seeds of both crops. Field and broad beans seedlings may emerge with tattered cotyledons and a blind apex to the stem between them. This is damage by the fly, which lays its white reticulated eggs on the soil in late spring and the larvae enter the soil and burrow into the bean where they fed undetected on the cotyledons. Pupation is in the soil, and in temperate climates this is the overwintering stage. There are two to five generations a year, depending on climate, and newly emerged females have a definite preoviposition period of a week or two before they are ready to oviposit.

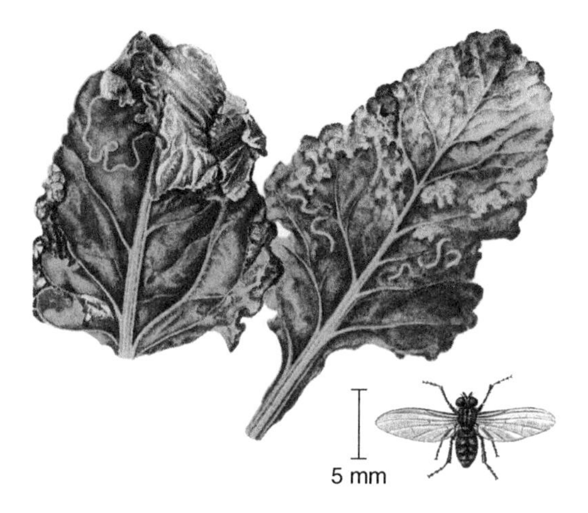

Fig. 10.47 Beet leaf miner (*Pegomya betae*) with blotch leaf mines (from Bayer 1968, with permission).

The Anthomyiidae also contain leaf miners. In contrast to the Agromyzidae (see Section 10.4.2.1, Family Agromyzidae) the mines tend not to be linear, but the larva eats in all directions to create an ever enlarging 'blotch' mine. The other contrast with agromyzids is that anthomyiid leaf miners leave the mine to pupate in the soil. *Pegomya betae* (beet leaf miner; Fig. 10.47) is such a leaf-mining anthomyiid pest of spinach and sugar beet in many temperate regions of Europe, Asia and North America. Large blotch mines appear in the leaves; there may be several larvae in one mine. Early attack is on the cotyledons, and this can kill seedlings whereas established plants can tolerate quite high infestations. The leaf miner has become more serious since sugar beet has been drilled to stand rather than oversown and later thinned (see also pygmy mangold beetle, Section 12.3.11.2, Family Cryptophagidae). In spinach, it is the leaves that are the marketed crop, and the presence of mines makes such leaves unsaleable. There are two to three generations a year.

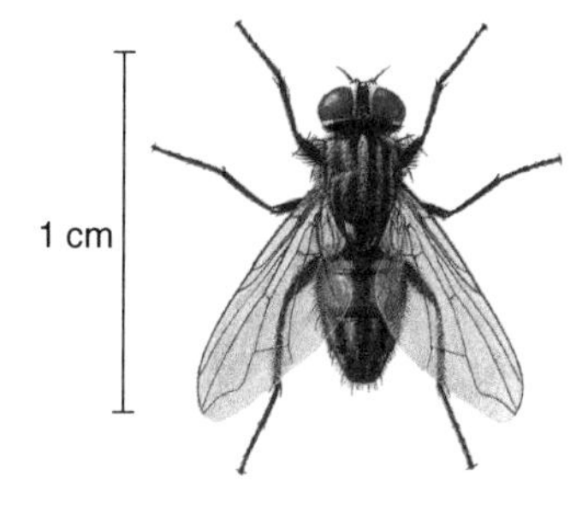

Fig. 10.48 House fly (*Musca domestica*) (from Zanetti 1977).

Family Muscidae

This Family includes the house fly (*Musca domestica*; Fig. 10.48), well known for contaminating food with fungi and bacteria carried on its feet. Most of the Muscidae are grey, rather hairy flies. The larvae are amphipneustic with the hind spiracles very large and

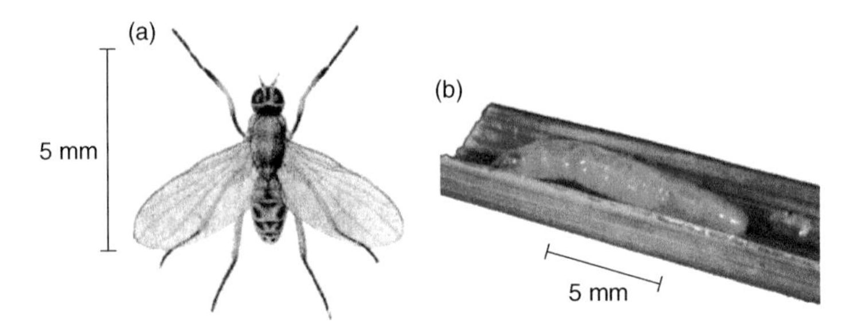

Fig. 10.49 Rice seedling fly (*Atherigona oryzae*): (a) adult (reproduced with permission of CSIRO, Australia); (b) larva tunnelling in rice stem (International Rice Research Institute).

on a backwardly facing plate (Fig. 10.2b). I once came across a student who thought these spiracles were eyes, and asked me 'Why do these maggots always wriggle backwards?'

Atherigona oryzae (rice seedling fly; Fig. 10.49a) causes 'deadhearts' in upland (non-irrigated) rice plants. The adult is mainly grey, but the abdomen and legs are largely yellow. The outer anterior edges of the compound eyes are rather angular in dorsal view. The problem with this fly is that it attacks early, when the plants are still tender and highly susceptible. Eggs are laid on leaves, and the larvae need a water film on these leaves to move down to the central growing point where they feed (Fig. 10.49b) and kill the shoot. Pupation is in the soil. The fly is a pest in tropical and subtropical Asia and Australia, and the life cycle takes 2–5 weeks, depending on temperature. The related sorghum shoot fly (*A. soccata*) attacks mainly sorghum, but also other cereals. It also feeds on the central shoot to cause 'deadhearts', and the tillers the plant produces to compensate for the loss of the central shoot may then in their turn be attacked as well. This fly is the most serious shoot fly pest of sorghum in many parts of the Old World, occurring throughout Africa, the eastern Mediterranean, and the near and middle-East as far as Burma and Malaysia. Pupation is in the shoot, though occasionally in the soil. Under hot and dry conditions, the pupae may not emerge till favourable conditions return (aestivation). The life cycle takes 2–3 weeks.

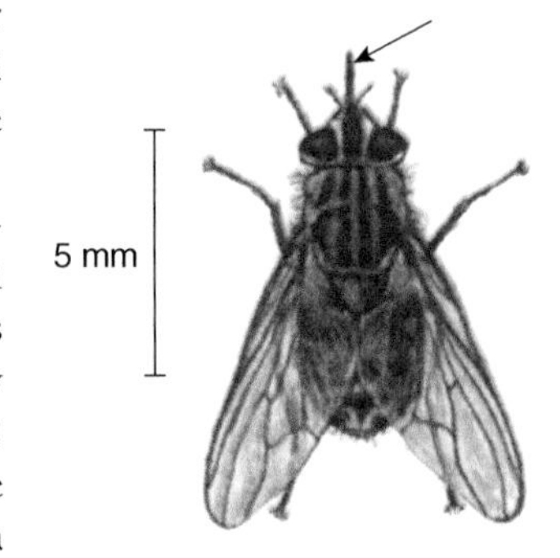

Fig. 10.50 Stable fly (*Stomoxys calcitrans*) (from Lyneborg 1968, with permission).

A cosmopolitan biting muscid is the stable fly, *Stomoxys calcitrans*, immediately recognisable by its rigid forward-projecting proboscis (Fig. 10.50). Both sexes feed on the blood of mammals, so in agriculture they are pests of horses and cows, and even of farmers! Three to four blood meals are necessary for the female to mature her eggs, and the larvae feed for up to a month on the microbial flora in moist decaying straw before they pupate there. The adults often roost in stables and cow sheds, and on sunny days will be found sitting on fences and walls near their hosts. Unfortunately, stable flies also vector diseases. In particular, they transmit a large number of trypanosome species (Protozoa) to domestic animals, as well as anthrax (transmitted also to humans).

Family Calliphoridae (bluebottles and greenbottles)

These largish flies (Fig. 10.51) with bright iridescent blue or green abdomens are familiar to many. Unlike the Muscidae, they have ***hypopleural*** bristles on the thorax (Fig. 10.52) but the postscutellum (see Section 10.4.2.2, Family Tachinidae) is slight or absent. Like house flies, they carry food contaminants on their feet and the larvae have the same large spiracles at the back of the abdomen. The larvae feed on carrion or other decaying organic matter. *Calliphora vomitoria* (blowfly) is the common bluebottle coming into houses and buzzing around dustbins.

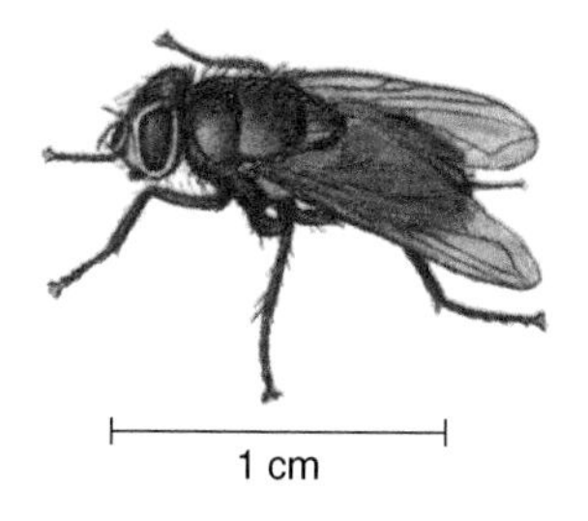

Fig. 10.51 Greenbottle, *Lucilia* sp. (Calliphoridae) (from Lyneborg 1968, with permission).

One of the greenbottles, *Lucilia sericata* (sheep maggot fly), is an important veterinary pest as the cause of 'sheep strike'. The eggs are laid in open wounds, and the larvae quickly erode the animal's flesh around the wound.

Another pest of farm animals in the southern USA and Central America is the screw-worm fly (*Cochliomyia hominivorax*), so-called because when the tapering larvae, widening out from a narrow anterior to the wide terminal plate with the large spiracles, have burrowed into the muscle of their host through any wound in the hide, they look not unlike a wood screw. The mature larvae drop to the soil to pupate. The adults are about twice house fly size, of a greenish-blue colour and large reddish-orange eyes. Screwworms literally eat their host alive, and can eventually kill it. The insect occupies a landmark position in the history of pest control, because it was the first example of the successful use of the sterile male release control method. This involved breeding up huge numbers of radiation-sterilised males in the southern USA and releasing them over several generations, increasingly to swamp the wild males, till virtually all matings were with sterile ones.

Family Sarcophagidae (flesh flies)

Very similar morphologically to the Calliphoridae, these flies are grey with a chequered abdomen. They are unusual in that the eggs hatch within the mother and it is the young larvae that are laid on or into the food (carrion for most species). A few are parasitic, feeding on the stored food in the nests of bees and wasps while others lay

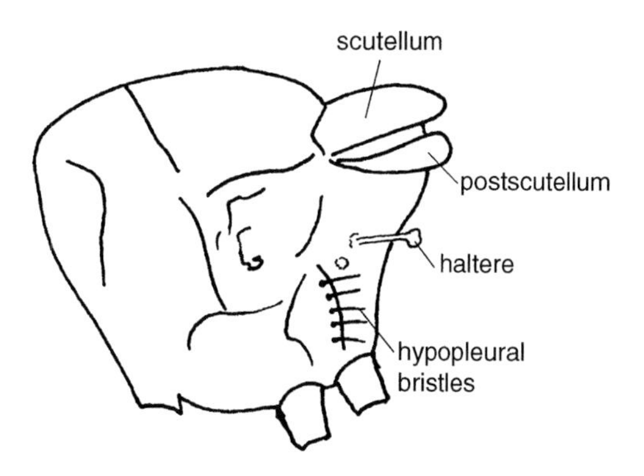

Fig. 10.52 Side view of tachinid thorax (from Alford 1999, with permission).

their eggs on the prey of solitary wasps after having tailed a prey-carrying wasp back to its nest.

Family Tachinidae

This is a large Family. Diagnostic (Fig. 10.52) is that the adults possess hypopleural bristles like the Calliphoridae, but there is also an obvious large ***postscutellum*** (in side view it looks as if the scutellum has a 'double chin'). As in some muscids and many calliphorids, the vein on the wing below the first cross vein is angled sharply forwards just beyond this cross vein, but in quite a lot of tachinids it actually makes a T-junction with the vein in front of it before the latter reaches the wing margin. The flies (Fig. 10.53) tend to be larger than house flies and are grey or brown, though some are green or have patches of red or orange.

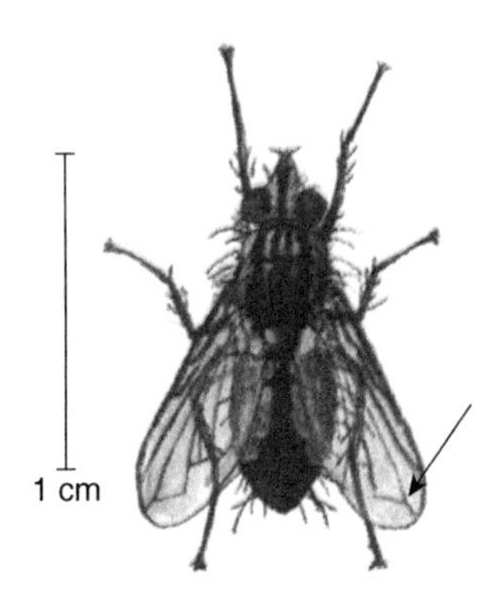

Fig. 10.53 Tachinidae (from Mandahl-Barth 1974, with permission).

The larvae are mainly internal parasitoids of other insects, and enter their host from eggs laid on or close to it, by the host swallowing eggs laid on its food, or by insertion into it through a cut made by the mother. Like other parasitoids in the Insecta (e.g. in the Hymenoptera, Chapter 12) the larva eats the host tissues beginning with less essential ones and only kills the host when the tachinid larva is ready to pupate. Compared with parasitoids in the Hymenoptera, tachinids have hardly been used for biological control of pests, though the species *Bessa remota* was used for biological control of coconut moth (*Levuana iridescens*) in Fiji as early as the 1930s.

Family Glossinidae (tsetse flies)

Species of *Glossina* are the biting tsetse flies, which are found across much of the middle of Africa between the Sahara and the Kalahari desert. They are relatively large flies, up to 1.5 cm in length, and are distinguishable by the hardened forward-pointing proboscis being attached to the bottom of the head by a distinct bulb, and by the fact that they fold their wings flat on top of each other over the abdomen when at rest (Fig. 10.54). A closer look will show that the arista of the antenna is plumose with feathery hairs. Tsetse flies feed on the blood of vertebrates and transmit trypanosomes (Protozoa). These parasites cause 'sleeping sickness' in humans and 'nagana' in cattle. 'Sleeping sickness' refers to the extreme lethargy of people once the infection reaches the central nervous system and brain. Death often results if the disease is not diagnosed sufficiently early, and in most years mortality of several hundred thousand people results. 'Nagana' causes reduced growth, lower milk productivity and weakness in cattle; the course of the trypanosomiasis is similar to that in humans so that again death may well result.

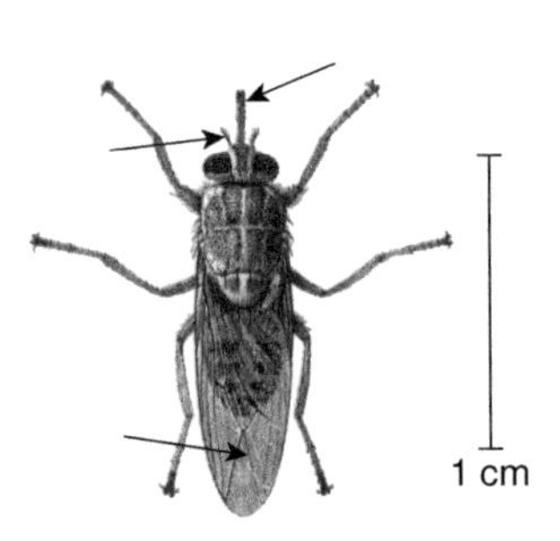

Fig. 10.54 Tsetse fly *(Glossina* sp.) (from Zanetti 1977).

There are several species of tsetse; probably the two most important ones are *G. morsitans* (an insect of the savannah) and *G. palpalis* (a riverine species). Like some tachinids (see Section 10.4.2.2, Family Tachinidae), tsetse larvae hatch inside the mother and are fed by her blood meals until they are 'laid' on the soil in their final instar. They then crawl into the soil to pupate inside their larval skin (puparium).

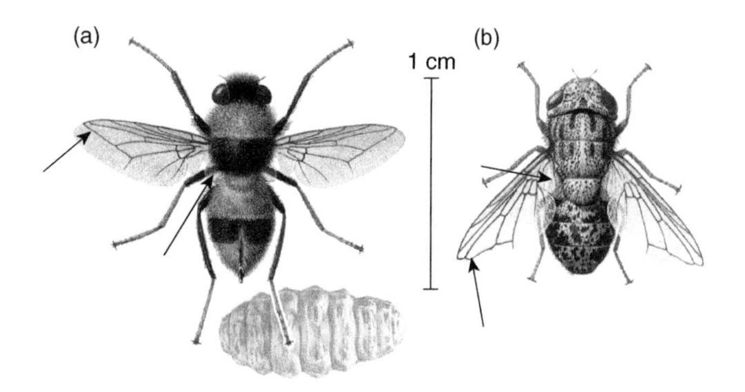

Fig. 10.55 Oestridae: (a) warble fly (*Hypoderma bovis*) and larva; (b) sheep-nostril fly (*Oestrus ovis*) (both from Zanetti 1977).

Family Oestridae

These are stout bristly flies with vestigial mouthparts and large calypters. As with so many other calyptrates, the veins on either side of the first cross-vein converge towards the tip of the wing. There is a strongly developed postscutellum (Fig. 10.52).

The larvae are parasites of mammals. Typical is *Hypoderma bovis* (warble fly; Fig. 10.55a). After hatching from eggs laid on the legs of cattle, the larvae bore into the skin, and then move through the body to end up under the skin on the back of the animal with the hind spiracles exposed through the skin. The inflamed swelling that develops around the larva is the 'warble'. The larvae finally fall to the ground and pupate.

Another veterinary pest in this Family is *Oestrus ovis* (sheep-nostril fly), with a warty head and thorax (Fig. 10.55b). Eggs are laid in the nostrils of sheep and goats and the larvae develop in the nose and sinuses. Infected animals display a characteristic giddiness. The larvae pupate on the ground after having been expelled from the host by a sneeze.

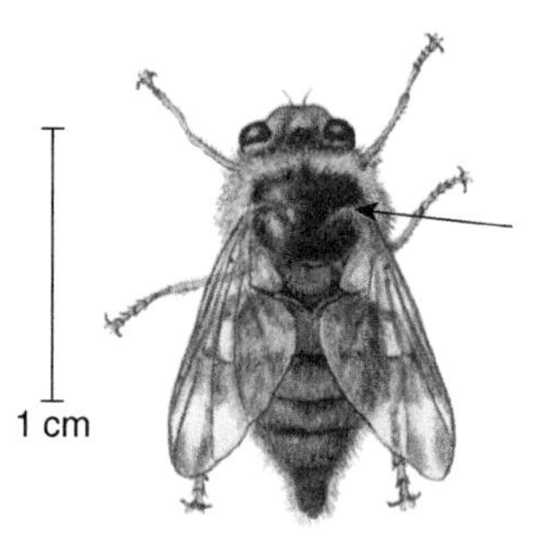

Fig. 10.56 Bot fly (*Gasterophilus intestinalis*) (from Lyneborg 1968, with permission).

The bot flies (*Gasterophilus* spp.) are regarded by some as a separate Family, the Gasterophilidae. Either way they are closely related to the other Oestridae, and therefore Calyptrata, in spite of having an incomplete dorsal suture on the thorax and small calypters. The adults are large and hairy and look not unlike bees (Fig. 10.56). The larvae are internal parasites of horses and their relatives. *Gasterophilus intestinalis* is the commonest species in the UK. Eggs laid on the body of the animal usually enter the mouth during licking and the hatched larvae use their mouth hooks to anchor to the wall of the stomach when they reach it. They feed there for about 9 months before letting go to be deposited on the ground (where they pupate) with the faeces. In other species, eggs may be laid on vegetation to be eaten by the host.

10.4.2.3 Division Pupipara (louse flies, keds)

These are completely different from other Diptera in being flattened, leathery and often wingless insects with the tarsal claws enlarged and modified for gripping the hairs or feathers of the mammals and birds on which both males and females are blood-sucking parasites. The labium (with its labellae) is stiff and sharp for piercing the host skin. The immature life cycle, similarly to that of the tsetse fly, takes place within the mother, and larvae develop one at a time to emerge from the mother fully grown and about to pupate. This of course is an adaptation to the parasitic habit and the lack of other food for the larvae on the host.

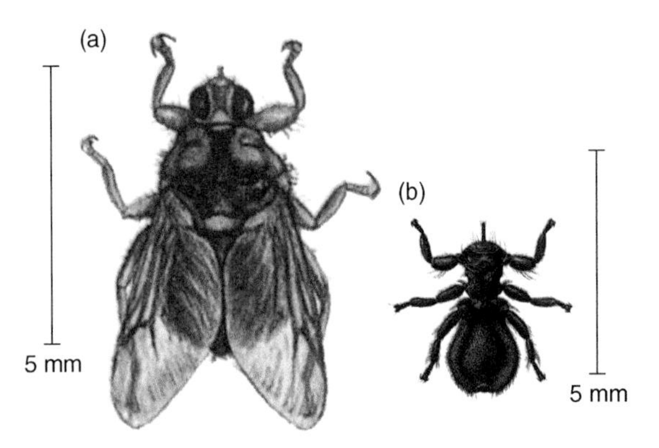

Fig. 10.57 Pupipara (Family Hippoboscidae): (a) forest fly (*Hippobosca equina*) (from Lyneborg 1968, with permission); (b) sheep ked (*Melophagus ovinus*) (from Edwards *et al.* 1939).

Family Hippoboscidae

Hippobosca equina (forest fly) is parasitic on deer, but also on horses and cattle in wooded areas (Fig. 10.57a). The hippoboscid of greatest agricultural importance in the UK is the wingless *Melophagus ovinus* (sheep ked; Fig. 10.57b). The amount of blood sucking is relatively harmless; more important is that the scratching of the sheep to relieve the irritation causes wounds, which allows the entry of diseases and *Lucilia sericata* (see Section 10.4.2.2, Family Calliphoridae). *Lipoptena cervi* (deer fly) is unusual in that although the adults emerge with wings, they shed these when they find a suitable host deer.

Another group of Hippoboscidae (e.g. *Nycteribia* spp.), which some place in a separate Family, the Nycteribiidae, are parasitic on bats. They are very small wingless hippoboscids with long legs, which may be more than twice the body length, and a distinctive groove along which the head is folded back.

11 Subclass Pterygota, Division Endopterygota, Order Hymenoptera (sawflies, ants, bees and wasps) – *c.* 120,000 described species

11.1 Introduction

This, like the Diptera, is one of the very large Orders of insects, with probably as many species yet to be described as already have been. Therefore again only a relevant selection of Families will be mentioned. The Hymenoptera have two pairs of wings coupled by a row of little hooks (***hamuli***) on the leading edge of the hind wing hooking into a fold on the trailing edge of the front wing (for those who know it, it is rather like a 'sailor's grip'; Fig. 2.17). Most species have a constriction (***petiole***) near the front of the abdomen. The petiole therefore divides the abdomen, with the anterior part (first abdominal segment) attached to the thorax known as the ***propodeum*** and the posterior part as the ***gaster*** (Fig. 11.1). The antennae are usually obvious, and this is an important contrast with most Diptera. Moreover, many Hymenoptera have a pigmented spot (***pterostigma*** or '***stigma***' for short) on the forewing near the tip (Figs 11.2, 11.4 and 11.39). Therefore, in sorting insects, beware of any 'fly' with obvious feelers, especially if there is a spot on the forewing – and check carefully that it has only one pair of wings before assigning it to the Diptera.

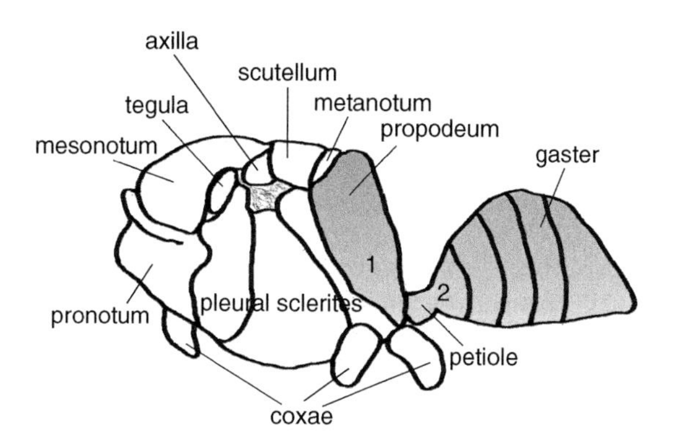

Fig. 11.1 Lateral view of thorax and abdomen of a hypothetical hymenopteran. The abdomen is shaded and the first two segments are numbered.

Antennae of the Hymenoptera are particularly likely to show a morphological distinction between the first (***scape***) and second segment (***pedicel***), with the remaining apical segments then rather similar to each other and forming the ***flagellum*** (Fig. 2.6). The early entomologists developed a distinctive terminology for characters on the

Handbook of Agricultural Entomology, First Edition. H. F. van Emden.

thorax in the higher Hymenoptera, as they failed to recognise how the plan of thoracic sclerites related to that in other Orders (see Section 2.5.1). However, today we can label (Fig. 11.1) the sclerites with their correct homologies, though in some taxa there is an extra demarcated area called the ***axilla***. Note also the oval sclerite (***tegula***) near the wing articulation, its position in relation to the back of the pronotum distinguishes 'bees' from 'wasps'. This tegula has been likened to 'shoulder pads'.

The adult mouthparts show fairly traditional maxillae and biting mandibles, but the glossae are often modified and elongated as a 'tongue' for lapping or sucking up nectar. In some bees visiting deep-throated flowers, this tongue can very long.

'Hymenoptera' means membrane-winged, a characteristic not unique to this Order. The wing coupling with hamuli and the frequent occurrence of a pterostigma have already been mentioned. Venation is often reduced and may virtually be absent in some small species; the larger ones, including the social species, tend to have more cross-veins, and therefore more cells, than the Diptera. Very often the hind wing and quite often the forewing have a little indentation in the hind margin. This marks the end of a fold in the wing which separates off the ***anal lobe*** (Fig. 11.2).

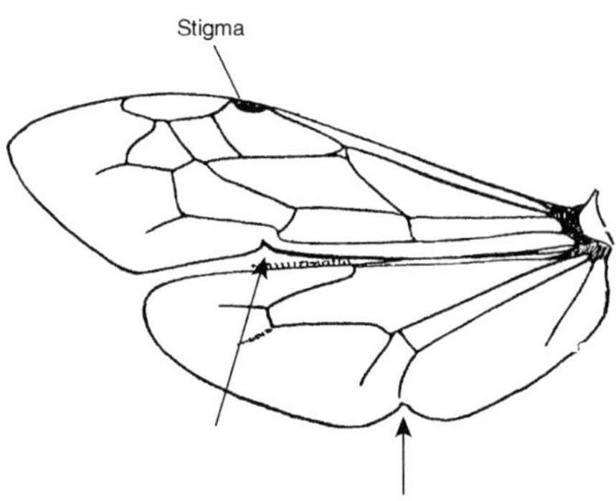

Fig. 11.2 Wings of a hymenopteran. The indentations of the rear wing margins that demarcate the anal lobe are arrowed (from Richards and Davies 1977).

Legs are usually normal, though it is common for the tibia to end in ***tibial spurs***. The hind legs of honey bees are specially adapted for carrying pollen (see Section 11.3.1.1, Family Apidae).

The females usually have a distinct ovipositor, which may be modified to form a sting.

Larvae are of one of three types – normal with only thoracic legs, a 'caterpillar' with prolegs, or a legless 'grub'. Pupae (e.g. Fig. 11.17b) have free appendages (exarate), but are usually protected by a cocoon of varying solidity.

11.2 Suborder Symphyta (sawflies)

This Suborder is very different from other Hymenoptera. There is no waist (petiole) in the abdomen (Figs 11.4 and 11.5), and the larvae (Fig. 11.3) are not legless, but have thoracic legs and also often even prolegs like the caterpillars of Lepidoptera. These prolegs are, however, devoid of crotchets and there are six to eight and not just five pairs. The ocelli are grouped close together, looking much more like an adult compound eye than do the ocelli of caterpillars of the Lepidoptera. The cuticle often has several folds within each segment, obscuring the underlying segmentation compared with Lepidoptera caterpillars. Adult and larval mouthparts are of the normal biting type and the wing venation tends to be more extensive than in other Hymenoptera. All but one Family (Cephidae) possess ***cenchri***, two small often light-coloured

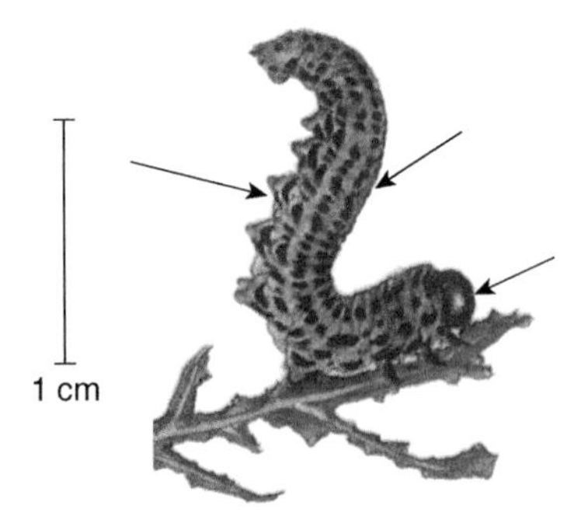

Fig. 11.3 Sawfly larva (from Mandahl-Barth 1974, with permission).

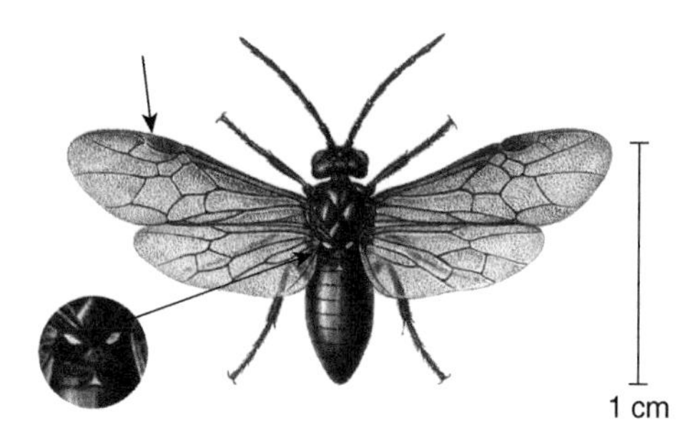

Fig. 11.4 Sawfly adult. The enlarged circle identifies the cenchri (from Zahradník and Chvála 1989).

headed studs on the metanotum (Fig. 11.4). These studs adhere like Velcro to rough patches on the underside of the forewings when the adult is at rest. The antennae are recognisably slender and filiform. The Suborder is nearly entirely phytophagous and the word 'sawfly' derives from the saw-toothed ovipositor of many Families, which is used to saw a slot into a leaf, stem or fruit into which an egg is then laid. This, as with leaf-mining Diptera (see Section 10.4.2.1, Family Agromyzidae) lessens the ability of the plant to shed the egg. You will now understand the paradox of a name 'sawfly' – the insect is a hymenopteran and not a dipteran, usually with lepidopteran-type larvae!

The larvae of many sawfly species feed parallel to the edge of the leaves. Often many can be found along the edges of the same tree leaf, and they will wave the rear half of their bodies in synchrony (the posture in Fig. 11.3) when a shadow (a predator?) is cast over them. Where the underside of the leaf of their host plant is conspicuously lighter than the upper surface, the larvae of certain species in the genus *Arge* (Family Argidae) show bilateral asymmetry in their colour, with the lighter side (left or right hand side according to species) matching the leaf surfaces. If a larva is placed on the leaf edge facing the wrong way, it will re-orientate itself.

11.2.1 Family Siricidae (wood wasps)

These are atypical Symphyta in that the larvae have no prolegs; indeed the thoracic legs are small and hardly visible. Also atypical is the ovipositor, which is not saw-toothed but long and hardened for drilling into wood, particularly of trees already under some stress. All wood wasps are well over 10 mm long, and the much larger black and yellow *Urocerus gigas* (Fig. 11.5) is often mistaken for a hornet (the ovipositor being misidentified as the sting) and thus killed unnecessarily in fear. The ovipositor carries spores of a fungus, which are left in the wood with the eggs. As the larvae tunnel in the wood, the fungus colonises the tunnel walls and the larvae get

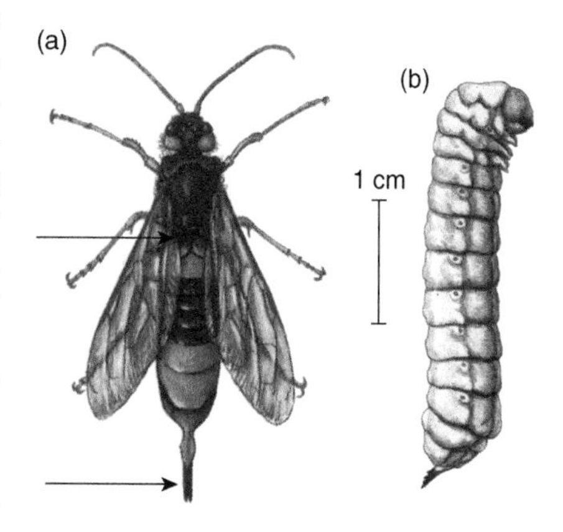

Fig. 11.5 Giant wood wasp (*Urocerus gigas*): (a) adult with (b) larva (from Mandahl-Barth 1974, with permission).

most of their nutrition by grazing the fungus. Siricids accidentally introduced from Europe became serious pests of conifers in Australia, but are now largely controlled there biologically with an adapted European parasitoid (*Rhyssa*; see Section 11.3.2.1, Family Ichneumonidae) and a nematode. The large larva is almost apodous, with just the three pairs of thoracic limbs very much reduced.

11.2.2 Family Cephidae

These sawflies can uniquely be recognised by the absence of cenchri. The adults are low-flying, and lay their eggs on the stems of grasses. The larvae (almost apodous like those of the wood wasps) bore into the stem and two species of the genus *Cephus*, *C. cinctus* and *C. pygmeus* (Fig. 11.6), are pests of cereals (wheat stem sawflies).

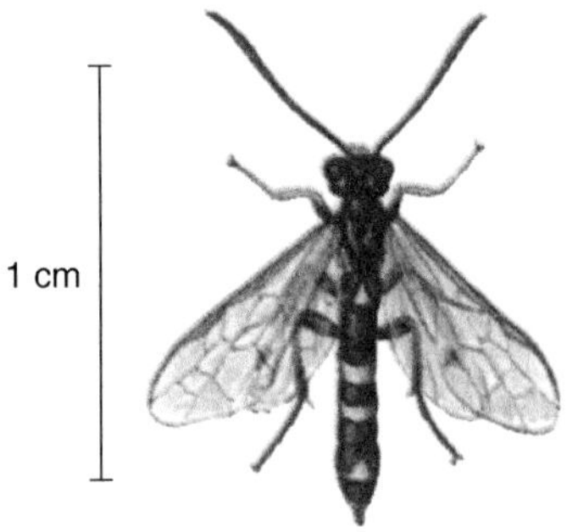

Fig. 11.6 The wheat stem sawfly *Cephus pygmeus* (from Lyneborg 1968, with permission).

11.2.3 Family Tenthredinidae

This is by far the largest sawfly Family. The size of adults (Fig. 11.4) varies widely (2.5–15 mm). The larvae are mainly exposed defoliators, and several species are occasionally minor pests of wheat. As in the other Families of sawflies below, the larvae have effective thoracic walking legs as well as prolegs (Fig. 11.3).

The apple sawfly, *Hoplocampa testudinea* (Fig. 11.7), uses its ovipositor to cut a slot in the surface of the receptacle of a flower and insert an egg; it does this around the time that the flowering of the apple tree is coming to an end and about 75% of the petals have fallen. The hatched larva first feeds on the soft edge of the oviposition slot and then tunnels into the receptacle, leaving a clearly visible hole in the surface of the fruit as it develops, just as the codling moth (see Section 9.2.3.7, Family Tortricidae) does. However, the apple sawfly's hole is accompanied by a crescent-shaped scar (the 'ribbon scar'), which is where the slot cut by the female and then enlarged by the young larva has healed but magnified as the fruit has grown. Infested fruit usually fall early while still relatively small and the larvae emerge to pupate in the soil. They

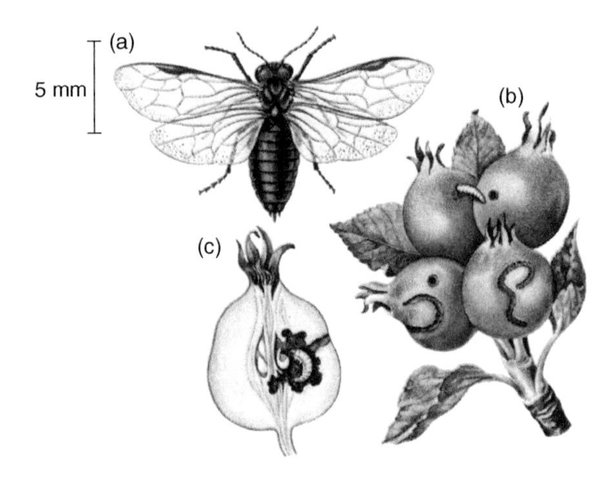

Fig. 11.7 Apple sawfly (*Hoplocampa testudinea*: (a) adult; (b) attacked fruitlet showing entry hole and ribbon scar; (c) larva feeding inside fruitlet (from Bayer 1968, with permission).

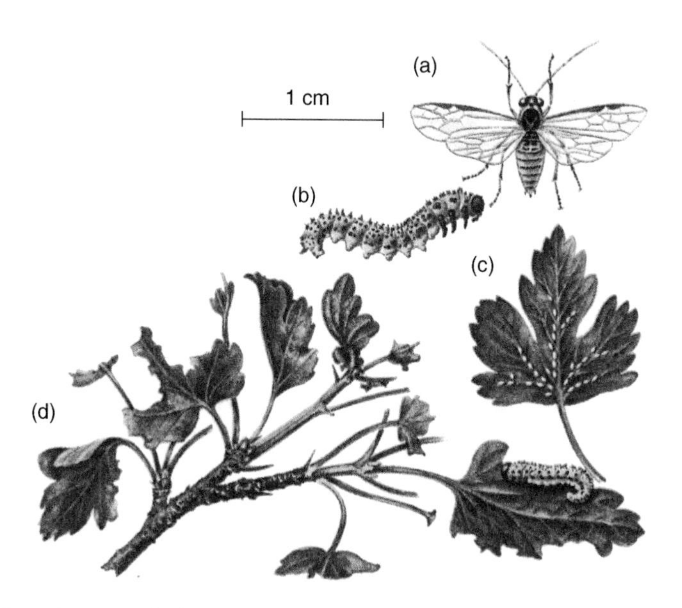

Fig. 11.8 Gooseberry sawfly (*Nematus ribesii*): (a) adult; (b) larva; (c) lines of eggs laid along leaf veins; (d) leaf damage (from Bayer 1968, with permission).

overwinter as immobile and rather shrunken larvae, and pupate in the spring. The adults are mainly black, but the legs and sternum are a reddish-yellow. The apple sawfly occurs in Europe and North America.

Dolerus haematodes is the sawfly most commonly found on cereals and grasses in Europe, where it can be a minor pest of wheat and barley. Pupation is in the soil, and there is one generation a year.

The gooseberry sawfly (*Nematus ribesii*) can cause serious yield losses on currants (except blackcurrant), but particularly on gooseberries in Europe and the USA. The adults are black with a yellow abdomen (Fig. 11.8).The white eggs are laid around April in rows in slits along the main veins, and the rows look like 'fingers' radiating from a 'palm'. The larvae are a bright green with many black spots and they initially feed together near where they hatched. Later they disperse and then cause serious damage with their voracious stripping of the leaves from the edges inwards to leave just the thickest veins. The larvae pupate in thin cocoons in the soil, and emerge soon thereafter to give a second generation in June, with a third generation in late summer. This last generation overwinters as a larva in its cocoon in the soil before pupating in the spring.

The turnip sawfly (*Athalia rosae*) is occasionally a serious pest of brassicas of all kinds in Europe, Asia across to China and Japan, and also occurs in parts of Africa and South America. The adults are black except for the bright orange abdomen; also the leading edge of both pairs of wings has dark veins. As occurs so frequently with sawflies, the eggs (which are white) are inserted singly into slots cut by the female's ovipositor. The small black larvae first mine into the leaf, but later feed on the leaf lamina, usually leaving just the midrib if populations are high, and characteristically easily fall off the plant if disturbed. Pupation is in a silk cocoon in the soil and there are three generations a year in the UK, with the last one overwintering as a pupa.

Caliroa cerasi is a small black sawfly (Fig. 11.9) whose larvae feed on the upper surface of leaves of fruit trees to leave a transparent window of the veins and lower epidermis. The insect is found in all temperate regions and pear, cherry, peach and plum are the most important hosts. The eggs are again inserted in slots cut in the leaf and the actually yellow larvae are very easily recognisable by their black shiny and slimy coating which gives the species the common names of 'pear slug sawfly' or 'cherry slugworm'. When mature after 2–4 weeks, the larvae descend to the soil where they pupate in a silk cocoon. There are two to three generations a year, with the last generation overwintering as a larva inside the cocoon before pupating in the spring.

Although of no economic importance, the fleshy lumps (which turn red) on willow leaves are a familiar sight. These lumps are galls inhabited by larvae of sawflies in the genus *Pontania*.

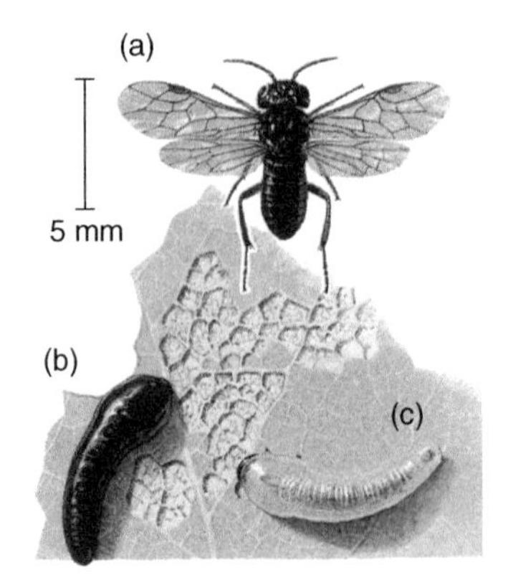

Fig. 11.9 Pear slug sawfly (*Caliroa cerasi*): (a) adult with damaged leaf below carrying (b) the black slug-like larva and (c) a larva with the black coating removed (from Zanetti 1977).

11.2.4 Family Diprionidae

These are the sawflies that feed on conifers, and so can be pests in forests. The adults can be recognised by the jagged (serrate in the female or pectinate in the male) outline of the antennae, and the larvae have prominent black spots. *Diprion pini* (pine sawfly) is common on pines, and *Gilpinia hercyniae* has been a serious defoliator of Canadian forests since its introduction there from Europe.

11.3 Suborder Apocrita (ants, bees, wasps and parasitic wasps)

These, in contrast to the Symphyta, have the 'wasp waist', that is the constriction (petiole) near the front of the abdomen (Figs 11.13 and 11.16). In many bees and wasps this may not immediately be obvious but can be revealed in a dorsal view by pushing the end of the abdomen downwards (e.g. Fig. 11.21). The Suborder has often been divided into two Divisions, the Aculeata and the Parasitica (or Terebrantia), but there is no easy distinction between these, and there are a number of Superfamilies which could almost fit into either Division. Most entomologists have now given up the two Divisions, and instead divide the Suborder directly into Superfamilies. I will follow this latter pattern, but begin by grouping together the social Hymenoptera (ants, bees and wasps – in which the female has a sting) and the Superfamilies of their close but solitary allies (i.e. the solitary bees and wasps).

Because the larvae of the Apocrita do not have to move to feed (e.g. fed by workers in social nests or being inside animal hosts or in plant galls), they are legless (Fig. 11.13d) and have very reduced heads (often with no visible head capsule).

11.3.1 The social Hymenoptera and their allies

These Hymenoptera are often regarded as comprising the most advanced insects, many with complex social behaviour. The Suborder contains the Division Aculeata of other classifications of the Hymenoptera. The ovipositor is modified as a short sting (Fig. 11.10) in contrast to the longer ovipositor of many of the other taxa of the Apocrita.

Fig. 11.10 Sting of a bee (from Zanetti 1977).

The structure of the colonies of social Hymenoptera, as with termites (see Section 6.11), revolves around nests founded by sexual castes that then raise sterile 'workers' to gather food and perform other services for the colony. In termites these workers can be either sterile males or females, but in the social Hymenoptera they are all female. The sexual females that founded the colony are 'queens'; the males in bees and wasps are known as 'drones'. The eggs laid by the queen are removed to brood areas and there are tended and fed by the workers. As with termites, the fate of the larvae as workers or sexuals is determined by pheromones emitted by the queen and the food they receive, but future sexuals are reared in special areas and their future is irrevocably determined by this. At certain times the new sexuals leave the nest to mate and found new colonies.

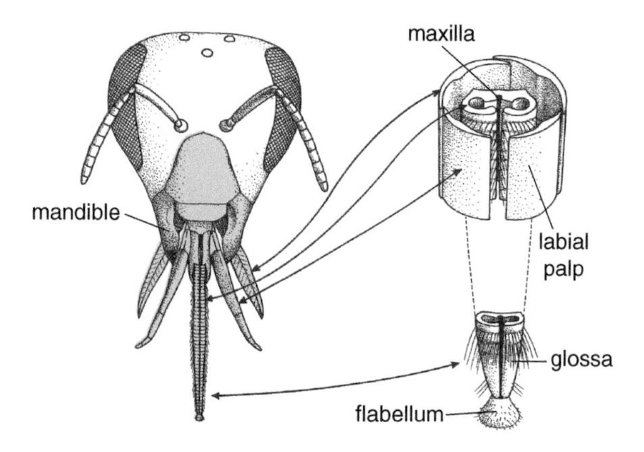

Fig. 11.11 Mouthparts of a bee (from Zanetti 1977).

11.3.1.1 Superfamily Apoidea

The hairs on the thorax are branched near the tip, and the pronotum does not extend back as far as the tegula. The mouthparts (Fig. 11.11) have a particularly long 'tongue' (glossa) for reaching nectaries deep within flowers (as in snapdragons, for example).

Family Andrenidae

The principle genus is *Andrena*. These are rather hairy solitary bees resembling honey bees, but with short tongues and no pollen basket on the hind tibia, although they also gather pollen and nectar with which to provision their burrows in the soil. Very often several females use a common entrance in the soil from which their individual burrows diverge to form chambers in which the eggs are laid. The mother probably has died before her young emerge. Different species are uni- or bivoltine.

Family Megachilidae (leaf-cutter and mason bees)

This is another Family of solitary bees. The leaf-cutter bees are species of the genus *Megachile* (Fig. 11.12) which cut out leaf circles (roses are a particular favourite) to construct the cells (Fig. 11.12b) in their nests, which are located in dead wood or in the soil. *Megachile* stores only pollen, not nectar. Each cell contains one egg and a ball of pollen as food for the larva.

The mason bees are megachilids of the genus *Osmia*. They construct their nests of sand grains glued together with saliva or chewed plant material in cavities (e.g. in the

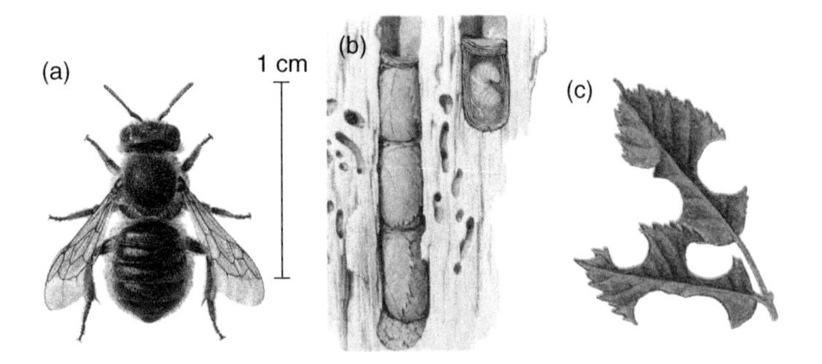

Fig. 11.12 Leaf-cutter bee (*Megachile* sp.): (a) adult and (b) cells in nest (from Zanetti 1977); (c) incisions made in rose leaves (from Lyneborg 1968, with permission).

mortar between bricks) already available or in ones they have themselves excavated. After several trips, the bee will have collected enough pollen and nectar to combine into a ball to provision one egg, which is then laid on top of this mass. The bee then makes a partition over the cell with mud to form the floor to separate off the next cell.

Fig. 11.13 Honey bee (*Apis mellifera*): (a) queen; (b) drone; (c) worker; (d) larva (from Zanetti 1977).

Family Apidae

Honey bees (*Apis mellifera*) are social bees (Fig. 11.13) which have been domesticated for centuries by man, though in many parts of the tropics local tribes set up hollow logs for wild bees to colonise. The nest is constructed of vertical sheets of hexagonal cells on which the adult bees live and move. The cells are made of wax and are used either to store pollen and honey (nectar collected from plants) or contain brood. When full of food or when brood have pupated, the workers seal the cell with a wax cap. The cells filled with pollen and honey are the familiar 'honeycomb'. The beekeeper buys man-made thin sheets of beeswax or plastic cells as 'foundation' for the bees to build upon. These racks of foundation are hung in wooden squares called 'supers', which can be mounted one on top of the other. These then form the familiar square tower of the modern beehive, though the image of the ancient 'skip' remains familiar as a 'logo' of a bee hive, in spite of its disappearance from the rural environment.

Many if not most crops that are harvested for their fruits, pods or seeds depend on pollination by insects, and so beekeepers provide an essential service to agriculture and horticulture. Periodically, a pandemic of bee disease or parasites sweeps across continents – a potential catastrophe if not checked.

Growers may arrange to have bee hives at adequately close spacing in their crops by hiring them or symbiotically offering free a rich source of nectar to beekeepers. The presence of hives in a crop limits the application of insecticides for the period they are there, but in some parts of the world insecticide use is so heavy that farmers have to buy the hives since there is little likelihood that the bees will survive the season to be returned to the beekeeper!

Honey is of course the main commercial output of the hive for beekeepers. In some countries, for example Australia, beekeeping is really big business and not the domain of the dedicated private beekeeper as it so often is in the UK. Large companies send out 'scouts' on motorcycles to find and establish 'squatter's rights' on good nectar flows (e.g. from eucalyptus trees in natural woodland). They will then phone back to base and hold the fort till large trucks carrying the supers arrive. Towers of nine or ten supers will be set up in contrast to the two or three usual in Europe. Even in Europe, a well-established hive may contain 50,000 bees.

However, liquid honey in jars is only one of the commercial products from a beehive. Honey in the form of honeycomb can be sold for rather more, and of course the bees have added more wax derived from pollen to the artificial foundation in the supers. This beeswax can be extracted and sold for various purposes such as the most expensive candles (often for religious use) and as a component of furniture polish.

Propolis, a resinous mixture that bees collect from tree buds and other plant sources to plug small gaps in the hive, is another valuable bee product claimed to have beneficial medical qualities.

Health-giving properties have also been claimed for 'royal jelly', the food with which the brood destined to become future queens are fed, and the substance is available commercially as a food supplement or added to a range of toiletries.

In the 1930s, the Swiss entomologist Karl von Frisch discovered what he called 'the language of the bees'. By using different colours to mark bees arriving at dishes of sugary liquid placed at various positions in relation to the hive, and then watching the behaviour of these marked bees when they returned to the hive, he established that bees could communicate to other bees information on the nature of a food source and its distance and direction from the hive. He observed that early in the day, worker bees acted as 'scouts' to locate good sources of food and that later other workers would go straight to these food sources for the rest of the day.

The 'language of the bees' begins with a scout bee fanning her wings and distributing the scent of the food source (e.g. apple or dandelion) on the comb. This scent and the dance the scout now performs attract the attention of other bees, which gather round and follow the scout mimicking and imprinting her movements on themselves. Eventually a veritable 'conga' chain of bees is performing the same dance. From the scent distributed by the scout, the bees learn what they are looking for. If the food source is close to the hive, no further information is communicated, and the scout dances a confused figure-of-eight. As distance from the hive increases beyond 15 m, the figure-of-eight becomes more defined with an ever longer 'waist'. While moving along this waist, the scout bee waggles her abdomen a number of times – no, I'm not kidding! The number of waggles increases with and is proportional to distance from the hive. How about direction? This is the really clever bit. The angle that the waist with the waggles makes to the vertical axis of the comb is the angle away from the sun that the food can be found. If the waggle-line is up the comb surface, the bees fly away from the sun in the direction it has come from; if downwards, the angle from the hive is ahead of the sun. What is amazing is that bees have translated a two-dimensional movement on the vertical aspect of a comb hanging inside the hive to the two-dimensional horizontal surface of the landscape outside the hive.

Bees pollinate most efficiently when collecting pollen, as they can often avoid contacting the anthers when collecting only nectar. The pollen that workers collect is packed into the specialised 'pollen baskets' on the back legs (Figs 2.15 and 11.13c) and it is only the pollen that they fail to recover from their body and wings which will cross-pollinate the crop. The workers try to maintain a fixed pollen : nectar ratio, and so beekeepers have developed some tricks to increase pollen-collecting by the workers. If the hive is artificially fed with nectar, then obviously the workers will try to restore the correct balance by collecting pollen. The other trick is to fit 'pollen traps' at the entrance to the hive. These are a bit like false eye-lashes and the bristles flip the pollen a bee is bringing back out of the pollen baskets. This prevents pollen building up in the hive and causes pollen collecting to be maintained.

Although honey bees are highly effective pollinators, they tend to be less active in poor weather, particularly if it is windy or cold. However, they move between flowers very quickly and it only needs rather less than 100 bees and 5 days of fine weather to achieve the necessary pollination on 1 hectare of an apple orchard. Some crops like cherries flower early, at a time when the weather deters honey bee activity in most years. Many insects, including solitary and bumble bees, beetles and flies are active in poor weather and can do the pollination needed, but so can wild honey bees (e.g. the Asiatic *Apis cerana*), which have already been mentioned as setting up combs with their colonies of workers in hollow logs.

There are several subspecies of *Apis mellifera*, and when the African subspecies hybridises with some European ones, such as the Italian subspecies, a very aggressive bee results. These so-called 'killer bees' invade and take over honey bee hives, and human deaths in the Americas have been attributed to them stinging in large numbers.

Bumble bees are wild bees with a larger and stouter body than honey bees, but like the latter they have pollen baskets on the hind tibiae. Bumble bees are mainly black, but often have whitish, reddish, orange or yellow hairs on or at the back of the abdomen (Fig. 11.14). Nests are built in the soil by the mated queen who is the only caste to survive the winter. Bumble bee nests have far fewer individuals than nests of honey bees; 200–300 workers would be a typical number. Bumble bees also collect pollen and nectar; they are therefore useful pollinators and are active in poorer weather than honeybees. *Bombus terrestris*, a common European species, is bred commercially, and occupied cardboard nests are sold to glasshouse growers to ensure adequate pollination of crops such as tomatoes.

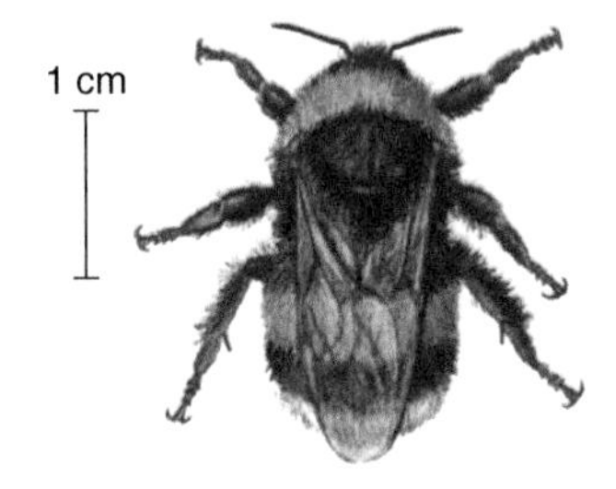

Fig. 11.14 Bumble-bee (*Bombus terrestris*) (from Lyneborg 1968, with permission).

Family Sphecidae (digger wasps)

These insects (Fig. 11.15) are called 'wasps' because of their superficial resemblance to true wasps, but are taxonomically much more closely related to the bees. They are easily distinguishable from wasps by the fact that, as in the bees, the pronotum does not reach the tegula. Also the wings are not folded longitudinally when at rest. Digger wasps provision their eggs in their burrows (not always divided into cells) in sandy places with all sorts of insect prey, such as aphids and leafhoppers as well as much larger insects; some species take spiders. The prey used varies with species, and an egg will be laid on a mass of several small victims before a cell is sealed. However, some species sting and paralyse quite large insects like grasshoppers or big caterpillars and lay eggs on them, often after the prey has been dragged back quite long distances to the tunnel. *Larra bicolor* has been used for the biological control of an orthopteran pest, the mole cricket *Scapteriscus* sp. When the tunnel is filled with cells, the mother usually abandons the nest for good. The larvae then eat the living but immobile prey. With some species using small prey, however, the female returns to re-provision the nest from time to time. As in some other wasp groups, the sphecids include 'cuckoo' species that lay their eggs on the prey collected by a different solitary bee or wasp.

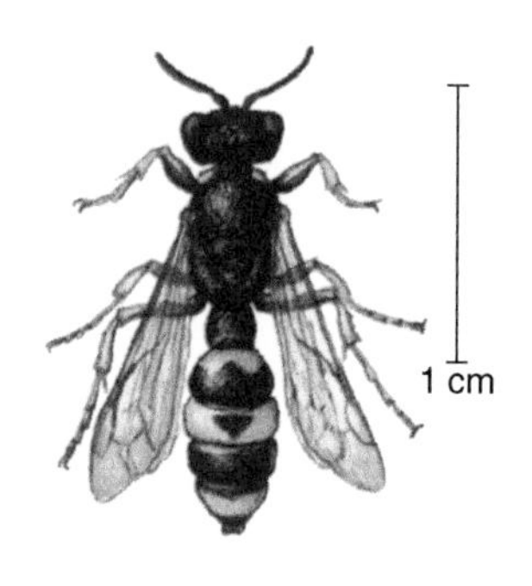

Fig. 11.15 Sphecidae (from Lyneborg 1968, with permission).

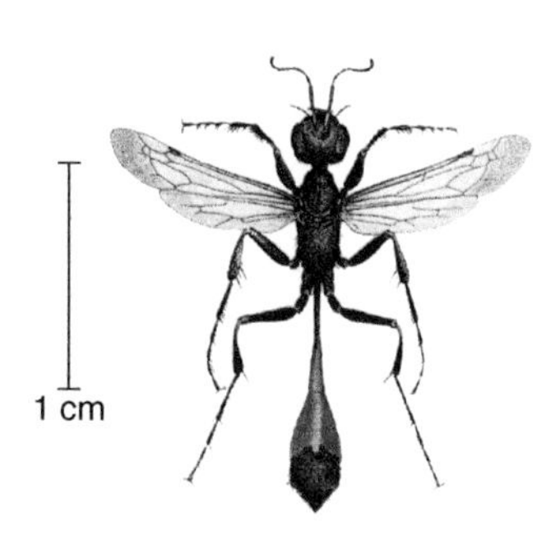

Fig. 11.16 *Ammophila* sp. (by Peter Skidmore, courtesy of Heather Skidmore).

Some sphecids (e.g. species of *Ammophila*; Fig. 11.16) have a quite different appearance and colouration from a wasp; they have a long waist of a reddish colour leading to a bulbous dark remainder of the gaster. Female *Ammophila* have the interesting behaviour of gripping a small stone between their mandibles and using it to tamp down the sand to seal the entrance to their burrow. So there you have it! Not long ago we thought man and chimpanzees were the only animals to use tools.

11.3.1.2 Superfamily Vespoidea (wasps)

These have unbranched hairs on the thorax and the pronotum reaches the tegula.

Family Vespidae

Wasps in this Family are social and include the familiar black and yellow wasps (e.g. *Vespula vulgaris*, Fig. 11.17, and *V. germanica*) with the painful sting that many readers will have experienced. Vespidae appear to have very narrow wings when seen at rest, as they are folded longitudinally, and the eyes are emarginated towards the centre of the head to appear kidney shaped, albeit rather asymmetrically (Fig. 11.18).

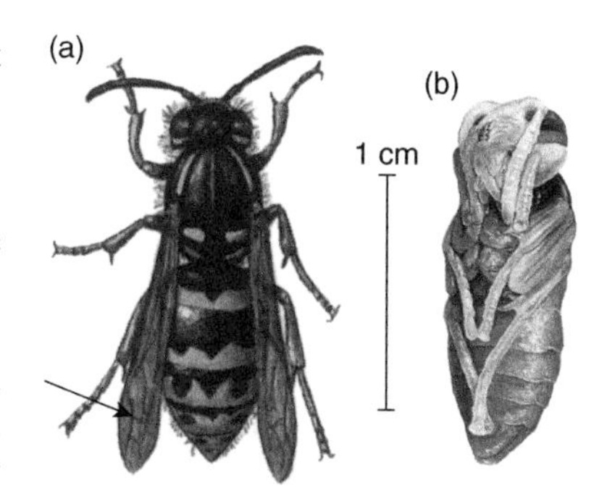

Fig. 11.17 Common wasp (*Vespula vulgaris*): (a) worker (from Mandahl-Barth 1974, with permission); (b) pupa (from Zanetti 1977).

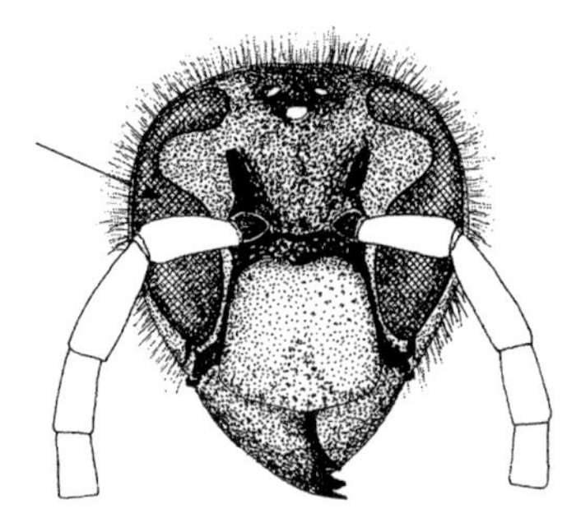

Fig. 11.18 Frontal view of the head of a wasp to show notched compound eye (from Alford 1999, with permission).

Other related and similar species of black and yellow wasps occur pretty widely in the world, but especially in the northern hemisphere. In contrast to those of honey bees, wasp nests are built to hang down from a pedicel in a free space. In nature this may be a cavity in the soil like a mouse hole, which the wasps further enlarge, or the nest may hang from some kind of anchor like a branch. Often the spherical nests are built in the loft space of dwellings, suspended from a rafter. The entrance to the nest is at the bottom, and there are usually six to eight tiered horizontal combs in the nest (Fig. 11.19). Also in contrast to honey bees, but as in bumble bees, nests are used for 1 year only and are vacated at the end of the season. Drones and new queens leave the nest in late summer to mate. The young queens find shelter for overwintering and, again as with bumble bees, are the only caste that survives the winter to start building a new nest and rear a first batch of workers the next spring. Wasp nests are built of a paper-like material made of wood particles cemented with saliva. A large number of workers scraping away with their mandibles gather the wood particles needed for nest building, especially as the number of horizontal combs grows, and this wood may come from timber of value to man, such as roof timbers, window sills and fence posts. The wasps tend to make an entry hole and then scrape away the wood from the inside, leaving a paper thin skin and no outward sign that the wood has been hollowed out!

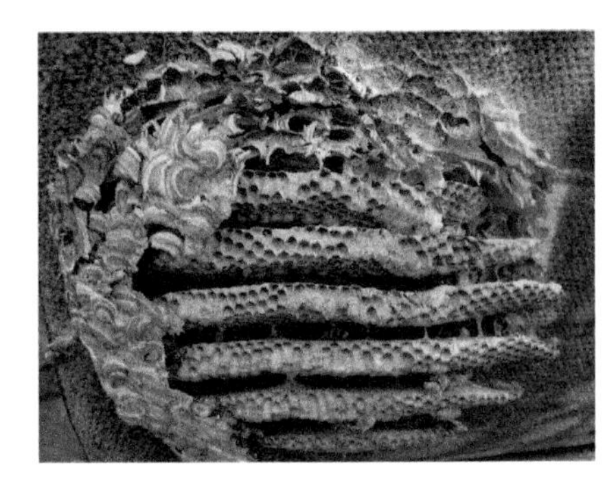

Fig. 11.19 Most of a wasp nest opened to show tiered structure. The outer casing still remains on the left side (photo by Richeman).

Adult wasps also damage fruit such as apples and pears by biting out pieces, but perhaps of more economic importance are their beneficial activities. Food is not stored in the nest, and the brood are mainly fed on animal food predigested by the workers; this food includes many species of insect pests such as small caterpillars and aphids. Wasps thus do have a beneficial aspect.

Paper wasps (*Polistes* spp.; Fig. 11.20) build open nests of a few hundred cells in bushes and small trees. These wasps are widely distributed, from southern Europe and the USA southwards to the whole of the southern hemisphere. They are again social wasps and carnivores, and thus prey on other insects that may be crop pests. Like European wasps they are attracted to ripening fruit and may cause some biting damage

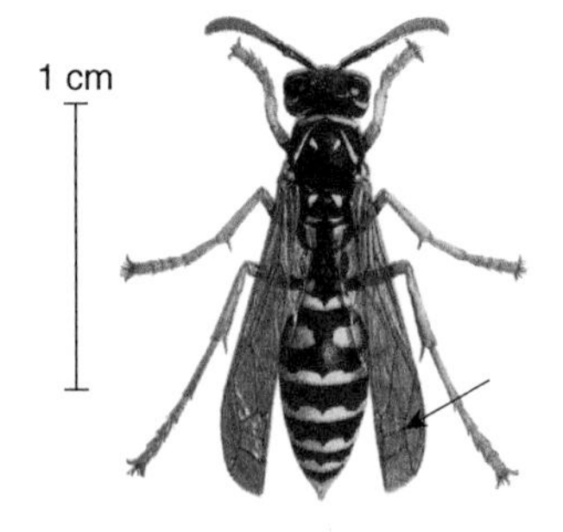

Fig. 11.20 Paper wasp (*Polistes* sp.) (from Zanetti 1977).

Fig. 11.21 Hornet (*Vespa crabro*) (from Mandahl-Barth 1974, with permission).

there. However, a more important impact on man is that they attack him particularly viciously (even for wasps) when they are disturbed!

The hornet (*Vespa crabro*; Fig. 11.21) is a much-feared large brown and yellow woodland wasp with a savage sting, which can be lethal to small children or the elderly. People annihilate anything they think may be a hornet, and many large flies, beetles and of course wood wasps (see Section 11.2.1) suffer as a result of human terror knee-jerk reactions.

Fig. 11.22 Velvet ants (Mutillidae). Left, male; right, female (by Gordon Riley from Chinery 1993, with permission).

Family Mutillidae (velvet ants)

Though not ants at all, the wingless females of these solitary wasps do look rather ant-like, and are covered with velvety hairs (Fig. 11.22). The males are nearly always winged, and both sexes are black or reddish in colour. Females invade the nests/burrows of bumble bees and some solitary bees and wasps, and lay their eggs near the larvae and pupae. The mutillid larvae then feed on their prey as ectoparasitoids, eventually killing it.

Family Pompilidae (spider-hunting wasps)

This Family of solitary wasps, with spiders as their prey, includes some of the largest Hymenoptera in the world, whose prey is large as well – tarantulas! The abdomen has no clear petiole and the hind legs tend to be rather long. In quite a lot of species, the antennae curl round on themselves like the beginnings of a spiral (Fig. 11.23). They are similar to the sphecid wasps in paralysing their prey and making their burrows in the ground. However, whereas the sphecids carry or drag their prey to a burrow already constructed, the Pompilidae wait till they have caught prey before excavating the soil, usually to create just one cell.

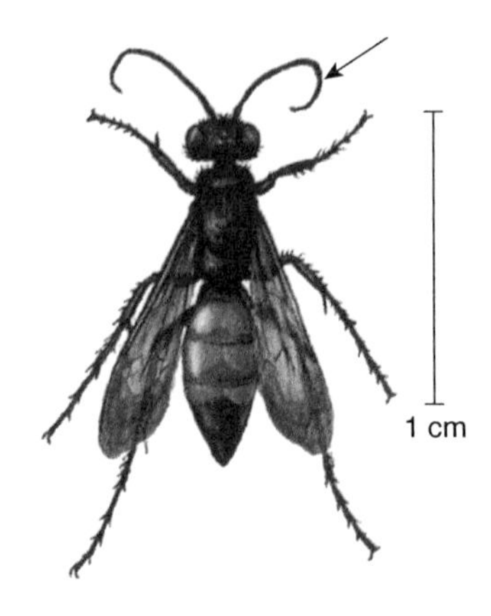

Fig. 11.23 Spider-hunting wasp (Pompilidae) (from Lyneborg 1968, with permission).

Family Scoliidae

These wasps are also among the largest Hymenoptera. They are hairy and black with bands or spots of brighter colours (Fig. 11.24), and the wings are often dark with a metallic sheen. They are ectoparasitic on some soil-inhabiting beetle larvae (Scarabaeidae or more rarely weevils). The female digs down to a larva she has located using her antennae while flying close to the ground, paralyses it and then excavates a cell around it before laying an egg upon it. Scoliids have been used successfully for the biological control of some chafers in Hawaii.

Family Tiphiidae

These are another Family of hunting wasps, smaller (often much smaller) than the Scoliidae. They have a more distinct petiole and clear wings, and in some species the females are wingless and are carried about by the winged male so that they can feed from flowers. Members of the Subfamily Thynninae are nearly the size of scoliids, and are again parasitoids of scarabaeid larvae, but *Diamma bicolor* attacks the mole crickets *Gryllotalpa* and *Scapteriscus*, which can be pests of cereals.

Fig. 11.24 Scoliidae (from Zahradník and Chvála 1989).

Family Formicidae (ants)

Given Superfamily status in many books, today the ants are mostly regarded as a Family within the Superfamily Vespoidea.

The 'Formic' in the name refers to formic acid, which ants can squirt at predators as a defence mechanism. The petiole of ants (see Figs 11.25b and 11.31) is very

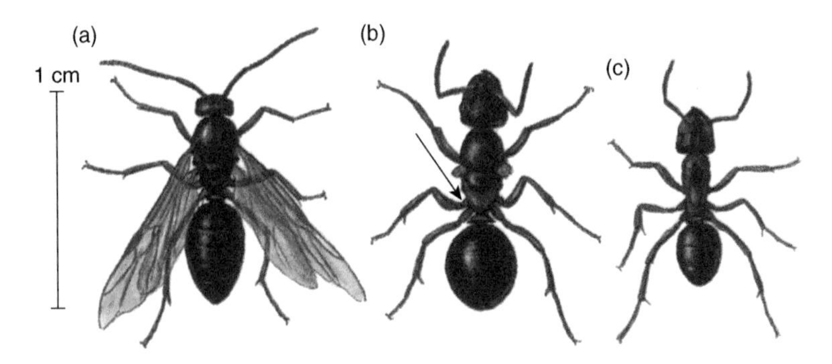

Fig. 11.25 Ant castes (the wood ant *Formica rufa*): (a) winged sexual (male); (b) queen in the nest (after losing her wings); (c) worker (all from Mandahl-Barth 1974, with permission).

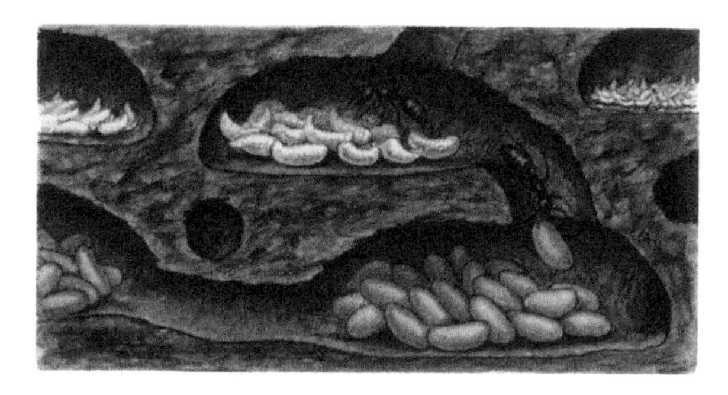

Fig. 11.26 Chambers and galleries in a section of a *Lasius niger* nest. Pupal cocoons are in the chamber at bottom right, with larvae at various stages of growth above (from Lyneborg 1968, with permission).

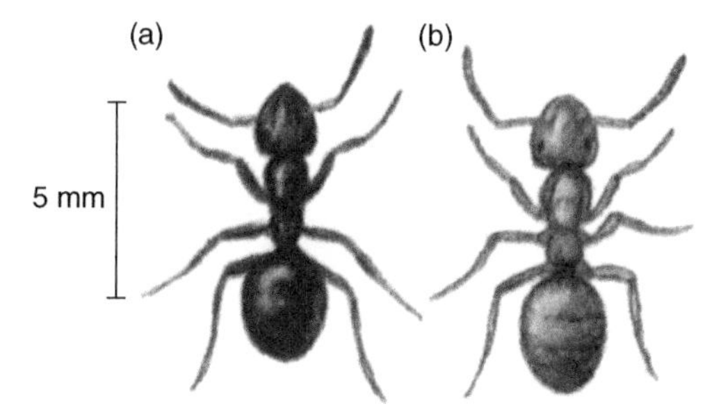

Fig. 11.27 Workers of (a) *Lasius niger* and (b) *Lasius flavus* (from Lyneborg 1968, with permission).

obvious compared to that of other Aprocrita, and may consist of several segments. In contrast to other social Hymenoptera, most adult ants are wingless, though the males and females are winged when they leave the nest (Fig. 11.25a). They mate on the wing and bite off their wings when they land to be wingless thereafter.

The nests of ants persist for many years, and involve a number of generations of queens 'ruling' the colony. The nest develops into quite a complicated structure, and epitomises the principle that many tarsi make light work! There are galleries devoted to specific purposes such as brood-rearing or storage of the pupa-containing cocoons (the fisherman's 'ants' eggs') linked by corridors (Fig. 11.26). Some species such as *Lasius niger* (Fig. 11.27a) build subterranean nests and little or nothing is visible from above ground. Ants like warmth, and in the UK it is common for *L. niger* to construct nests in the sand underneath patio slabs, which absorb heat and conduct it to the sand beneath. The untidy piles of excavated sand at the junctions between the slabs often cause the householder to try to kill the ants. Many other species (e.g. *Lasius flavus*; Fig. 11.27b) build their nests as roughly circular earth mounds (Fig. 11.28). These increase in size with time and can reach 50–60 cm. The mounds are colonised by plants, especially grasses, but often low-growing flowering plants like thyme grow on the flattish tops, which absorb the warmth of the sun. The ants move their pupae daily from lower in the nest, even below ground, where it is warmer at night, to near the top of the mound during the day. The much larger wood ants (*Formica* spp.; Fig.

Fig. 11.28 Nest of *Lasius flavus* with a 30-cm ruler.

Fig. 11.29 Nest of *Formica* sp. with a 30-cm ruler.

11.25) make even larger mounds of plant debris on the soil surface (Fig. 11.29). Such nests are particularly associated with pine woods where fallen needles become the main ingredient of the structure, which may be several metres across. Good drainage is important for ants and is one reason why so many species have nests elevated above soil level. I can recall once finding mounds obviously built by *L. flavus*, but occupied by a different species with subterranean nest-building behaviour. What had happened was that the farmer had built a road at the bottom of the field. This road had blocked the original drainage through the soil, which had become very wet to the point of being waterlogged. The subterranean species had solved their problem by driving out the mound-building species and occupying the mounds themselves.

As with other social insects, it is the workers that forage for food and bring it back to the nest. For their size, ants are immensely strong, and can drag food much larger than themselves for long distances. Some species (e.g. *Formica sanguinea*) avoid such chores by raiding the nests of other species and capturing the occupants to act as their

Fig. 11.30 Honey-pot ants (from Zanetti 1977).

slaves. Indeed, ants are pretty ferocious with strangers (see the example in the paragraph above concerning drainage), and battles to the death even occur when conspecifics, but from different nests, meet when foraging.

The food of ants is varied. Many species are carnivorous, and drag any 'meat' they find (such as caterpillars) back to the nest. Other species (as mentioned in Section 7.5.3.2, Family Aphididae) 'farm' aphids and scale insects for their honeydew as well as collecting nectar. Thus species of the genus *Oecophylla* (red tree ants) are often associated with scale insect outbreaks, as they protect the scales against predators and parasitoids. However, the same ants are primarily carnivorous, and can be useful biological control agents of other pests on trees. Some ant species store honeydew and nectar in the abdomens of some of their colleagues, which hang like bags from the roof of special galleries (Fig. 11.30). Some Australian Aborigines seek out these 'honey-pot' ants as a source of sweet food. Often mound-building and subterranean species colonise the roots of grasses at the edges of their nests with root-feeding aphids and so have a source of honeydew very close by and protected from rain.

Leaf-cutting ants are familiar in the tropics, and may present a serious pest problem on crops such as citrus. The economically important leaf-cutters are found in two genera, *Atta* and *Acromyrmex*. Both species occur in the tropical regions of the new World. *Atta* spp. (Fig. 11.31) are primarily forest insects and so crops near forest edges are most likely to suffer attack, while *Acromyrmex* spp. occur more in urban and cultivated landscapes. *Atta* spp. make huge nests, perhaps more than 10 m in diameter and 4 m in depth, harbouring perhaps 2 million ants. In contrast, the nests of *Acromyrmex* are very much smaller and are usually less than 1 m in diameter. Like termites, both these ant genera develop 'fungus gardens' on the cut leaf portions and graze the fungus developing on this substrate. Both genera are polyphagous, though most damage seems to be done to citrus, cocoa, coffee and maize. Trees/plants can be so totally defoliated that they die.

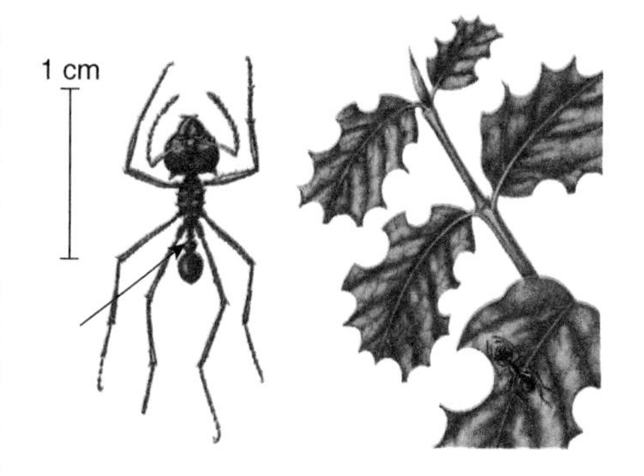

Fig. 11.31 Leaf-cutting ant (*Atta* sp.) and damage to coffee leaves (from Bayer 1968, with permission).

The harvester ants (e.g. *Messor barbarus*) feeds on many species of grasses and cereals, but is mainly a pest of grassland. It is found in the drier regions of east Africa. The ants cut pieces from the plants in order to build their nest mounds above their large underground nests, to the extent that the area around the nests becomes completely bare. The food of the ants is grass seeds, which are stored in the nests in great quantities.

Solenopsis geminata (fire ant) damages citrus, avocado, and many other fruits and seedlings (especially of tobacco). The trees are damaged by the ants biting into the bark to collect the sap that exudes; the damage may result in girdling of the trunk/branch. They also attend aphids. The ants live in burrows in the soil around the trees in many tropical regions of the world.

Where common, the fire ant can be a nuisance by their aggressive behaviour towards farm workers; the bite is rather painful. This nuisance problem also makes other ant species 'pests'; *Oecophylla* was mentioned earlier in this Section. Fire ants live in small nests made by sewing a few leaves together. The colonies are very sensitive to vibration, and rush out and bite workers tending the trees. The biting ant *Tetramorium aculeatum* on coffee builds delicate nests (each containing a few hundred small brown ants) in the bushes. This does virtually no direct damage to the crop; neither does the ant promote honeydew-producing insects. However, its bite again causes such severe pain to workers that they may have to abandon the task they had embarked on.

11.3.1.3 Superfamily Chrysidoidea

This Superfamily of solitary wasps has little economic importance, although they are parasitoids of other insects. I mention it only because two of the several Families are easily recognised by spot characters. Thus female wasps in the Family Dryinidae, whose larvae are endoparasites of the auchenorrynchan Hemiptera, have the most bizarre chelate (i.e. with an articulated claw) tarsi on the front legs. The larvae of the Family Chrysididae (ruby-tailed wasps), where the adults are usually of a brilliant metallic colour (Fig. 11.32), are ectoparasitic on the eggs and larvae of some Families of solitary bees and wasps.

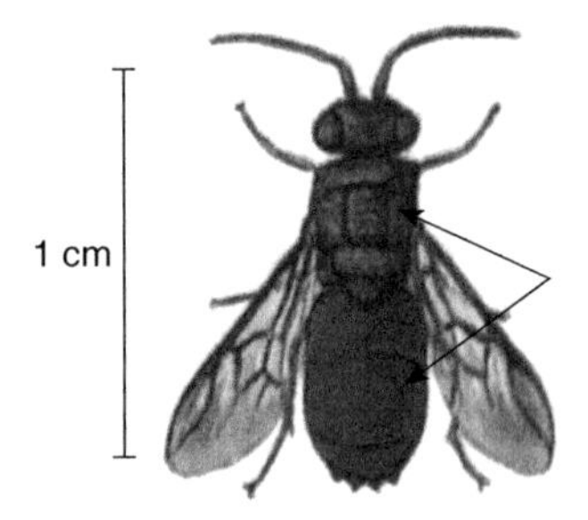

Fig. 11.32 Ruby-tailed wasp (Chrysididae) (from Lyneborg 1968, with permission).

11.3.2 Other Superfamilies

These do not have a sting in place of the ovipositor, and many species use the ovipositor to insert eggs into living organisms (see below in this Section); another difference from the social Hymenoptera is that the forewing has no anal fold. The hind wings are usually very small and, as pointed out in Section 11.1, these may be so hard to distinguish in specimens at rest or dead that the mistake may be made of thinking they have only one pair of wings (i.e. they are Diptera). However, they usually have quite obvious filiform antennae as well as that pigmented area (pterostigma) near the apex of the front wing so common in the Order Hymenoptera. The advice given earlier is therefore repeated here: look very closely at any 'fly' that has obvious antennae and a pterostigma before placing it in the Diptera.

Most of the wasps in these other Superfamilies of the Apocrita are parasitoids of other insects or have larvae which develop is such close association with plants as to cause the latter to respond with very specific nutritionally rich and protective structures (galls). The parasitoids have been used extensively for the biological control of pests, and thus have enormous economic importance. Although the parasitoid habit was briefly described in Chapter 1, this seems an appropriate place to give a fuller treatment.

Although the terms 'parasitoid', 'endoparasitoid' and 'ectoparasitoid' have already been defined in Chapter 1 and have been used above in describing solitary wasps, it is the group (often given the umbrella name 'Parasitica') of the Apocrita we have now reached which contain many such carnivores, frequently used in biological control. I therefore propose to define the terms again at this point. 'Parasitoids' complete their development on or in one prey individual (the 'host'), which is eventually killed as a result. Most pierce the cuticle of their host (an egg, larva or pupa) and lay an egg inside it. Others lay their eggs on the host and the larvae that hatch may either enter the host or feed through the cuticle, but from the outside. The latter are 'ectoparasitoids'; those that develop inside their host are 'endoparasitoids'. One female may lay several eggs in a host, but one egg per host is more usual. Then if another female later oviposits in the same host, 'superparasitism' has occurred. Usually contest will occur, and the smaller later larva will usually not survive. Avoiding superparasitisation is thus in the interest of a parasitoid species, and many females leave chemical markers on hosts they have attacked, which females arriving later will recognise.

The larvae are thus often enclosed in host fluid. They show no distinct head capsule and respire by exchanging oxygen between the exterior fluid and their own body fluids through their skin. The parasitoid larvae first feed on non-vital structures such as the fat body, and the host normally survives until the parasitoid larva is ready to pupate. Pupation is often within the dead skin of the host, but the larvae of many other species emerge from the dead host before they pupate.

There are very many Superfamiles and Families comprising a multitude of species, many of which are still undiscovered. This section of the Apocrita is one of the very largest insect 'taxa'; therefore only a sample of the Superfamilies and Families can be presented here. The adults range from quite robust insects to the smallest insects there are, with the larvae completing their development in the eggs of other insects. The wasps may be primary or secondary parasitoids. The latter are known as 'hyperparasitoids' – they develop in or on the larvae of primary parasitoids. Some species may be both primary parasitoids in their own right as well as hyperparasitoids of another species of primary parasitoid. There is even the example of the genus *Aphelinus* (see Section 11.3.2.2, Family Aphelinidae) where the male is a hyperparasitoid of the female of its own species (a primary parasitoid of aphids).

The taxon includes wasps with a variety of unusual features. Commonly, the female (which as in other insects stores the sperm received at mating in her ***spermatheca***) decides whether or not to fertilise eggs as they pass down the oviduct. Fertilised eggs develop into females, but unfertilised ones produce the males. Females may also develop parthenogenetically, and in some species males are virtually unknown. Another unusual phenomenon is the occurrence of polyembryony, where a single egg may divide to become many (up to 2000) embryos and thus many young may stem from a single egg. There are also cases where male and female eggs are laid in different host species, and then mated and unmated females show quite different host selection behaviour. With over 40 Families in the taxon, I will refer to those of economic importance, but also to a few others with particularly interesting features or the activities of which you are very likely to come across.

11.3.2.1 Superfamily Ichneumonoidea

This Superfamily contains the majority of the species important for natural or managed biological control. Many are parasitoids of Lepidoptera, but they also parasitise many other Orders that include pest species, including Hemiptera, Diptera and Coleoptera. Wing venation tends to be more complex than in the other Superfamilies, and the antennae are clearly simple and filiform. There is often a pterostigma. The two Families in the Superfamily are most easily distinguished on wing venation characters (Fig. 11.33): the Family Ichneumonidae has a second m-cu cross vein (2m-cu) on the forewing and in the hind wing the r-m cross vein is much further along the wing than the half way mark. In the Family Braconidae there is no 2m-cu cross vein on the forewing, and the r-m cross vein on the hind wing is not more than half way along.

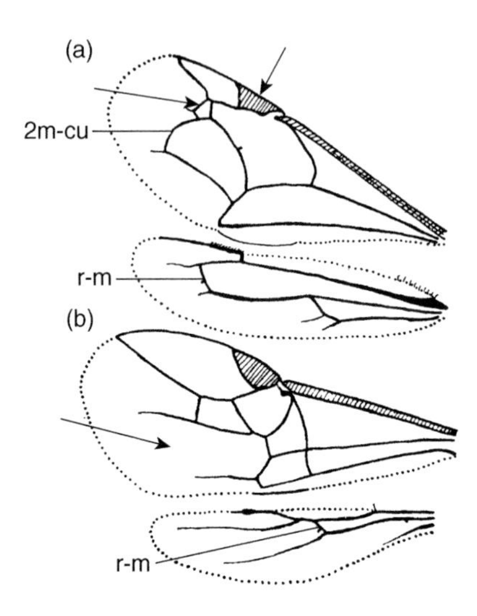

Fig. 11.33 Wings in the Ichneumonoidea: (a) Family Ichneumonidae; (b) Family Braconidae (from Richards and Davies 1977).

Family Ichneumonidae

This Family includes the largest of the parasitic Hymenoptera; the common name is 'ichneumon flies'. This is etymologically incorrect, since they are not flies, like for example, hover flies, and thus 'ichneumonflies' should really be one word, like for example butterflies, but it never is! Some species have the diagnostic character that the r-m cross veins forms a much reduced and therefore small polygonal cell near the apex of the front wing at the top of the r-m cross veins (Fig.11.33).

Ichneumonidae are mainly parasitoids of butterfly and moth caterpillars, though other hosts are larvae of sawflies and beetles. Some are hyperparasitoids. Commonly attracted to light in the UK are two yellow-brown parasitoids of noctuid moth caterpillars, *Ophion luteus* (Fig. 11.34) and *Netelia testacea*. A very dramatic-looking black species with an extremely long ovipositor is *Rhyssa persuasoria* (Fig. 11.35), which parasitises wood wasp larvae (see Section 11.2.1) deep in the timber of pine trees. The female parasitoid lays its antenna on the tree bark to detect the small body movements

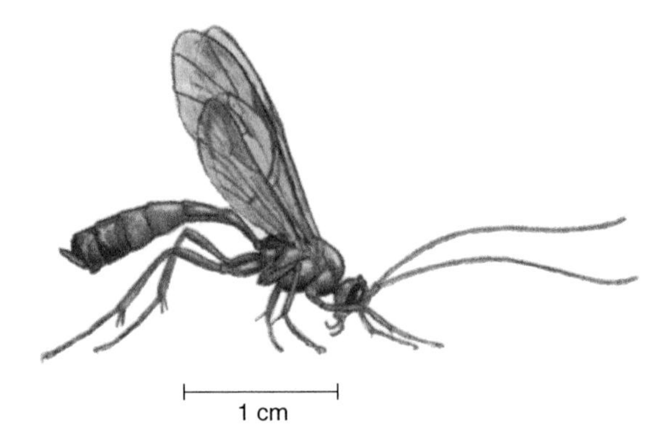

Fig. 11.34 *Ophion luteus* (from Mandahl-Barth 1974, with permission).

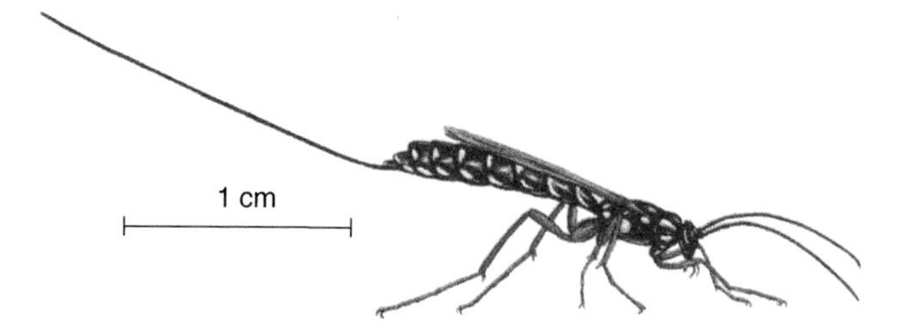

Fig. 11.35 *Rhyssa persuasoria* (from Mandahl-Barth 1974, with permission).

of its prey, and then uses its ovipositor first as a drill to bore down to the host before then squeezing an egg down the tube.

Of course, predators of insects have their own parasitoids, which then can reduce the impact of biological control. Thus the ichneumonid species of *Diplazon* parasitise hover fly larvae, which are rather useful predators of aphids (see Section 10.4.1.1).

Family Braconidae

The Braconidae share the parasitoid habit of the ichneumonids, but are usually somewhat smaller. An important genus is *Cotesia. Cotesia flavipes* is a common and regular occurring parasitoid of the legume pod borer *Maruca vitrata, C. plutellae* is an internationally important biological control agent for the diamond back moth (see Section 9.2.3.3, Family Plutellidae), a very serious pest of brassicas, and *C. glomerata* (Fig. 11.36) is the major parasitoid of cabbage white butterflies (*Pieris* spp.). More than 100 larvae may emerge from a single *Pieris* caterpillar to spin their yellow silk cocoons and pupate (Fig. 11.36b).

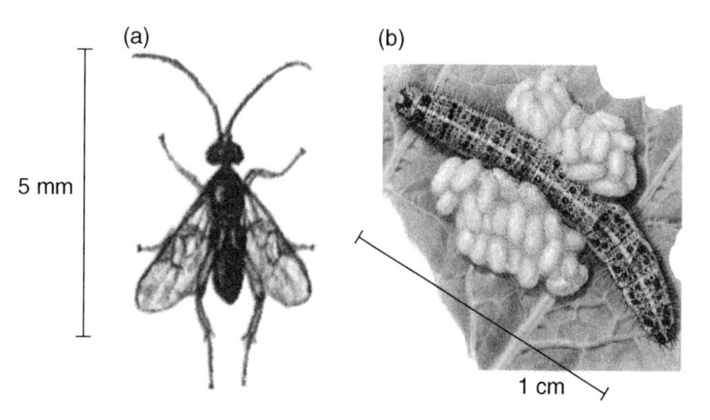

Fig. 11.36 *Cotesia glomerata*: (a) adult (from Lyneborg 1968, with permission); (b) *Pieris brassicae* caterpillar with cocoons (from Zanetti 1977).

As just pointed out above in Section 11.3.2.1, predators of insects have their own parasitoids, which then can damage biological control, and the braconid *Dinocampus coccinellae* has ladybird larvae as its prey.

Very important as natural enemies of aphids are members of the Subfamily Aphidiinae (Fig. 11.37). The largest genus is *Aphidius*. In common with most other parasitoids of aphids, the larva cuts a slot in the bottom of the aphid and attaches it to the leaf or stem with silk, with which it then goes on to spin its cocoon within the skin of the now dead aphid. The dead aphid skin is brown and is known as the 'mummy' (Fig. 11.38a). The adult parasitoid emerges through a circular hole it cuts in the cuticle of

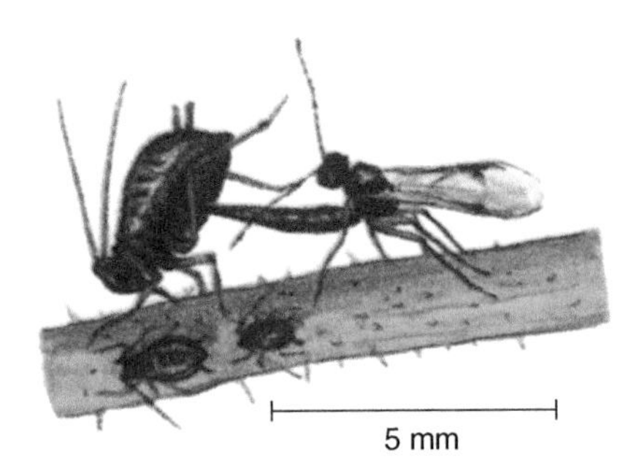

Fig. 11.37 An aphidiine parasitoid (*Lysiphlebus testaceipes*) ovipositing in an aphid (from Lyneborg 1968, with permission).

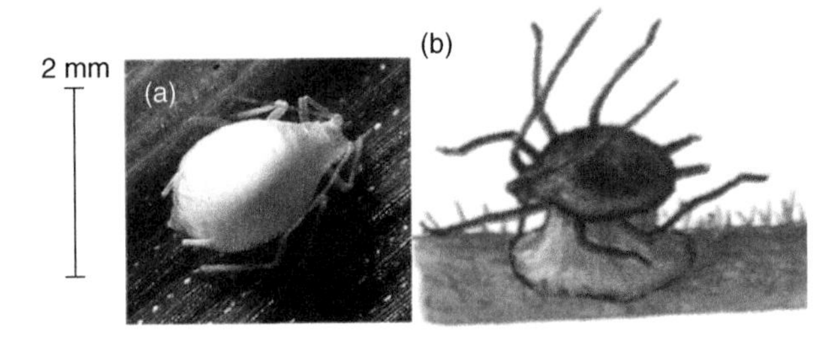

Fig. 11.38 Aphid mummies formed by the parasitoids (a) *Aphidius* sp. and (b) *Praon* sp. ((a) photo courtesy of G. R. Gowling; (b) from Mandahl-Barth 1974, with permission).

the mummy. *Aphidius colemani* is reared commercially in many countries for biological control of aphids, especially in greenhouses. Most Aphidiidae are rather or very host specific; the commercial advantage of *A. colemani* is that it is very much a generalist.

The genus *Praon*, also a generalist, is another common genus of parasitoids of aphids. It differs from the other Aphidiidae in that it does not pupate within the mummy, but in a silken tent it constructs underneath with the mummy forming the roof (Fig. 11.38b). Emergence is then through a hole cut in the side of the tent.

In contrast to these two generalists, *Trioxys pallidus* is a specialist parasitoid of the walnut aphid (*Chromaphis juglandicola*), and release in California of strains from Europe and Iran have given good control of the pest.

11.3.2.2 Superfamily Chalcidoidea

This is probably the largest of all insect Superfamilies, with many species still awaiting discovery. The majority are parasitoids, but many are hyperparasitoids, either solely so or in addition to being primary parasitoids. Some, however, are herbivorous; they are seed feeders or form galls on plants. They are mostly small to minute, with a much reduced wing venation. The forewing is either devoid of veins or has a single anterior vein leading to a small pterostigma (Fig. 11.39). Either way, there are no cells. The adults are often very pretty with bright metallic colours, especially green and blue. Antennae are often small and geniculate or filiform.

Family Agaonidae (fig wasps)

The pollination of most varieties of fig (admittedly a minor crop) depends on the activity of these minute wasps with most unusual biology. The females (Fig. 11.39) are fully winged and mate with the males that are mostly wingless (Fig. 11.40) and never leave the fig in which they developed. The male then bites a hole in the fig wall to enable the female, which has now acquired some pollen, to escape and search for another fig in which to oviposit. The female only lives for 2 days, and has to enter a new fig through a very narrow opening (the ostiole) at the apex of the fruit. Squeezing in usually causes her wings to break off, and thus she remains trapped inside this fig where she leaves pollen and lays eggs. The inflorescences of the tree are within the fig, and most have flowers of three kinds: male (producing pollen), short female flowers (where the female can reach the ovaries to oviposit and which are then eaten by the larvae to produce more wasps) and long female flowers (which are pollinated but where the ovaries cannot be reached for oviposition). These are the flowers that successfully produce seeds.

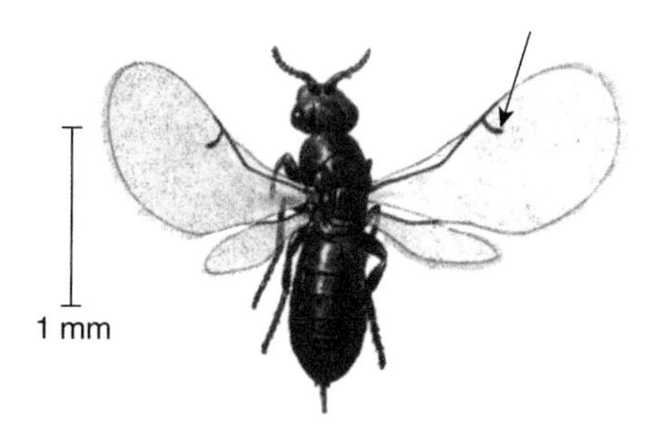

Fig. 11.39 Chalcidoidea (female fig wasp, *Blastophaga psenes*) (from Zanetti 1977).

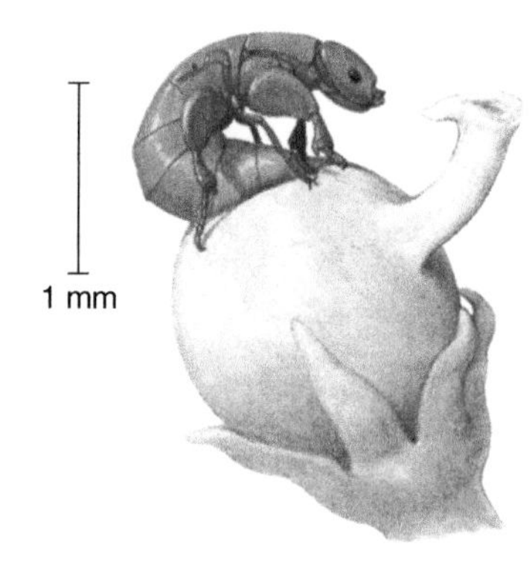

Fig. 11.40 Male fig wasp, *Blastophaga psenes*) (from Zanetti 1977).

Family Aphelinidae

The Family includes endoparasitoids of aphids and whiteflies, and both endo- and ectoparasites of scale insects. The Aphelinidae show the bizarre biology that male larvae may develop as hyperparasites on the female larvae of their own species!

The parasitoids of aphids in this Family are very similar in appearance and habits to the Aphidiidae, though the adult females initially act as aphid predators, and feed on the juices of young aphids to obtain the nutrition for the eggs to mature before they switch to normal parasitoid behaviour and oviposit into their host. *Aphelinus* is the main genus, and *A. mali* has been released for biological control of the woolly apple aphid (*Eriosoma lanigerum*).

Species of the genus *Aphytis* are ecto- or endoparasitoids of scale insects in the Family Diaspidae (e.g. *A. mytilaspidis* parasitising the mussel scale *Lepidosaphes ulmi*). There have been several instances where use of insecticides has caused an outbreak of scales because it destroyed *Aphytis* parasitoids.

Encarsia formosa (Fig. 11.41) is a parasitoid of whiteflies (Hemiptera: Aleyrodidae). It appeared in Hertford-

Fig. 11.41 *Encarsia formosa* (courtesy of Ward Stepman, BCP Certis).

shire in the UK in 1926 and was noticed because the normally pale scales of whiteflies become black after parasitisation. *Encarsia* proved such an efficient parasitoid that it was bred at a government research station for release on tomato and cucumber crops in greenhouses. The availability of DDT as an easier solution halted this biological control practice in the 1940s, but *Encarsia* became important again in the 1960s as part of control systems to minimise insecticide use under glass. For many years now, *E. formosa* has been reared commercially, and the black parasitised whitefly scales have been purchased by many growers. The genus *Encarsia* gives us another example of the weird biologies we find scattered in the Aphelinidae; the sexes of *E. porteri* develop in hosts from different Orders: females develop in scale insects and males in the eggs of Lepidoptera!

Family Chalcidae
This Family includes a huge number of species, but few of economic importance as biological control agents.

Fig. 11.42 Pteromalidae (courtesy of Andrew Polaszek).

Family Pteromalidae
This again includes very many species, but of limited economic importance. I mention it mainly because it has so many species and is commonly encountered. Most species are easily recognised by the bright green metallic or bronze colours and the pointed very triangular abdomen (Fig. 11.42). However, species of the genus *Pachyneuron* are hyperparasitoids of *Anagyrus* (see Section 11.3.2.2, Family Encyrtidae) and have damaged the biological control of Kenya mealybug (*Planococcus kenyae*) by *Anagyrus*.

Family Torymidae
These are very similar in appearance to the Pteromalidae, with bright metallic colours and a triangular abdomen. However, the females have an obvious long ovipositor, adapted to reach gall-inhabiting insects. Also the coxae of the hind legs are noticeably swollen.

Family Encyrtidae
A spur at the middle of the tibia is commonly found. The wasps are parasitoids of eggs, larvae and pupae of a wide range of insects, especially Hemiptera and Lepidoptera. *Anagyrus* (Fig. 11.43) is one of the genera that has featured in successful biological control. The cassava mealybug (*Phenacoccus manihoti*) was successfully controlled in the 1980s over much of Africa by *A. lopezi* introduced from South America.

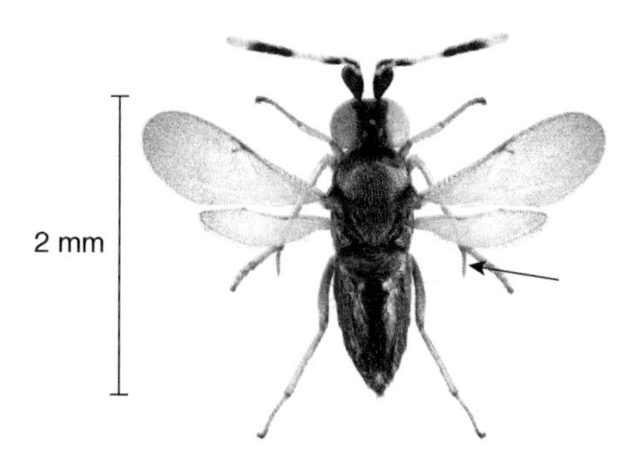

Fig. 11.43 *Anagyrus lopezi* (courtesy of G. Goergen, Biodiversity Centre, IITA).

Family Trichogrammatidae

These are parasitoids of insect eggs. Understandably, they are therefore rather small wasps (Fig. 11.44). Many are black, but some are lighter coloured. A spot character for the Family is that the otherwise featureless forewing has lines of hairs on its surface.

These egg parasitoids have proved extremely valuable and widely used biological control agents, especially of Lepidoptera. They are currently used in the biological control of close on 30 pests. *Trichogramma* has therefore been produced commercially on eggs of *Ephestia kuehniella*, the Mediterranean flour moth which is easy to rear in quantity. The eggs are sold as cards of about 1000 parasitised eggs which are black in colour, signifying that the *Trichogramma* within will soon emerge as an adult. These cards are then hung in trees or on plants, though over large fields in South America, the parasitised eggs are sometimes broadcast from a microlight aircraft. Several species of *Trichogramma* with different host preferences are available for different pests, but *T. evanescens* is very polyphagous and probably the most frequently used species.

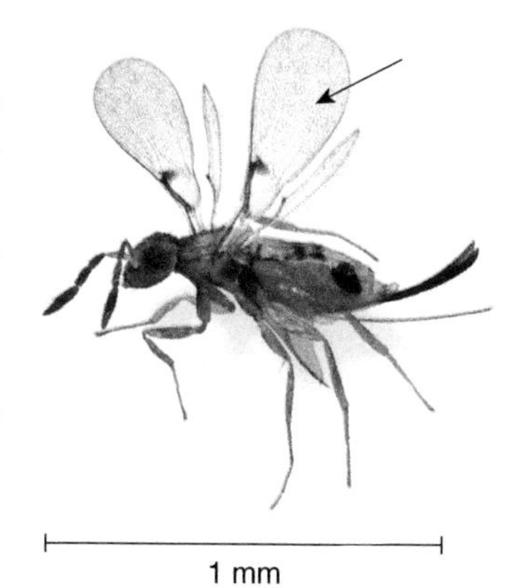

Fig. 11.44 Trichogrammatidae (courtesy of H. Negri Oliviera).

Family Mymaridae (fairy flies)

Clearly these are not true flies, so 'fairyflies' should really be the common name of this Family. These are again parasitoids of eggs. Some of these are probably the tiniest insects (see Section 1.1) at under 0.25 mm long. As an adaptation to avoid being wetted, their wings begin as a narrow stalk, expanding into a slightly wider paddle fringed with long hairs (Fig. 11.45). They are brown insects with obvious filiform antennae. The biological control of grape leafhopper (*Erythroneura elegantula*) in California depends on a fairy fly (*Anagrus epos*), but then also depends on the presence of other leafhoppers (e.g. *Dikrella californica* on blackberries) as this overwinters as an egg, whereas *E. elegantula* overwinters as an adult. The grape leafhopper therefore provides no hosts for an egg parasitoid's overwintering generation.

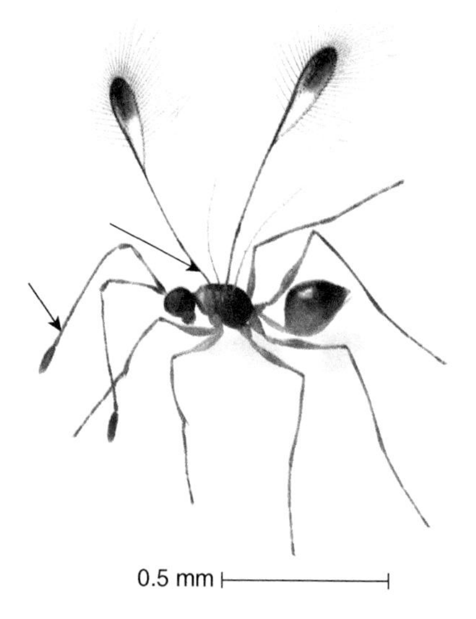

Fig. 11.45 Mymaridae (courtesy of John Huber).

Family Tanaostigmatidae

This Family is unusual for the Chalcidoidea in being almost exclusively phytophagous. Most form galls in plant stems, leaves or seeds. The mesonotum is often so arched that the pronotum appears vertical.

In India, I came across *Tanaostigmodes cajaninae*, the pod wasp, which feeds in the pods of pigeon pea and retards their development. This pest had become a problem for a bizarre reason. The insecticide endosulfan had been used against lepidopteran pod borers because it is relatively non-toxic to the hymenopteran parasitoids, giving a useful degree of biological control. Of course, as a hymenopteran, the pod wasp was equally spared and thrived in the absence of competition from the lepidopteran pod pests.

11.3.2.3 Superfamily Proctotrupoidea

These are yet more small or tiny parasitoids, some of which parasitise eggs but the Superfamily also includes larval and pupal parasitoids. They are brown or black wasps, with very reduced wing venation, but the hind wing has a demarcated anal lobe. The forewing may have only a few faint longitudinal veins with or without some strongly sclerotised veins near the leading edge; some species are totally wingless. The front tibia has one spur.

This Superfamily is also the classic example of polyembryony in the Hymenoptera, whereby many larvae develop from a single egg (see Section 11.3.2).

Members of the Family Proctotrupidae are common, but in economic terms the most important Family is the Platygasteridae which specialises on gall midges (Cecidomyiidae). This includes *Platygaster zosine,* which can prevent the hessian fly (see Section 10.2.3) from reaching pest status on wheat. The Platygasteridae have an almost complete absence of wing veins (Fig. 11.46).

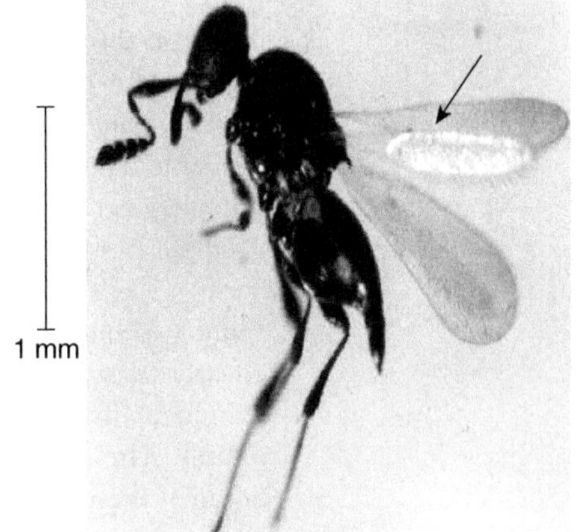

Fig. 11.46 Platygasteridae (www.fernandoriazs.galeon.com).

11.3.2.4 Superfamily Evanioidea

These are black, but may have a red abdomen; they are of medium size. The Superfamily is easily identified from the fact that the slender petiole is attached right at the dorsal tip of the propodeum.

Family Evaniidae

These have a very large thorax in relation to the small laterally flattened gaster, which thus looks like a downwardly pointing flag at the end of the petiole (hence the common name 'ensign wasps' sometimes given to the Family). The Evaniidae are parasitoids of the eggs of cockroaches carried in the ootheca (see Section 6.10.1).

Family Gasteruptiidae

These have a very long slender gaster and the ovipositor of the females is also very long. The Family are predators rather than parasitoids, eating the larvae and stored food in the nests of solitary bees and wasps.

11.3.2.5 Superfamily Cynipoidea

Cynipoidea are small and black or chestnut-brown, and many have a rather globular thorax and abdomen looking a bit like two beads stuck together. They have filiform antennae and very diagnostic wing venation (Fig. 11.47). The strongest veins on the forewing are all in the front half, and form a zig-zag pattern. To me these veins often look like a distorted and laterally stretched letter 'w' with the ends of the strokes pulled out sideways like the serifs on a printer font like 'Times New Roman'. This pattern includes a very triangular radial cell (the cell under where the stigma would be in other Hymenoptera).

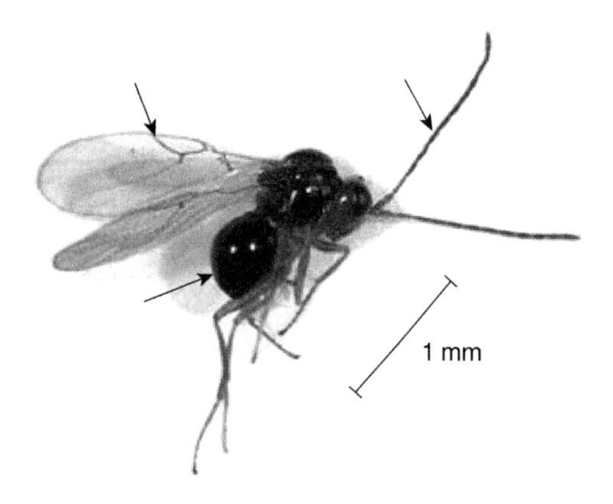

Fig. 11.47 Cynipidae (photo by S. E. Thorpe).

Family Charipidae

These are common hyperparasitoids of those Braconidae (e.g. *Aphidius* spp.) that parasitise aphids, and more rarely of those (e.g. *Aphytis* spp.) that parasitise coccids.

Family Cynipidae (gall wasps)

These have an elongated gaster more often than the Charipidae. Many cynipid galls are common and well-known, such as the 'robin's pin cushion' (Fig. 11.48) on roses (more scientifically called the Bedeguar gall) caused by *Diplolepis rosae*. These galls vary enormously in size; they may reach a diameter of as much as 10 cm. There are also cynipid species which form different galls and/or galls in different places in their alternating unisexual (all parthenogentic females) and bisexual generations. The 'oak apple' is a familiar gall found on the petioles of oak leaves and on oak twigs. Oak apples (Fig. 11.49) are caused by the unisexual generation of *Biorhiza pallida*, but the bisexual generation of the same insect induces not dissimilar swellings on the roots of the tree.

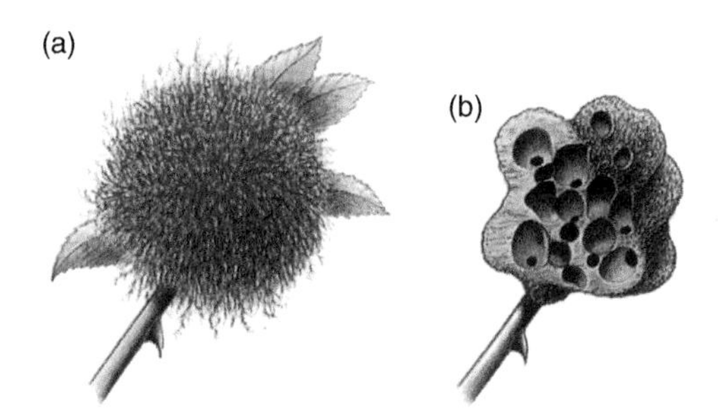

Fig. 11.48 (a) Bedeguar gall ('robin's pin cushion) on rose and (b) section of gall (by Gordon Riley from Chinery 1993, with permission).

Fig. 11.49 Oak apples (from Zanetti 1977).

The genus *Neuroterus* forms purplish spherical buds that look like red currants (currant galls) on the leaves and on the catkins of oak in the spring (Fig. 11.50). The larvae that form these galls hatched from eggs produced asexually (parthenogenetically), but from these galls males and females emerge to lay eggs into the leaf tissue. Quite different galls are then produced, in which larvae develop that will all emerge as asexually-reproducing females. These galls are the 'spangle galls', rather flat discs on the leaf which have different appearances (Fig. 11.51) depending on the species of *Neuroterus* involved. Some are uniformly green, some are red, some are hairy, some have a central 'pimple' and others are like doughnuts. Although the larvae of the different species produce such varied galls which are easy to identify to cynipid species, the adults look identical and are very hard to identify. One expert at the Natural History Museum in London identified them by 'the way they walked', but had great difficulty if presented with dead specimens!

Fig. 11.50 Currant galls on the catkins of oak (by Gordon Riley from Chinery 1993 with permission).

Fig. 11.51 Spangle galls on oak leaf (from Mandahl-Barth 1974, with permission).

12 Subclass Pterygota, Division Endopterygota, Order Coleoptera (beetles) – *c.* 350,000 described species

12.1 Introduction

With over 300,000 described species, the Coleoptera is the largest Order of insects (about 40% of all described insect species) – and indeed the largest Order in the Animal Kingdom. Some people have estimated that there may actually be between 5 and 8 million species of beetle! Probably about 1 in 4 of all the species of animals there are is a beetle. I told a relevant story about the late Professor Haldane and beetles at the beginning of this book (see Section 1.1).

Not surprisingly, so many species show tremendous diversity. Life styles include carrion feeders, pests of museum insect collections, scavengers on dead vegetable material including textiles and carpets, herbivores (feeding on roots, stems, leaves, flowers, seeds and fruits; Fig. 12.1), pests of stored products such as cereals and flour, and carnivores. Beetles are also well represented in the aquatic environment.

The size range of beetles is vast. At one end is the Goliath beetle – probably the largest extant insect (see Section 1.1); at the other are some of the smallest insects such tiny beetles avoiding wetting by living in the caps of mushrooms.

12.1.1 Adult morphology

Beetles have very varied antennae. All the types of antennae shown in Fig. 2.7, except for the cyclorrhaphous, pectinate and plumose types, could have been drawn from beetles. Also,the number of segments can vary over the huge range of one to 27!

By contrast, there is less variation in mouthpart design. Mouthparts are of the basic biting type (see Section 2.4.2) with mandibles, maxillae and the labium clearly identifiable. Herbivores and scavengers have robust mandibles with an incisor and a large grooved molar area, while carnivores tend to have longer, curved and pointed jaws.

The pronotum (Fig. 12.2) is usually the only obvious part of the thorax visible from above, since the rest of the thorax and the abdomen are usually concealed by the front wings modified as wing cases (each is an ***elytron***, plural ***elytra***). However, a small triangular ***mesoscutellum*** from the mesothorax may show between the elytra at the boundary with the pronotum.

The junction of the elytra makes a ***dorsal suture*** since the elytra abut tightly and are indeed fused together in some flightless beetles. Dorsally, the elytra have several characters used in identification. They may have longitudinal grooves (***striations***), depressions like tiny craters of pinhole dimensions (***punctures*** – their pattern is called ***puncturing***) and normal hairs or ones flattened into scales. The elytra project beyond the sides of the abdomen and are reflexed back, so that a strip (the ***epipleuron(a)***) is visible in a ventral view (Fig. 12.11). This enables the abdominal spiracles to open

Handbook of Agricultural Entomology, First Edition. H. F. van Emden.

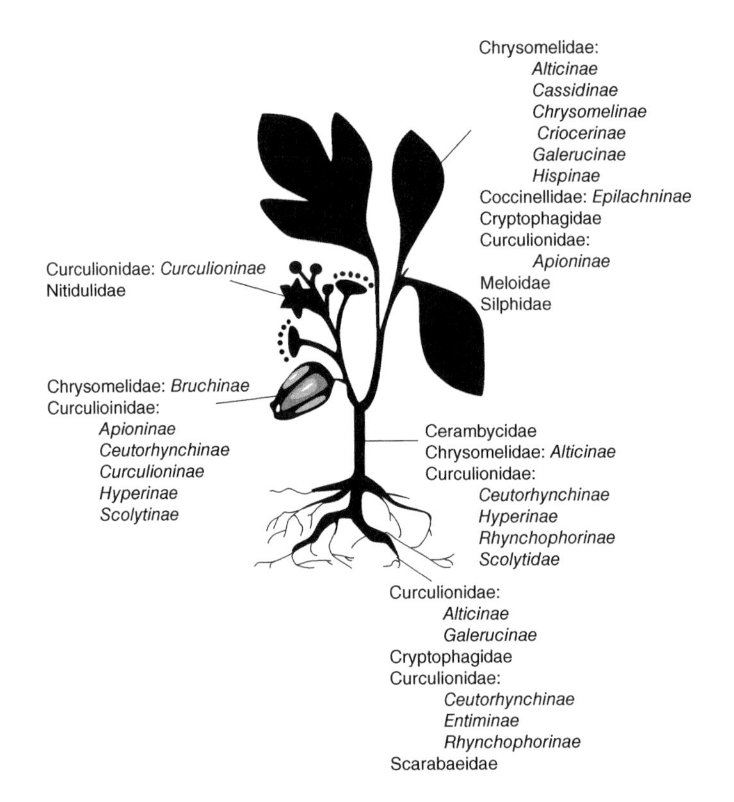

Fig. 12.1 Coleoptera which are pests on different parts of a plant.

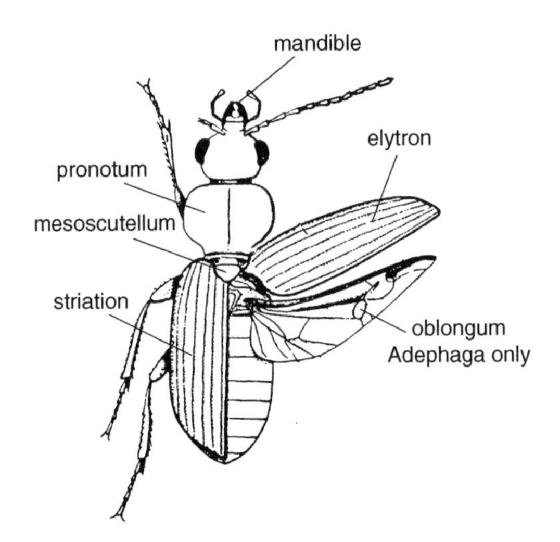

Fig. 12.2 External features of Coleoptera (dorsal view).

under the elytra, which prevents them becoming clogged with, for example, soil and is also important in the respiration of aquatic beetles (see Section 12.2.3).

In some Families, the adult beetles may stridulate by rubbing together ridges on the head and on the ventral surface of the thorax.

Legs are of the walking type, except where they are adapted for swimming in aquatic beetles. However, there is wide variation in the number of tarsal segments. This can vary from three to five, and may differ between the three pairs of legs. This leads to the concept of the ***tarsal formula*** (e.g. 3–4–5 respectively for front, middle and hind legs), frequently used to separate Families. There is also the phenomenon of the ***pseudotetramerous*** tarsus – one with five segments, but with the fourth so small that it is almost invisible. How can one tell if such tarsi are truly or only pseudo-four-segmented? I'll leave this conundrum till Section 12.3.12, when we reach the taxa where it is relevant.

The forewings are of course hardened to form the elytra. These are held out sideways in flight, which is the function of the hind wings. These may have pigmented markings, but are devoid of clearly identifiable veins.

Now to the abdomen. In using an identification key, the first requirement is to distinguish between the two Suborders, the Adephaga and Polyphaga. So the first couplet usually goes something like 'Hind coxae immovably fused to the metasternum, completely dividing the first visible abdominal sternite . . .' Not everyone may immediately assimilate what this means. What it means is that, if you carefully dissected the abdomen of a beetle in the Adephaga away from the thorax, the first abdominal segment would appear as an incomplete ring of cuticle, with a gap fitting against the hind coxae of the thorax (Fig. 12.3a). Moreover, the hind coxae are fused to the thorax. They are articulated with it and therefore moveable in the Polyphaga, and the first abdominal segment forms a complete ring (Fig. 12.3b). The easiest way to check out the segmentation on an undissected beetle is to view it from underneath. Identify the hind legs, and look between them for the first horizontal line marking the end of an abdominal segment. In the Polyphaga, this will be the hind edge of the first visible abdominal segment, but of the second in the Adephaga. So follow this line sideways to near the edge of the abdomen, and then look carefully at the chitin there forwards to where the abdomen joins the thorax. If you cross another line on the way, you have crossed the hind edge of that incomplete first abdominal segment of the Adephaga. However, even this may not always be easy to be sure of, and under 'Adephaga' (Section 12.2) I suggest another way of separating many of the members of the two Suborders.

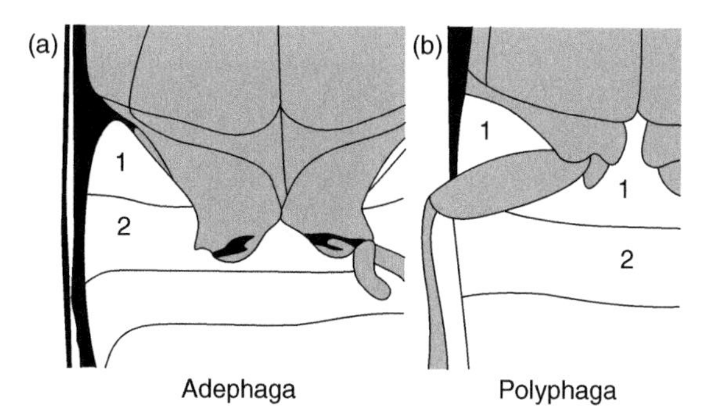

Fig. 12.3 Difference in ventral view of beginning of abdomen between (a). Adephaga and (b), Polyphaga.

12.1.2 Larvae and pupae

Beetle larvae, like the adults, are quite variable and equally highly evolved. They of course have often evolved with different selection pressures from those affecting the adults. Sometimes, therefore, the larvae are more easily identifiable than the adults and *vice versa*. It is actually quite a difficult and lengthy process to write an unambiguous morphological definition of a beetle larva, and larvae of different Families have many characters shared with some Neuroptera or some Hymenoptera.

The eggs of Coleoptera are simple, and the first instar larva breaks out of the egg shell using raised structures (***egg bursters***) situated on the head or abdomen. Larvae are classed as 'campodeiform' or 'eruciform', a division basically between carnivores and herbivores/scavengers. The cartoon (Fig. 12.4) demonstrates that the differences between the two types listed in Table 12.1 have more general application than simply to beetle larvae!

Within the eruciform category, there are four main distinct types of larvae in terms of their external appearance:

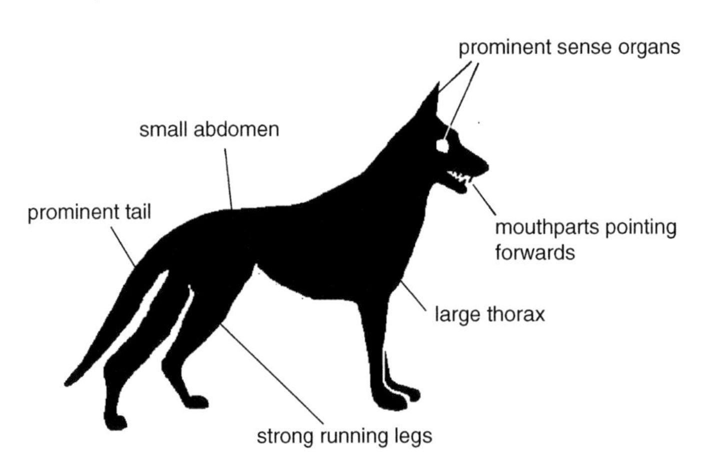

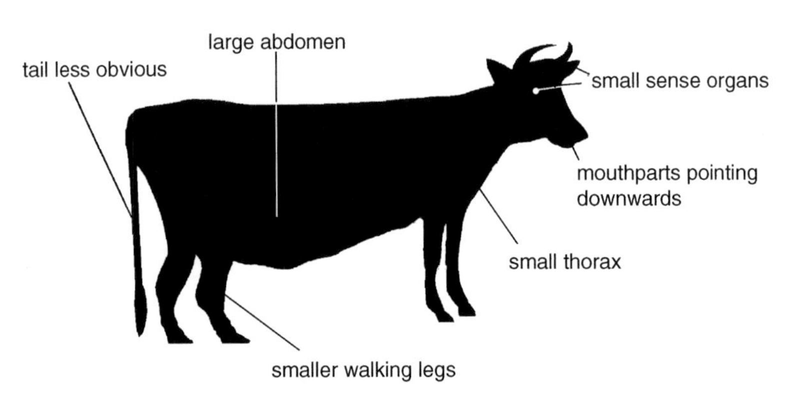

Fig. 12.4 Cartoon relating differences between campodeiform and eruciform beetle larvae to generalised differences between carnivores and herbivores.

	CAMPODEIFORM	ERUCIFORM
Feeding habit	Carnivorous	Herbivorous/scavenging, some carnivores on mainly sessile prey
Sense organs	Long filiform antennae and cerci, prominent ocelli	Short antennae often requiring magnification to see, cerci single segmented or absent, ocelli small
Mouthparts	Prognathous (i.e. pointing forwards)	Hypognathous (i.e. pointing downwards)
Legs	Long, for fast movement	Short
Thorax	Broader than abdomen	Often narrower than abdomen
Abdomen	Slender (food easily digested)	Broad

Table 12.1 Differences between campodeiform and eruciform larvae.

- straight larvae with short horny cerci (e.g. some Elateridae; Fig. 12.5b);
- straight larvae with no cerci (e.g. Coccinellidae and Chrysomelidae; Fig. 12.5c);
- highly curved larvae (scarabaeiform type) with the end of the abdomen curled round to reach close to the head; no cerci (e.g. stag beetle and cockchafer larvae; Fig. 12.5d);
- legless (apodous) with no cerci (e.g. Curculionidae; Fig. 12.5e).

Campodeiform larvae (Fig. 12.5a) are characteristic of the Suborder Adephaga and eruciform larvae of the Suborder Polyphaga. However, the key criterion for separating larvae of the two Suborders is that those of the Adephaga have normal insect legs finishing with a tarsus at the end of the tibia, whereas those of the Polyphaga (which also includes all apodous larvae) have only one segment (the ***tibiotarsus***) distal of the femur (Fig. 12.6). Pupae of beetles are exarate and often formed in hollow cells in the soil. Beetle larvae do not produce silk, and so the kind of cocoons found in many Lepidoptera and Hymenoptera are absent.

The enormous diversity of beetles makes it necessary perhaps to be even more selective than for the other Orders about which Superfamilies and Families to include in this book.

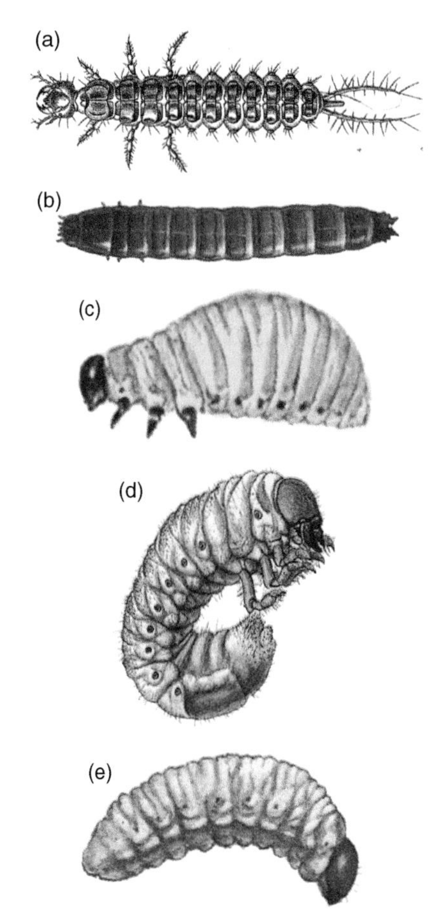

Fig. 12.5 Types of beetle larvae (see text): (a), Carabidae; (b), wood-boring Elateridae; (c), Chrysomelidae; (d), Scarabaeidae; (e), Curculionidae. (a, b, e) from Mandahl-Barth 1974, with permission; (c, d) from Lyneborg 1968, with permission).

12.2 Suborder Adephaga

These are mainly some water beetles (though there are also water beetles in the other Suborder) and the ground beetles.

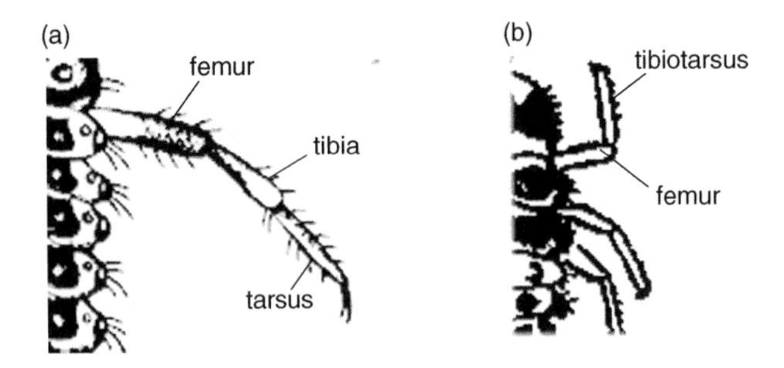

Fig. 12.6 Distinguishing larvae of (a), Adephaga from (b), Polyphaga on their leg structure.

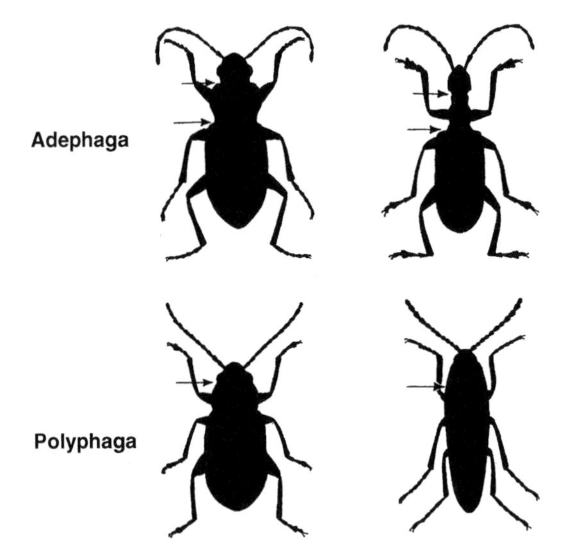

Fig. 12.7 Silhouettes of some terrestrial Adephaga and Polyphaga (from Lewis and Taylor 1967, with permission).

They are characterised by that incomplete first abdominal segment mentioned above. However, if we exclude the water beetles, then the terrestrial Adephaga have some other ways of distinguishing them from the Polyphaga. Firstly, the majority have a characteristic silhouette when viewed from above (Fig. 12.7). The silhouette narrows behind the head and behind the pronotum, so that 'notches' in the silhouette identify the divisions between head, pronotum and elytra. Also, the terrestrial Adephaga are the only beetles to have both antennae and tarsi simply filiform, AND the antenna is composed of 11 segments AND the tarsi of five segments on all three pairs of legs. However, the antennal and tarsal segments are not always that easy to count. So these less clear-cut differences with the Polyphaga can be used as follows.

If at least one of the 'notches' in the smooth outline of the silhouette is lost AND the antennae are not a simple 'string' (e.g. perhaps clubbed) **or** the mid and hind tarsi have a lobed segment (Fig. 12.56) near the tip (the expanded tarsal segments on the forelegs of some Carabidae could be mistaken for 'lobed'), then the specimen must be Polyphaga. This should leave rather few specimens for which it is then necessary to check the abdominal segment character.

The Orders comprising the aquatic Adephaga are often referred to under the blanket term *Hydradephaga*.

12.2.1 Family Gyrinidae (whirligig beetles)

Their common name derives from the rapid rushing about in dizzy circles that these small, shiny, black beetles (Fig. 12.8) do at the water surface of ponds and slow-moving streams. They normally occur in a group of many individuals. Each compound eye is divided by a cuticular bar into a lower area that sees below, and an upper area that sees above, the water surface. The meso- and metathoracic legs are flattened for swimming and covered in hairs, which trap air as does the underside of the abdomen; thus wetting is prevented and the beetles do not sink. The adults feed on small insects which have fallen on the water, while the larvae have abdominal gills and are fully aquatic. They are also carnivorous.

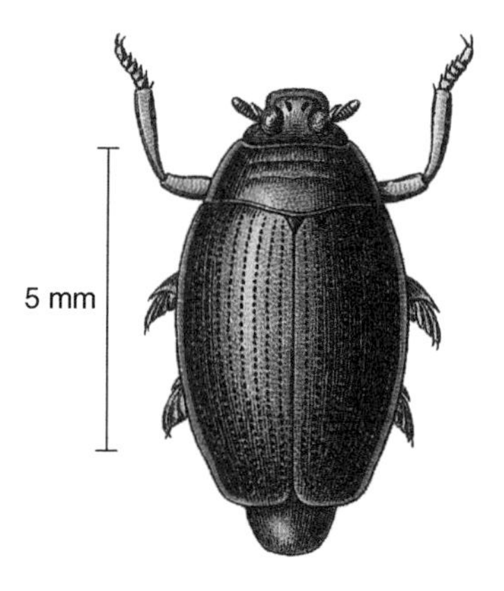

Fig. 12.8 Whirligig *Gyrinus natator* (from Reitter 1908–1916).

12.2.2 Families Haliplidae and Hygrobiidae

The Haliplidae are small beetles recognisable by the elongated punctures on the elytra. The adults are more often found walking on the algae on which they feed than actively swimming. The Hygrobiidae are larger (10–12 mm long) and *Hygrobia hermanni*, the only UK species, is known as the 'screech beetle' because of the sound it can emit by rubbing the dorsal side of the tip of the abdomen against the underside of the elytra. Larvae of the Hygrobiidae can be identified by their three long tails at the tip of the abdomen.

12.2.3 Family Dytiscidae

These very smooth ovoid (even the head and eyes hardly break the oval outline) water beetles have long hairy hind legs for swimming and can vary in size from a few millimetres to the 4 cm or so of *Dytiscus marginalis* (Fig. 12.9a) with its orange rim to

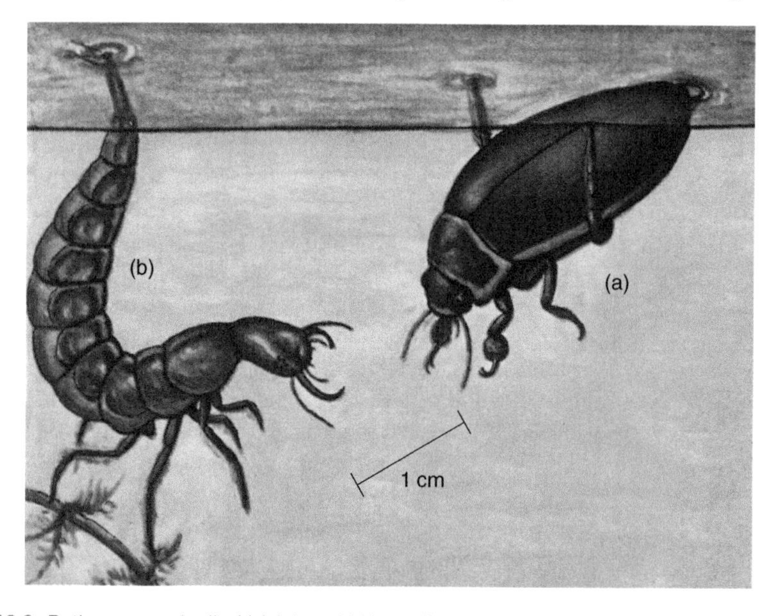

Fig. 12.9 *Dytiscus marginalis*: (a) Adult and (b) larva (from Mandahl-Barth 1973, with permission).

the elytra. Both adults and larvae of the Dytiscidae are fierce predators, and ones as large as *D. marginalis* will even tackle small frogs and sticklebacks.

Both adults and larvae breathe air. The larva (Fig. 12.9b) comes to the surface and inhales air with its terminal siphon (a bit akin to a snorkel tube, but at the rear end), while the adults expose the tip of their abdomen and replenish the air bubble under the elytra from which the dorsally pointing abdominal spiracles obtain their air. The meniscus of the air bubble is held by hairs, so that oxygen diffuses from the water into the air bubble as the hairs resist any shrinking of the bubble (plaston respiration, see Section 7.5.1.1, water boatmen). This enables the beetles to remain submerged for a long time.

The larvae have long sickle-shaped mandibles, with which they suck out the juices of their prey.

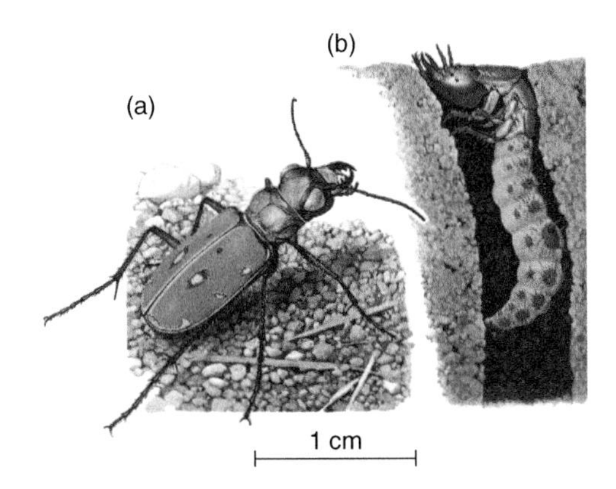

Fig. 12.10 Tiger beetle *Cicindela campestris*: (a) Adult and (b) larva (from Zanetti 1977).

12.2.4 Family Cicindelidae (tiger beetles)

This is the first Family of terrestrial Adephaga to receive mention. They are actively running beetles with smooth (i.e. non-striated elytra – contrast the Carabidae, see Section 12.2.5) coloured elytra with small circles of a contrasting colour. Thus the common *Cicindela campestris* (Fig. 12.10) is green with yellow circles. Tiger beetles tend to occur in light sandy soil with sparse vegetation. There the larva digs a burrow in which it sits with its armoured pronotum making a lid to the tunnel. When this lid is touched by passing prey, the larva shoots up and ambushes it.

12.2.5 Family Carabidae (ground beetles)

This Family demonstrates the facies of the terrestrial Adephaga of having a division in silhouette between both head and thorax and thorax and abdomen, as well as having simple filiform antennae and tarsi. The elytra are strongly striated, in contrast to the Cicindelidae (see Section 12.2.4). The trochanter is 'bean-shaped' and lies out of line with the adjacent segments of the leg (Fig. 12.11), though this character is also found in a few other Families. Quite a number of species are flightless, with the elytra fused together at the midline. The common name 'ground beetle' applies to the familiar, often black or brown beetles of 1–2 cm in length (Fig. 12.13), which hunt over the soil surface, usually at night, and hide under stones and other shelter by day. The larvae are campodeiform

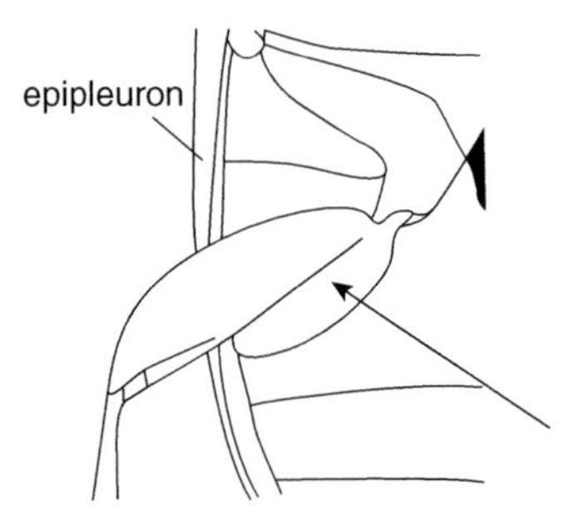

Fig. 12.11 Trochanter of Carabidae.

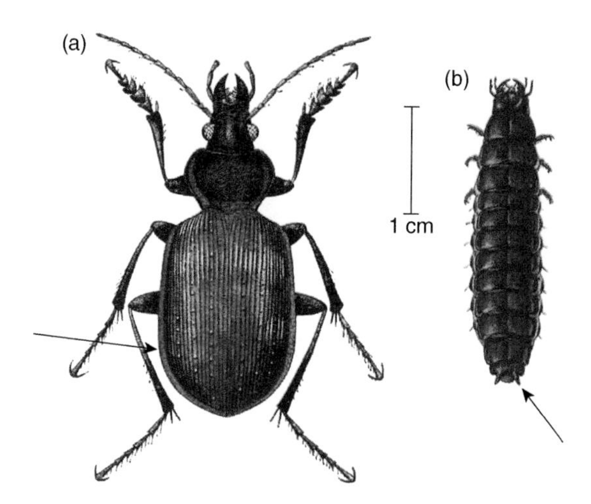

Fig. 12.12 (a) Adult *Calosoma sycophanta* (from Becheyne 1956); (b) larva of the related *Carabus nemoralis* (from Lyneborg 1968, with permission).

(see Section 12.1.2); apart from the short cerci of the larger genera like *Carabus* (Fig. 12.12b) and *Calosoma*, the cerci are relatively long (Fig. 12.5a). The larvae of most genera are carnivorous and hunt in the soil or on the surface, mainly at night.

Both adults and larvae of Carabidae are regarded as important generalist predators of many crop pests. Thus they consume the eggs of flies such as wheat bulb fly and cabbage root fly, which lay their eggs on the soil. Often their activity renders further control of such pests unnecessary. The guts of these predators have also been found to contain the remains of many aphids, and the adults may climb plants at night to feed on the aphids there. There is some controversy about how far their predation of aphids relies on such climbing and how far it reflects consumption of the many aphids that fall off plants onto the ground.

Carabidae feature greatly in the literature on selectivity of insecticides and the effect of cultural systems and floral diversity on the impact of biological control of crop pests. This overemphasises the importance of carabids as biological control agents, and arises from the ease with which their populations can be assessed by pitfall trapping; many other predators live on plants and are much harder to sample and count. So, in applied entomological research, carabids are often the 'representative' predators in experiments, especially as they can be excluded from areas of the field by barriers so that pest populations in areas 'with and without' these predators can be compared.

The controversy about how frequently carabids climb plants has been mentioned. However, there is no doubt that adult carabids can and do climb. One evening after dark, I was waiting for transport on the University of Oklahoma campus next to a street lamp, the metal post of which must have been 5 m high. I became aware of the continual procession of beetles climbing up the post to catch insects coming to the light high above; the majority of these beetles were carabids. Also, some tropical carabids are arboreal as their normal habit. They are also very different in appearance from typical 'ground beetles' in that they may be perhaps 6 cm long, their heads may be greatly prolonged and the elytra may be flattened at the sides to expand greatly sideways, giving the beetle a large almost circular surface area, as well as being brightly coloured.

The first rudiments of such an expansion of the sides of the elytra (Fig. 12.12a) are seen in the largest (about 25 mm long) European carabids, in the genera *Carabus* and

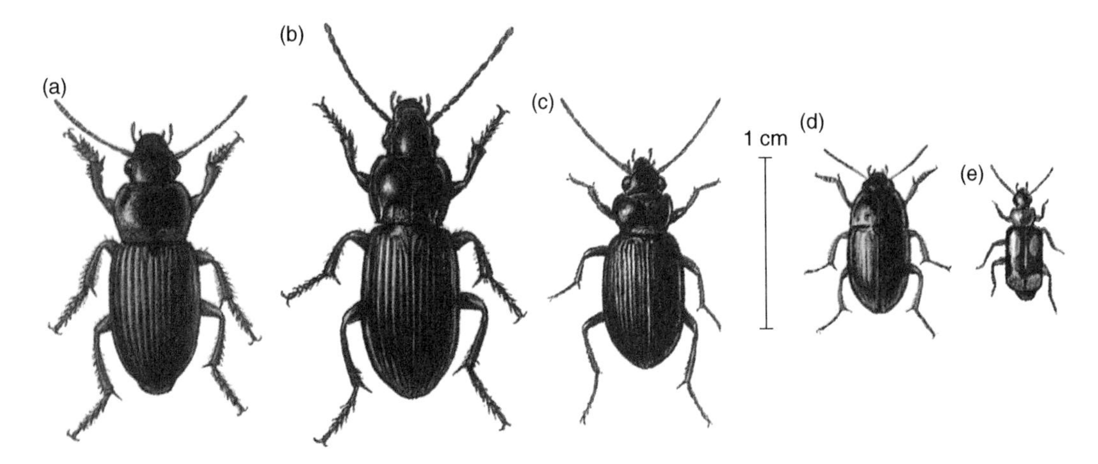

Fig. 12.13 Representatives of some common ground beetle genera: (a) *Pseudoophonus rufipes* and (b) *Pterostichus melanarius*; (c) *Nebria brevicollis*; (d) *Amara aenea*; (e) *Dromius quadrimaculatus* (a, b, d) from Lyneborg 1968, with permission; (c, e) from Mandahl-Barth 1974, with permission).

Calosoma. *Carabus violaceus* is a common species with the rim of the elytra a vivid violet, and other species also have colouration in the shiny elytra. Those of *Carabus nemoralis* are coppery, and those of *Calosoma sycophanta* (Fig. 12.12a) are iridescent with most of the colours of the rainbow!

Many of the Carabidae in the next size range (around 20 mm long) are considered the most important predators in the Family that are found in arable crops. They are mainly in three genera, *Pterostichus, Harpalus* and *Pseudoophonus* (previously included under *Harpalus*), though at least larvae of the latter also take plant food. Important predators that are smaller still are in the genera *Nebria* (an important contributor as both adults and larvae to pitfall trap catches) and *Leistus*. Three easily recognised genera of small carabids are *Amara* (Fig. 12.13d) with a much more ovoid smooth silhouette than other genera, *Elaphrus* with protruding eyes and *Dromius* (Fig. 12.13e) with a red pronotum and lightish-brown elytra with a large black patch.

Several of the common generalist predators, especially *Pseudoophonus rufipes* (Fig. 12.13a), *Pterostichus madidus*, *P. melanarius* (Fig. 12.13b) and *Nebria brevicollis* (Fig. 12.13c), can become a pest problem in strawberry fields. Adults visit the crop during the time the strawberries are ripening and bite chunks out of the fruit. An often used common name for *Ps. rufipes* is indeed 'strawberry seed beetle'.

Finally, of entomological if not of applied interest, is the genus *Brachinus*, the bombardier beetles (Fig. 12.14). These have a red pronotum and metallic bluish-green elytra and defend them-

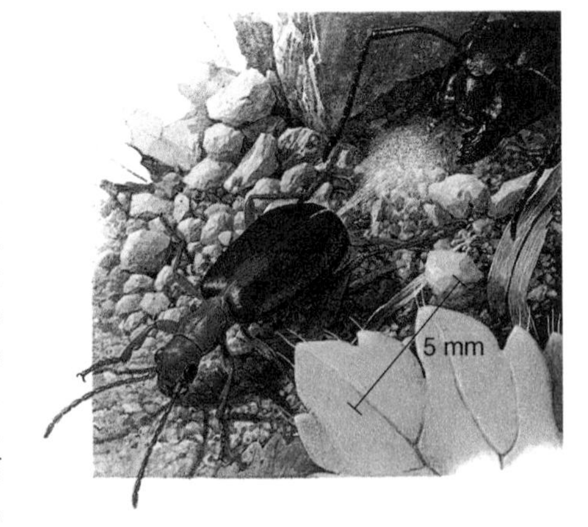

Fig. 12.14 Bombardier beetle (*Brachinus* sp.) 'firing' at an approaching predator (from Zanetti 1977).

selves against attack by producing a puff of smoke (accompanied by an explosive 'pop') from their rear. The 'smoke' is actually fine droplets of an irritant volatile produced by the mixing of two chemicals from special glands at the rear of the abdomen.

12.3 Suborder Polyphaga

These, as explained earlier, are beetles with the first abdominal segment complete ventrally. Antennae are very variable and are often not filiform as in the terrestrial Adephaga, and either the division between head and pronotum, or between pronotum and elytra, or both, are often not distinguishable in silhouette. There are many Superfamilies, which between them contain very many Families, and only those of economic importance, ease of recognition or of special entomological interest can be included in this account. Some Families with similar characters but not necessarily phylogenetically related, are grouped into 'Series' rather than Superfamilies.

12.3.1 Superfamily Staphylinoidea

The character common to members of this Superfamily is that the abdomen is at least partially longer than the elytra, so that the terminal segments are exposed in dorsal view.

12.3.1.1 Family Staphylinidae (rove beetles)

Here the elytra are square and very short, so that most of the abdomen is exposed (Figs 12.15 and 12.16). Indeed, the beetles can look very much like small earwigs, but with no terminal forceps. Like earwigs, the wings folded under the short elytra are quite large when unfolded, and the beetles can fly well. The antennae are simple and filiform. One large black staphylinid, the devil's coach horse (*Ocypus olens*) is not far off earwig size (Fig. 12.15). Another easily recognised genus is *Stenus*, the pronotum and elytra of which are strongly punctured and the eyes are large and protrude noticeably from the outline of the head (Fig. 12.16a).

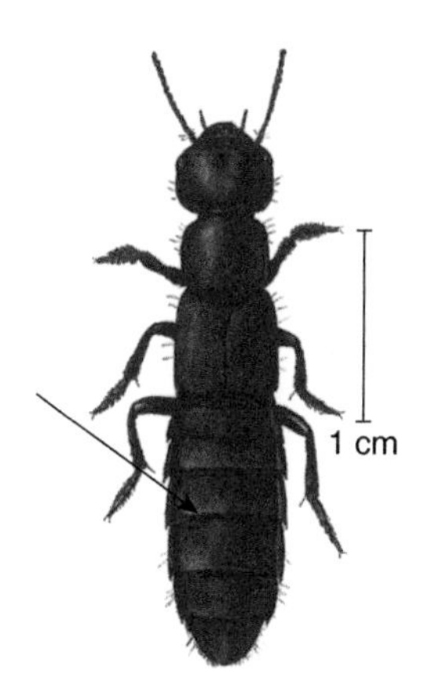

Fig. 12.15 Devil's coach horse (*Ocypus olens*) (from Mandahl-Barth 1974, with permission).

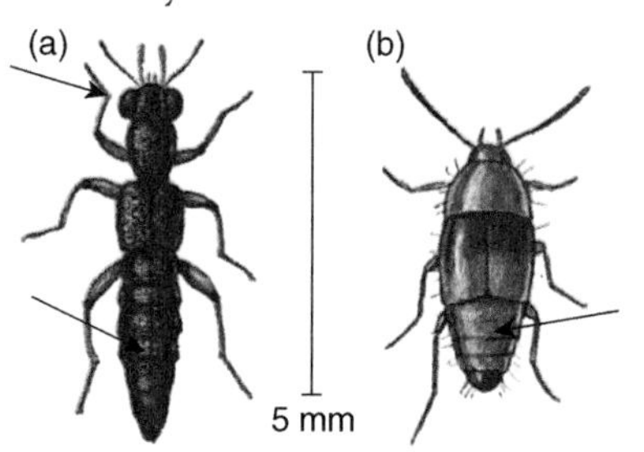

Fig. 12.16 Rove beetles (Staphylinidae); (a) *Stenus* sp. (from Lyneborg 1968, with permission); (b) *Tachyporus* sp. (from Mandahl-Barth 1974, with permission).

The larvae (Fig. 12.17) are prognathous, and look rather like carabid larvae. However, being Polyphaga, staphylinid larvae have a single tibiotarsus rather than the separate tibia and tarsus of Adephaga larvae. Feeding habits are variable, and individuals may take a mixture of animal and plant food. Many are scavengers or feed on fungal spores. Thus they are common contributors to the breakdown of cattle dung.

Members of the genus *Tachyporus* (Fig. 12.16b) are recognisable by their pronotum and elytra being wider than the tapering abdomen and the yellow colour of the elytra compared with the black of many of the other small rove beetles. They are known to feed on aphids as well as fungal spores, and are soil-surface dwellers that can be trapped in pitfall traps. Thus they tend to be considered along with carabids as 'representative generalist predators' in arable crops (see Section 12.2.5).

Fig. 12.17 Staphylinid larva (from Zanetti 1977).

12.3.1.2 Family Silphidae (burying, sexton or carrion beetles)

The antennae are usually clubbed, and only the last few segments of the abdomen protrude past the end of the elytra in dorsal view. Burying beetles (Fig. 12.18) include the large and totally black *Nicrophorus humator* and *N. vespilloides* with its bright red elytra with black bands. The adults are capable of burying the corpses of animals as large as mice by using their mouthparts and strong tibial spines to excavate the earth away from under the corpse so that it slowly sinks. Adults are attracted from a long distance by the odour of a corpse and beetles work alone or in pairs, driving off any other beetles that arrive subsequently. The female then lays eggs in a chamber excavated adjacent to the corpse and will stay there and feed the hatched larvae with food she regurgitates until they are strong enough to fend for themselves. Some species are more carnivorous than scavenging and feed on the other insects, mainly Diptera larvae, which soon colonise a new corpse.

Fig. 12.18 Burying beetle (*Nicrophorus* sp.) (from Zanetti 1977).

Burying beetles have a strange life history phenomenon known as 'hypermetamorphosis', which takes the form of larvae undergoing up to three distinct morphological stages. The first stage is a campodeiform larva, the second eruciform and a third stage is an apodous grub. Each larva finally pupates in its own cell hollowed out in the soil near the burial site.

The beet carrion beetle, *Aclypea opaca* (Fig. 12.19), is unusual in being a herbivore, the adults and larvae feeding on leaves of Chenopodiaceae, but also on potato and cereals. Especially on the continent of Europe, they can be a pest problem on sugar beet when the overwintering adults move from the weed hosts they initially feed on in the spring until the leaf tissues become mature and tough. The one generation per year begins with eggs laid in the soil in early summer. Larvae then feed on the foliage of their host plants, but may also attack the roots. Pupation is in the soil, and the adults emerge after about 2 weeks to overwinter in shelter like leaf litter outside the crop.

12.3.1.3 Family Pselaphidae

These are interesting tiny beetles with a narrow pronotum and head, the latter bearing long clubbed antennae and commonly also an unusually long clubbed palp. Many live in damp vegetable refuse, but others (e.g. the genus *Claviger*) inhabit ants' nests and

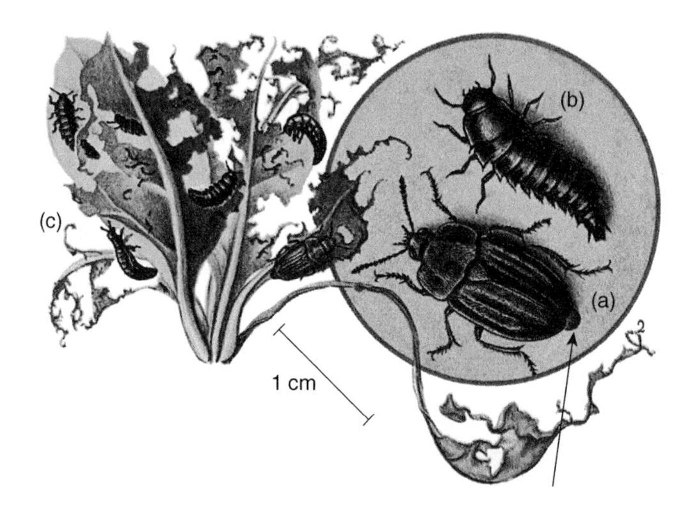

Fig. 12.19 Beet carrion beetle, *Aclypea opaca*: (a) adult; (b) larva; (c) damage to beet (all from Bayer 1968, with permission).

the adults move around unhindered, but whether the larvae are also such inquilines in ant nests is not known. As yet, no larvae of a *Claviger* species have been found, and it is possible that they feed in a totally different environment.

12.3.1.4 Family Ptiliidae

These are included here as being among the smallest beetles, easily recognised by the strap-like and hair-fringed wings typical of tiny insects (see Section 2.6.4) and often protruding from under the elytra (Fig. 12.20). The Family is found world-wide, and both adults and larvae live in all kinds of decaying plant material, in fungi and under bark.

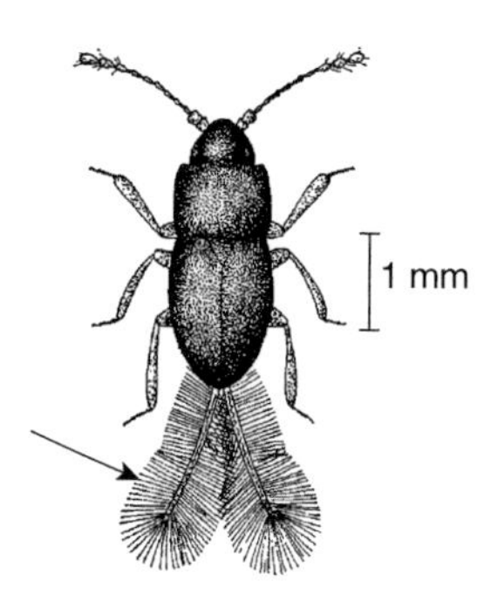

Fig. 12.20 Ptiliidae (from Lewis and Taylor 1967, with permission).

12.3.2 Superfamily Hydrophiloidea – Family Hydrophilidae

These interesting beetles do not have legs adapted for swimming, yet many are aquatic. However, they tend to remain on water plants in stagnant water. In contrast to the Hydradephaga (see Section 12.2.3) the plastron respiration system involves hairs in a groove on the antennae leading to a continuation groove on the thorax and eventually to spiracles. It is through this channel that the air bubble under the elytra is replenished with oxygen, but it does mean the beetles come up head first, rather than tail first, to take in air. The whole body is clothed in water-repellent hairs, so that submerged beetles look silver rather than their natural black or brown. With the antennae dedicated to respiration, the prominent palps by which the Family is recognised assume the role of antennae. The adults are herbivorous, but many species have carnivorous larvae.

The Family contains by far the largest European water beetle, the 5 cm long *Hydrophilus piceus* (great silver water beetle – a black beetle, the 'silver' of course referring to its air-clad appearance under water – see previous paragraph). Though a herbivore, it can give you a nasty nip, but the weapon turns out to be a ventral spine on the underside of the body, not the mandibles.

12.3.3 Superfamily Histeroidea – Family Histeridae

These are shiny black beetles with two abdominal segments exposed and short elbowed and clubbed antennae. They have very small heads and a curved and globular shape, and so have a strong similarity in appearance to scarab dung beetles. They are also found in dung, and both adults and larvae are probably mainly carnivorous on other insects there.

12.3.4 Superfamily Scarabaeoidea

These beetles are all of substantial size and are characterised by their elbowed and lamellate antennae (Figs 2.7 and 12.24a), the tip of which may look like a club when folded, but actually consists of several 'sheets' which can be opened out rather like a lady's fan. The larvae are also distinctive. They are thick bodied and curled into an almost complete circle (Fig. 12.5d). As the end of the abdomen is close to the mouth, the anus is not open, and excreta are contained at the end of the abdomen inside the cuticle which is shed like a tied-off rubbish sack at each moult. The larvae have legs, but remain on their side in cells usually excavated in the soil adjacent to their food, and finally pupate in these cells.

12.3.4.1 Family Lucanidae (stag beetles)

The largest beetle in Europe is the stag beetle *Lucanus cervus* (Fig. 12.21), with the large antler-like mandibles of the male. These are used for fighting other males, particularly over a female. Though looking fierce, the male has little leverage on these jaws, and they are quite harmless in comparison with the small jaws of the female. These can close with force and inflict a painful bite. The larvae feed, usually underground, on dead wood and roots, particularly of oak. They are thus a 'pest' of oak fence posts, and when these were abundant in towns and cities and sunk into the ground without being embedded in concrete, stag beetles were a common sight in late summer. Males would often be seen flying slowly and rather vertically with a deep whirring sound. Lately, oak fence posts have been replaced by other timbers or post and strand fencing, and stag beetles have become so rare that they are now a protected species. A male *L. cervus* may look impressive to those in temperate countries, but tropical stag beetles can be up to 12 cm

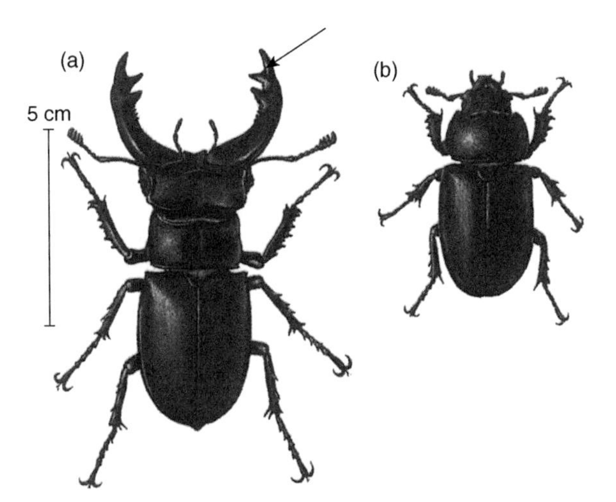

Fig. 12.21 Stag beetle, *Lucanus cervus*: (a) male; (b), female (from Mandahl-Barth 1974, with permission).

long; some genera (e.g. *Prosopocoilus* sp.) have dramatic mandibles almost as long as the entire body!

The lesser stag beetles (*Dorcus parallelopipedus* and *Sinodendron cylindricum*) have no antler-like mandibles and both sexes look rather like small female *L. cervus*. They also feed in rotten wood, but more in fallen logs than underground.

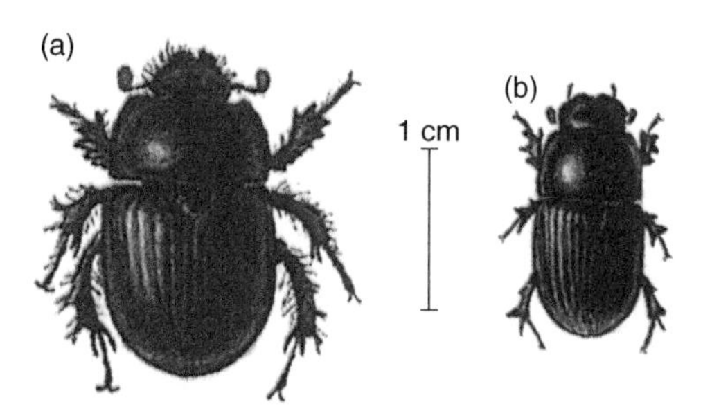

Fig. 12.22 Dung beetles: (a) *Geotrupes stercorarius* (Geotrupidae) (from Mandahl-Barth 1974, with permission); (b) *Aphodius fossor* (Scarabaeidae) (from Lyneborg 1968, with permission).

12.3.4.2 Family Geotrupidae (dung beetles)

These beetles feed on the dung of cattle and other mammals, and are important beneficial insects in accelerating the disappearance of such deposits. As well as eating the dung themselves, the adults remove and bury dung on which they lay their eggs and the larvae develop. A common genus is *Geotrupes* (Fig. 12.22a).

When settlers brought European livestock to Australia, the local dung beetles (which had evolved with marsupials) were unable to deal with cattle dung. The amount of grazing land covered by such dung became a serious problem, especially as irritant bush flies (*Musca vetustissima*) bred in huge numbers in the dung, leading to those corks on strings around the rim of bush hats which today identify Australians in cartoons. The problem was dealt with by importing European (both geotrupid and scarabaeid; see Section 12.3.4.3) dung beetles.

12.3.4.3 Family Scarabaeidae

This large family includes another group of dung beetles, many of which feed on the dung as adults and larvae without burying it. The genus *Aphodius* (Fig. 12.22b) is particularly widespread (there are many species). The dung rollers (*Scarabaeus* spp.) are well known for their Herculean efforts in scraping together a large ball of dung, akin to making a giant snowball, and then rolling it with their back legs (Fig. 12.23) for some distance to a place suitable for burying it.

Fig. 12.23 The sacred scarab, *Scarabaeus sacer*, with dung ball (from Zanetti 1977).

Many other scarabaeids are in the group known as the 'chafers'. They

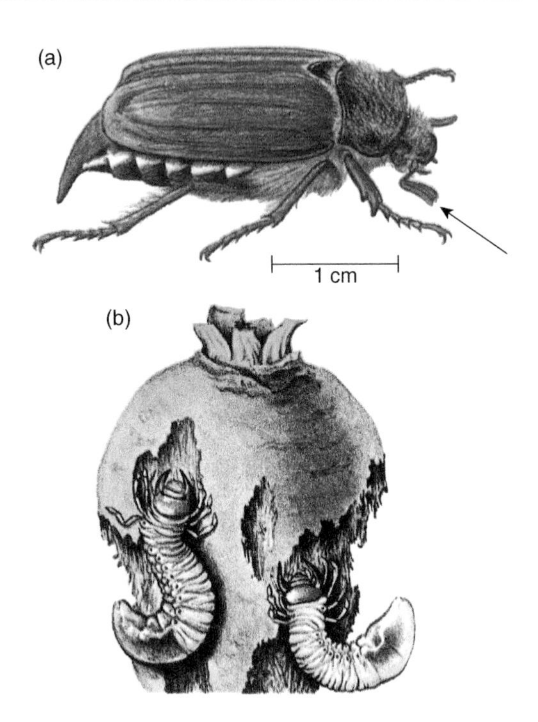

Fig. 12.24 Cockchafer, *Melolontha melolontha*: (a) adult (from Lyneborg 1968, with permission); (b) larvae attacking root of beet (from Bayer 1968, with permission).

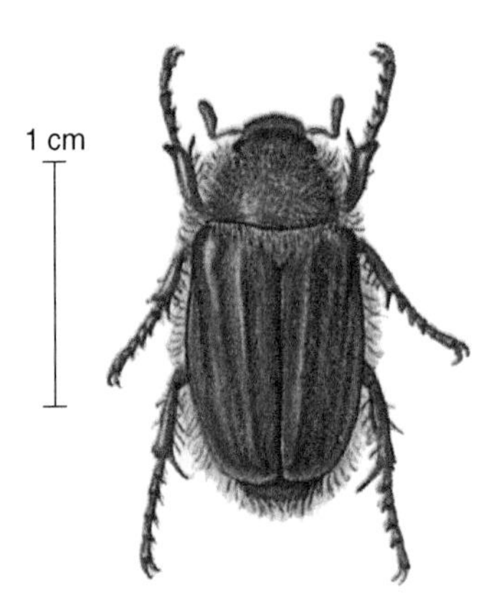

Fig. 12.25 Junebug, *Amphimallon solstitiale* (from Lyneborg 1968, with permission).

have very obvious elbowed lamellate antennae, particularly the males. Familiar in Europe as adults are the cockchafer *Melolontha melolontha* (Fig. 12.24) – also known as the Maybug – and the Junebug (*Amphimallon solstitiale*; Fig. 12.25). The former is quite large at more than 3 cm long, and both fly to light in early summer and frighten people when they land in their hair. However, adult chafers feed on flowers and leaves and are harmless in contrast to the larvae, which in many species are serious pests, living for 1–4 years in the soil curled up in earthen cells (where they also pupate) and feeding on plant roots. The larvae of chafers are known as 'white grubs' and are pests world-wide. Thus the larvae of both *M. melolontha* (which may feed for up to 3 years) and *A. solstitiale* (which has one generation a year) can become pests of grassland if populations build up to large numbers, and can especially become pests of cereals or other crops planted into recently ploughed pasture, given the much smaller amount of root there then is per area of land. Chafer larvae may stridulate by rubbing their legs together but, since they are almost immobile and isolated in their cells, one has to ask 'Why?'.

Several white grub species attack sugar cane in different parts of the world. Important in East Africa is the sugar cane white grub *Cochliotis melolonthoides*. Feeding by the larvae, of which populations can reach 50,000 per hectare, causes sugar cane leaves

Fig. 12.26 Rose chafer, *Cetonia aurata* (from Zanetti 1977).

to yellow and wilt; sometimes the roots are so reduced by the feeding that the cane topples over. The life cycle takes about 1 year. White grubs of the genus *Schizonycha* also attack sugar cane in East Africa, Sudan and Egypt, but also other cereals such as sorghum and maize as well as dicotyledonous crops such as sunflower and groundnut. The life cycle takes about 7 months.

A not uncommon sight feeding at flowers is the emerald green rose chafer (*Cetonia aurata*; Fig. 12.26).The larvae feed in rotting wood and do not present a pest problem. Related are several genera of other emerald green or brown beetles (including *Anomala* and *Popillia*), whose larvae feed on the roots of many crops and ornamentals while the adults damage both foliage and flowers. *Popillia japonica* (Fig. 12.27) is the Japanese beetle of the Far East, though it has been introduced to North America, and other similar species of *Popillia* occur in China, and some also in India and Africa. They can be serious pests on the foliage and flowers of many cultivated plants while the larvae are also potentially serious pasture pests. Thus different chafer species occur in most parts of the world but, taken together, the biggest impact of chafers is probably as root grazers in grassland. Other related and brightly coloured but much smaller chafers, only about 1 cm long, are *Hoplia* spp. and the garden chafer *Phyllopertha horticola*. These also feed on grass roots, and are pests of valuable short turf such as bowling greens and golf tees. Here small circular brown patches of grass appear defining the cell beneath in which the larva lives. There is one generation a year.

Fig. 12.27 Japanese beetle, *Popillia japonica* (courtesy of Krista Hamilton).

The rhinoceros beetles (*Oryctes* spp.) are largish shiny black beetles, 3–5 cm long and with the male having an obvious dorsal horn arising from the frontal region of the head (Fig. 12.28); this horn is present but not so immediately obvious in the female. In Europe, the genus is represented by *Oryctes nasicornis* (Fig. 12.28), with the large larvae found feeding in rotting wood. However, elsewhere rhinoceros beetles can be serious pests. It is not the larvae that do the damage, but the adults feeding on

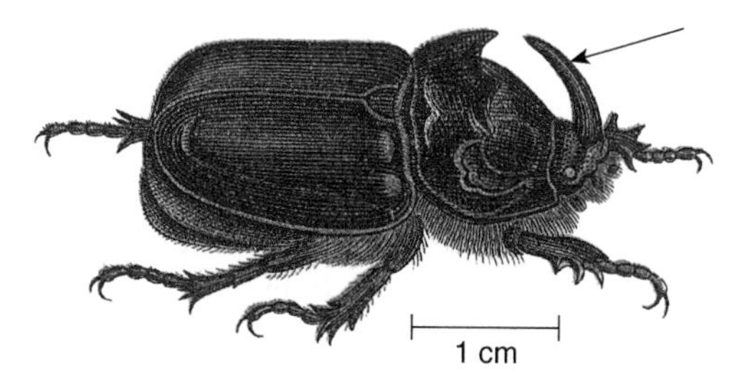

Fig. 12.28 Male rhinoceros beetle, *Oryctes nasicornis* (from Reitter 1908–1916).

the growing points of the plant. *Oryctes boas* occurs in most of central and southern Africa, with coconut palm as the principal host and oil and date palm as other hosts. *Oryctes monoceros* has a more restricted distribution in tropical Africa; it also has coconut palm as its main host, but also feeds on oil and date palm. *Oryctes rhinoceros* is, by contrast, the Asiatic rhinoceros beetle. It occurs in Southeast Asia from India and Pakistan to Indonesia and Taiwan. The life history and damage symptoms of these three species are essentially similar. Eggs are laid in rotting plant material, especially dead palm trunks, though *O. rhinoceros* also oviposits in other decaying vegetable matter in compost heaps and rubbish dumps. The larvae feed and pupate here, though in *O. monoceros* the immature stages may also be found in the soil. Adults are active fliers at night. The damage they do to the growing point of the palm results in the emergence of fronds with all or some V-shaped sections of the leaflets missing. If the growing point itself is destroyed, the palm dies. Generations take 3–5 months, and there are one or two generations a year.

The black maize beetle (*Heteronychus* spp.) is related to the rhinoceros beetles, and is a black oval beetle 15–20 mm long. Eggs are laid in the soil at the base of the host plant, and the larvae eat the roots of maize, wheat, other cereals, sugar cane, tobacco and some vegetables. However, this damage is of minor importance unless numbers are very large, when young sugar cane and maize may be killed. By contrast, one adult beetle may destroy several seedlings by chewing through the stems just below soil level. The pest is found in many parts of Africa; there is just one generation a year.

12.3.5 Superfamily Buprestoidea – Family Buprestidae

This Superfamily contains just this one Family of shiny insects with metallic colours (Fig. 12.29). The eyes are very prominent on the head, which is otherwise largely sunk into the thorax, and the antennae are serrate. The larvae of most species are borers in healthy timber, and so these insects are pests in forestry, particularly in the tropics. Timber is low in nutrients and larvae of all timber borers can take a long time to develop to pupation; 30 years has been suggested in connection with a buprestid. *Agrilus* spp. (citrus jewel beetles) are green or bronze beetles, which are important pests of citrus in the Far East and pests of ornamental trees, pear and soft fruit canes in the USA. One species (*A. acutus*) is a stem borer of Jute in India, while *A. rubicola* bores currant branches in the USA. Eggs are laid in cracks in the bark, and the larvae make long galleries in the bark as they feed. Unlike most other boring beetle larvae (such as the *Cerambycidae*, see Section 12.3.12.1), the frass is not expelled but is packed tightly in the gallery behind the feeding larva as it progresses. The larva pupates at the end of the gallery just under the bark surface and the emerging adults cut a hole to escape. Depending on climate, a generation may take 1 or 2 years.

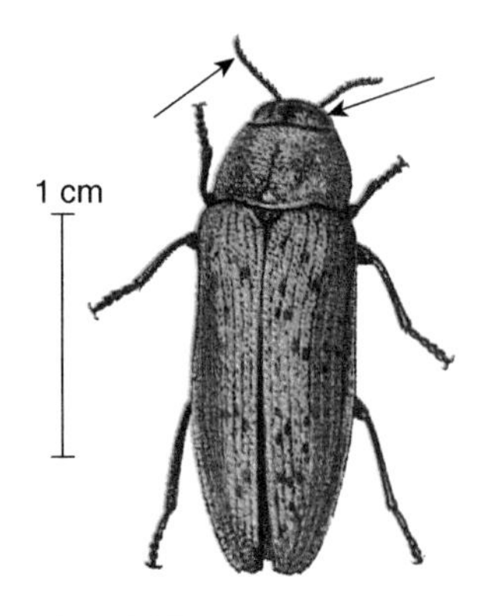

Fig. 12.29 Buprestidae (from Zanetti 1977).

12.3.6 Superfamily Elateroidea – Family Elateridae

The most important Family in this Superfamily is the Elateridae (click beetles). The common name comes from the loud click the adults make when they jump in the air from an upside-down position to right themselves. The trick is based on a mobile articulation between the meso- and metathorax, and a ventral process on the former. This process forms a spring when it catches (through the beetle arching its back) on the edges of a cavity on the metathorax (Fig. 12.30).

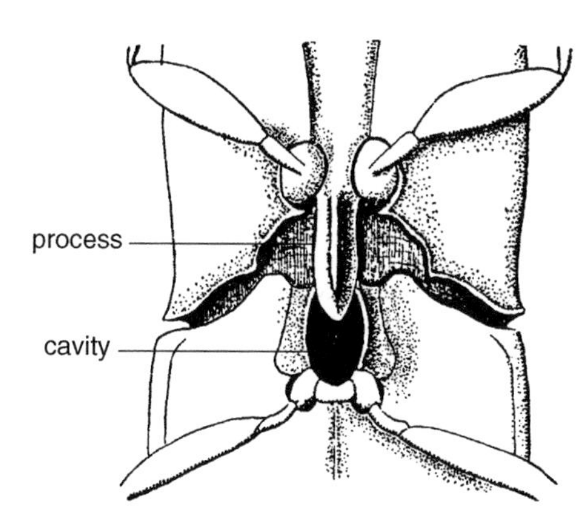

Fig. 12.30 'Click' mechanism of Elateridae, showing the process on the prosternum and the cavity in the mesosternum on the edges of which the process catches (from Richards and Davies 1977).

The adults (Fig. 12.31a) are rather spindle-shaped with a totally smooth silhouette since the head is sunk into the thorax. The 'click' mechanism is visible dorsally at the hind edge of the pronotum, which has little backwardly pointing projections in the centre and at the edges giving the pronotum the shape of an inverted shield. These projections form the 'hinge' which enables the beetle to arch its back. The antennae are serrate, and the beetles are usually a dark brown though a few are more brightly coloured, and then usually red.

The larvae are heavily chitinised and thus a yellowish-brown. The abdomen often ends in two stiff curved horns forming a U-shape between them, but the abdomen of one genus (*Agriotes* spp.) ends in a rounded cone with a dark spot (***lateral pit***) on each side (Fig. 12.31b). These larvae are known as 'wireworms'. Wireworms live in the soil for 3–4 years, grazing on the roots of a wide range of plants, including grasses. Populations rarely build up to show symptoms in grassland, but when old grassland is ploughed and reseeded to cereals, the wireworms' food supply is dramatically reduced and the many larvae have to 'home in' on the few roots of the seedlings that become available (as with chafers, Section 12.3.4.3). Thus wireworms were perhaps the most serious UK agricultural pest in World War II, when large areas of previously permanent grassland were ploughed up to increase home cereal production. The main species are *Agriotes lineatus, A. obscurus* and *A. sputator*. Another crop that can suffer badly from wireworm is potato, where the larvae tunnel into the tubers. Pupation in the soil lasts 3–4 weeks. The first keys to separate the larvae of these three *Agriotes* species were based on the shape of the mandibles. Much later it was realised that soil quickly wears down the mandibles, and that all the keys were doing was identifying how long it had been since the larva last moulted, regardless of species!

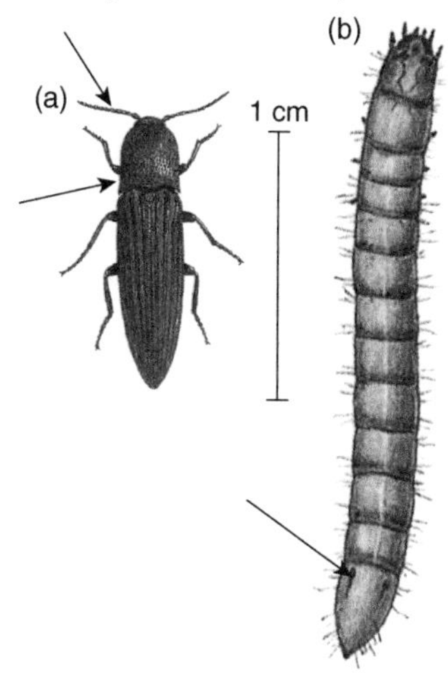

Fig. 12.31 (a) Adult (from Zanetti 1977) and (b) larva (from Lyneborg 1968, with permission) of the agricultural wireworm *Agriotes lineatus*.

The larvae of many of the other elaterid genera feed in wood of broad-leaved trees, particularly fallen logs, and are if anything beneficial in accelerating nutrient recycling in forests.

12.3.7 Superfamily Cantharoidea

Adults are distinguished by their rather slim bodies with soft slightly hairy elytra verging on the papery, often rather darker at the tip.

12.3.7.1 Family Cantharidae (soldier beetles)

These are commonly found feeding on flowers, particularly on the white platforms of umbellifers. Although they do take pollen and nectar, they also predate other insects visiting the flowers and so, like their larvae, are carnivorous. The latter are dark and velvety in appearance, and are predatory soil-surface dwellers. The term 'soldier beetles' derives from the fact that their elytra are often red like guardsmen's dress uniforms; *Rhagonycha fulva* (Fig. 12.32) is a common flower visitor in the UK.

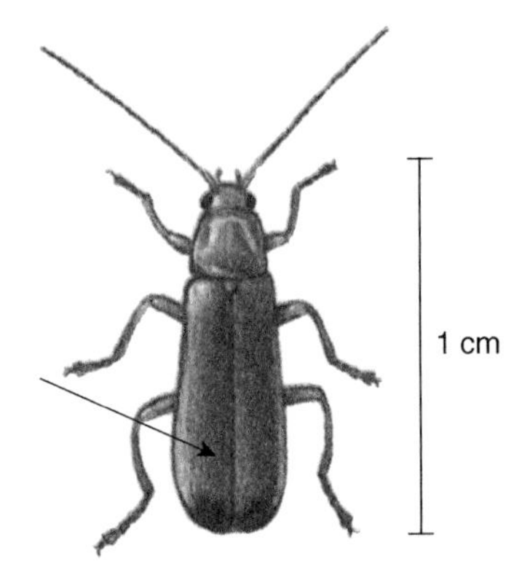

Fig. 12.32 Soldier beetle (*Rhagonycha fulva*) (from Lyneborg 1968, with permission).

Soldier beetles are often found in copulating pairs, and many people erroneously link this with the chemical cantharidin, once thought to be an aphrodisiac. However cantharidin, in spite of its name, comes from the totally unrelated blister beetles (Meloidae, see Section 12.3.11.1, Family Meloidae).

12.3.7.2 Family Lampyridae (glow-worms)

Male lampyrids look like normal beetles capable of flight. By contrast (Fig. 12.33), the female is a strange larva-like insect and it is only the female, the eggs and the larvae that 'glow in the dark'. On the females and larvae the light-emitting organs, which can be switched on and off, are at the rear of the abdomen. The emission of light depends on the oxidation of a compound called 'luciferin', a process which emits light but virtually no heat. The purpose of the female glowing is to make her position visible to the nocturnally flying males. I'm afraid I cannot enlighten (pardon the pun!) you as to why the larvae and eggs also glow.

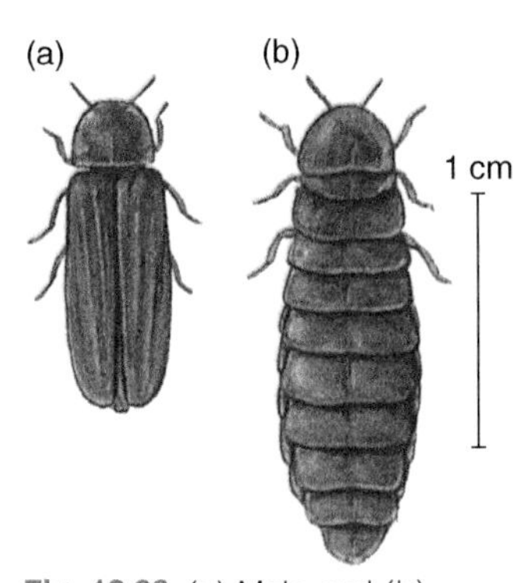

Fig. 12.33 (a) Male and (b) female glow-worm (*Lampyris noctiluca*) (from Mandahl-Barth 1974, with permission).

The larvae are predatory on small snails and slugs, which they liquidise by injecting an enzyme through their jaws prior to ingestion. The need of snails for calcium (for the shells) means that glow-worms are mostly found on chalk or limestone.

Lampyrids have an unusual economic importance based on their production of luciferin. Laboratory apparatus measures the light produced by luciferin to quantify microbial activity in soil samples since the strength of the light correlates with the amount of free oxygen in the soil air spaces. The technique is, for example, used to quantify the decline with time in the toxicity of pesticide residues in the soil to microorganisms.

12.3.8 Superfamily Dermestoidea

The main Family is the Dermestidae. The adults are 1 cm or less in length, brown with usually some darker or lighter patches on the elytra (Figs 12.34 and 12.35), which are downy or bear flat hairs (scales). The antennae are clubbed, and most genera have a single and large ocellus on the top of the head. Often the larvae are covered with long hairs and are known as 'woolly bears' (Fig. 12.35b). They are scavengers on dried

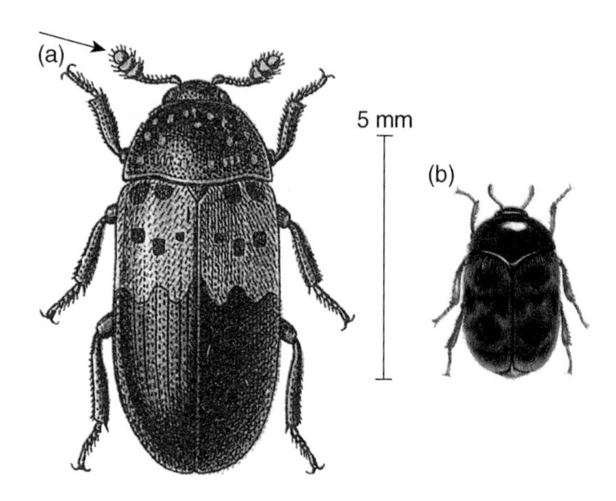

Fig. 12.34 Dermestidae: (a) *Dermestes lardarius* (from Reitter 1908–1916); (b) Khapra beetle (*Trogoderma granarium*) (courtesy of United States Department of Agriculture).

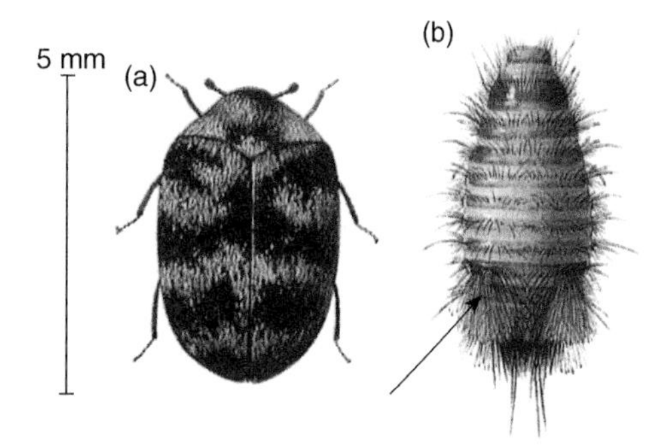

Fig. 12.35 (a) Adult and (b) larva of *Anthrenus museorum* (from Zanetti 1977).

organic matter, and so in nature they are found in bird nests and on dried carcasses. The problem with dermestids is that we humans value a whole range of dried organic matter – which is all prone to dermestid attack. This range includes carpets, fur, natural fabrics, dried meat, stored grain and museum specimens.

Thus *Dermestes lardarius* (Fig. 12.34a) attacks stored meat. The Khapra beetle (*Trogoderma granarium*) (Fig. 12.34b) is a widespread pest, the very hairy larvae hollowing out the grains of stored cereals, groundnut and pulses, especially in warmer climates. There the life cycle can take only 3 weeks, though if food is absent the larvae congregate in large numbers in crevices in the store, and can survive there for many months. *Anthrenus* spp. are the carpet beetles and the museum beetles (e.g. *A. museorum*; Fig. 12.35).

12.3.9 Superfamily Bostrychoidea

The pronotum of the adults covers the head completely (not in the Lyctidae, however) and the head is then usually referred to as 'hooded'. The larvae are soft, white and curved like the larvae of chafers. Their economic importance rests in the considerable damage they can do as timber borers.

12.3.9.1 Family Lyctidae (powder-post beetles)

The pronotum does not cover the head completely, and the antennae are clubbed (Fig. 12.36a) but unusually for the Superfamily the club is two- and not three-segmented.

Also the last segment of the tarsus is disproportionately long. The insects (e.g. *Lyctus linearis*) bore in wood, including furniture, expelling a very fine wood powder as evidence of their activity.

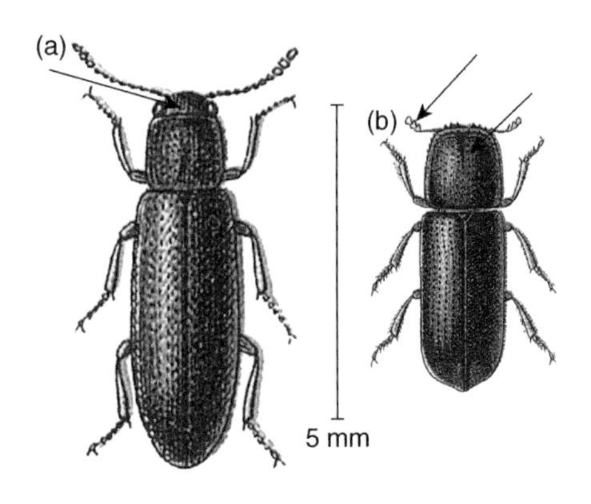

Fig. 12.36 Bostrychoidea: (a) Lyctidae (*Lyctus* sp.) and (b) Bostrychidae (*Rhizopertha dominica*) (from Reitter 1908–1916).

12.3.9.2 Family Bostrychidae

The Family can be identified by the asymmetry of the antennal club. The important pest species is the lesser grain borer, *Rhizopertha dominica* (Fig. 12.36b), now a cosmopolitan pest of stored grain and some other stored foodstuffs including flour and cassava. The larvae feed on grains from the outside and pupate in those that they have hollowed out; the adult can thus emerge without leaving an emergence hole. In warm conditions the life cycle takes about 4 weeks.

If there is a lesser grain borer, there has also to be a greater grain borer. This is *Prostephanus truncatus*, a native of South America where it does little damage in stores. However, its arrival in Tanzania showed its potential as a very serious pest, now reproduced in other parts of Africa to which it has spread.

12.3.9.3 Family Anobiidae

Most anobiids have the prothorax prolonged forwards over much of the head like a hood. The antennae are toothed and often have both well-separated insertions on the head and the antennae are lightly toothed. The Family includes two particularly serious wood-boring pests. 'Woodworm' is the larval stage of the furniture beetle (*Anobium punctatum*; Fig. 12.37a), but the word is also used for the symptom – when we see the small holes through which the adults leave, we say that the furniture has 'woodworm'. I have seen a wicker basket in a loft which looked just like a woodworm-riddled wicker basket until I touched it to discover it was a shape in sawdust!

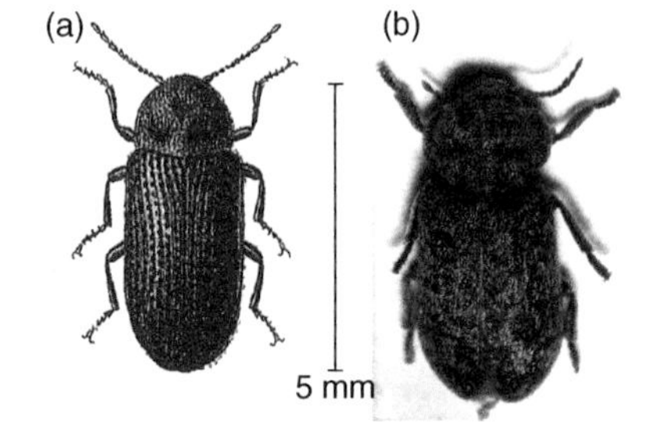

Fig. 12.37 Bostrychoidea, Anobiidae: (a) *Anobium punctatum* (from Reitter 1908–1916); (b) death-watch beetle (*Xestobium rufivillosum*) (photo by Sarefo; this file is licensed under the Creative Commons Attribution-Share Alike 3.0 Unported license).

The other really important wood-borer is the notorious death-watch beetle (*Xestobium rufivillosum*; Fig. 12.37b), so-called because of the staccato knocking sound the adults make by striking their heads against wood, commonly the roof timbers of churches at night. This tapping is a mating signal. Such infestation of really old timbers may not be such a serious problem, as the larvae cannot penetrate the rock-hard parts of the beams and so their feeding is often confined to the outer layers of the wood.

The Family also includes some cosmopolitan pests of stored products. The drugstore beetle (*Stegobium paniceum*) and the cigarette beetle (*Lasioderma serricorne*) attack various types of dry products, including tobacco leaves in storage after drying and cocoa beans, groundnut, peas and beans, grain (where the germ is preferentially eaten) and flour. Attack is characterised by the emergence holes of the adults from the grain etc., but both larvae and adults will also make holes in the cardboard of cigarette packets. The life cycle takes 4–7 weeks.

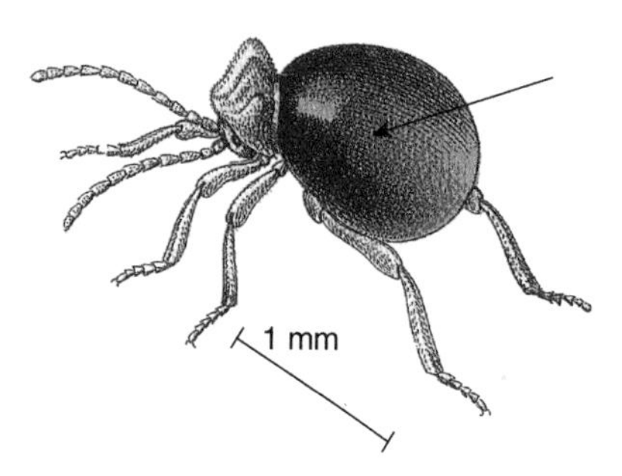

Fig. 12.38 A spider beetle (Ptinidae) (from Reitter 1908–1916).

12.3.9.4 Family Ptinidae (spider beetles)

These small beetles, with their spherical abdomen (Fig. 12.38) and long legs and long filiform antennae (whose insertions on the head are close together), move fairly quickly and do rather look like tiny spiders. In nature they scavenge in the nests of birds, mammals and bees, but adults and larvae are also found in flour, dry seeds, dried fruit and textiles as well as in other stored products.

12.3.10 Superfamily Cleroidea

12.3.10.1 Family Cleridae

These beetles are hairy and often brightly coloured (Fig. 12.39), but another feature is the presence of floppy lobes on some of the tarsal segments. The antennae are clubbed or strongly serrate. Many are found in association with wood-boring beetles, since their larvae are predators on the borer larvae. I was once called to a house in a poor part of town because the owners had complained of large numbers of grubs falling onto their bed from the bedroom ceiling! The grubs were indeed clerid larvae, escaping the roof timbers which the local council had just had fumigated against woodworm. One species (*Necrobia rufipes*) attacks stored sides of bacon and hams.

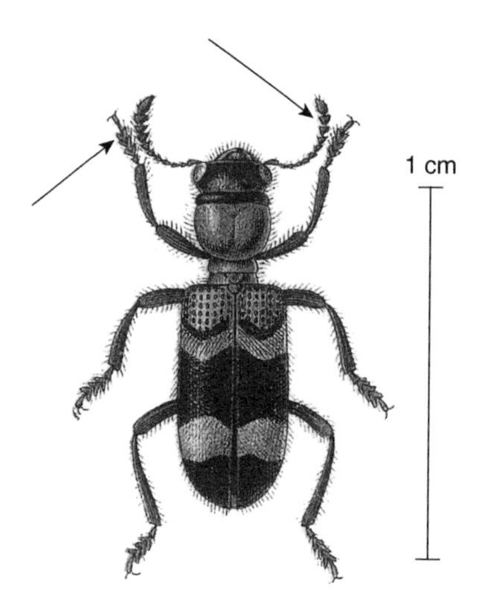

Fig. 12.39 Cleridae (*Thanasimus formicarius*) (from Reitter 1908–1916).

12.3.10.2 Family Trogossitidae

Trogossitidae have a rather long last tarsal segment with a small lobe between the claws. The Family includes the Cadelle beetle (Fig. 12.40), which has the misleading Latin name *Tenebroides mauritanicus*) – misleading because the Family Tenebrionidae (see Section 12.3.11.1, Family Tenebrionidae) is a Family of quite unrelated beetles. The Cadelle beetle is found in food and grain stores; it both feeds on other insect larvae and itself does some feeding on the stored material.

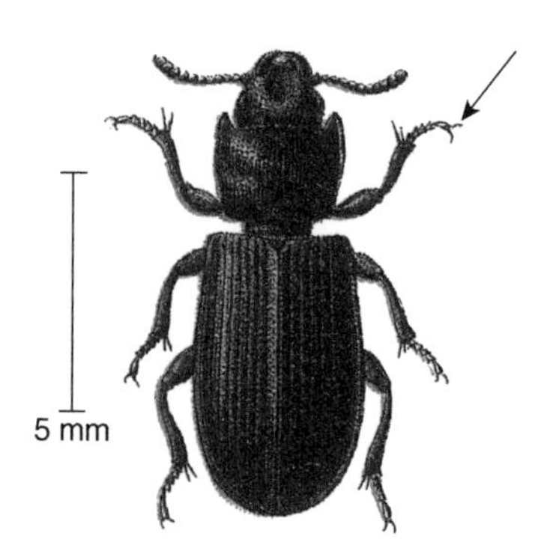

Fig. 12.40 The Cadelle beetle (*Tenebroides mauritanicus*) (from Reitter 1908–1916).

12.3.10.3 Family Melyridae

Melyrids are often brightly coloured beetles with a fringe of long hairs at the side of the thorax. They have rather delicate elytra, and are often found at flowers feeding on pollen. Otherwise both adults and larvae are predacious on small caterpillars and other small insects.

12.3.11 Superfamily Cucujoidea

This huge Superfamily is conveniently divided into two series, the Heteromera with the tarsal formula (see Section 2.5.2) 5–5–4 and the Clavicornia. The latter usually have an obvious antennal club and the tarsi of females is never 5–5–4. I shall also give (in smaller italic orange typeface) a more recent classification, which distributes the Families in this Superfamily into three Superfamilies (Melooidea, Tenebrionoidea and Cucujoidea), which are not as clearly separable as the division into two Series followed here.

12.3.11.1 Series Heteromera (tarsal formula of both sexes 5-5-4)

The tarsal formula of both sexes is 5–5–4.

Superfamily Melooidea

Family Anthicidae

I have included this Family because the adults are often encountered and are easily recognised. They are only a few millimetres long and have non-striated elytra with an exceptionally strong constriction between the head and thorax. They live in a variety of decaying vegetable material.

Family Meloidae (oil beetles, blister beetles)

These soft-bodied beetles have a downwardly pointing head, which is sharply demarcated from the thorax. The tarsal claws are usually saw-toothed. The elytra (hind wings are absent) are often papery and brightly coloured with broad bands of black or another contrasting colour, and they may (e.g. *Meloe*) have a gap between them at their posterior end (Fig. 12.41) and only cover part of the back of the abdomen. The common names for the Family reflect the oily and sometimes caustic defensive secretions that the adults can produce.

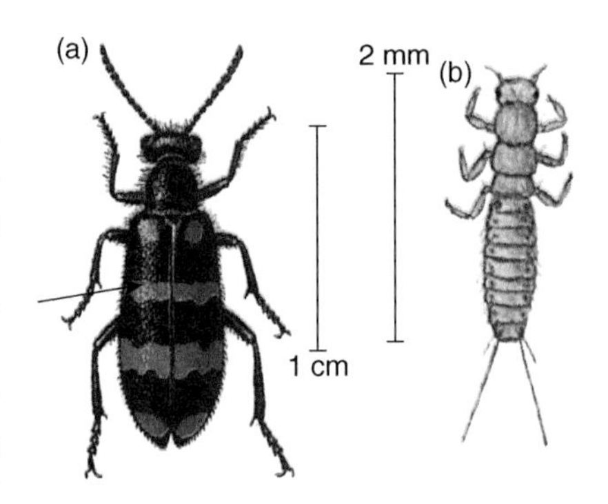

Fig. 12.41 Meloidae: (a) *Mylabris polymorpha* (from Chinery 1986); (b) triungulin larva (from Lyneborg 1968, with permission).

The Family is noteworthy particularly for its unique life cycle, which involves the phenomenon of 'hypermetamorphosis' (cf. burying beetles, see Section 12.3.12), that is the occurrence of an additional metamorphosis during the immature stage. The larvae predate the eggs of Orthoptera and bees, particularly the non-social species. Adult meloids lay many eggs (usually 2000–10,000) on the soil. This allows for a huge mortality of the first instars, since locating a host in time is an unlikely event. The first instar larvae are known as 'triungulins'; they are small, active and tough larvae with strong claws. If they encounter their appropriate host they attach themselves with claws; species that predate the eggs of bees will climb onto flowers and await the arrival of a host. At their next moult, most change their appearance to a shorter-legged eruciform and rather inactive grub. These feed on the eggs of their host. With bee hosts, the triungulin will move from the bee's body onto the egg and be incarcerated with the egg into a sealed brood cell. The subsequent life cycle still springs other surprises. There is a resting stage (the pseudopupa) of the larva within an unshed skin, from which the larva later emerges to complete its development to pupation. Larvae may move between cells and predate several eggs. Towards the end of the life cycle they may move to feed on the pollen and nectar stored in the nest. There is some variation between species in the detail of the life cycle.

Adult Meloidae are flower visitors feeding on pollen and nectar. Several species of the genus *Mylabris* do considerable damage on cowpeas and other pulses as adults, both to the flowers (resulting in pod loss) and to the foliage. Eggs are laid in batches in the soil. The larvae are not herbivores, but feed on the egg pods of Orthoptera. If the orthopteran species is a pest, the *Mylabris* larvae, in contrast to the adults, are actually beneficial. Pupation also takes place in the soil. The adults are quite large at around 3 cm in length and the insects are found throughout Africa and in the Indian subcontinent.

Striped blister beetles (*Epicauta* spp.) feed as adults on the foliage and fruits of legumes, particularly pulses and alfalfa, but also of most vegetables including solanaceous crops. The beetles are 1–2 cm long, and usually black with longitudinal stripes. In the USA, an important species is *Epicauta vittata* with yellow stripes, while the main species in East Africa is *E. albovittata* with white stripes. Damage may completely defoliate the plants.

The blister beetles contain the pharmaceutical cantharidin, so named because *Cantharis* (now a genus in the unrelated Family Cantharidae see Section 12.3.7.1) is an old synonym of the blister beetle genus *Lytta*, 'Spanish fly'). So don't be misled (many people are!) – the soldier beetles you often see sitting on flowers don't contain that chemical. It is a toxin that blisters human skin and accounts for the name 'blister beetles'. Crushed beetles may be present on alfalfa hay when it is harvested and baled for animal fodder, and horses have been known to die as a result. The elytra contain much of the toxin and are ground to extract the drug, which is used to remove warts and tattoos and (very risky medically) to induce male arousal.

Family Ripiphoridae (previously Rhipiphoridae)

Like some Meloidae, the Ripiphoridae (Fig. 12.42) are soft bodied with short gaping elytra and no hind wings, but in addition the antennae, at least in the males, are flabellate (i.e. some segments have projections which almost make them appear branched). Such reduced front wings and flabellate antennae are characters also found in another Order of insects, the Strepsiptera. Like some Strepsiptera, some Ripiphoridae are parasitoids of aculeate Hymenoptera and have a hypermetamorphosis with triungulin first instar larvae. The Strepsiptera have therefore been regarded as taxonomically allied to the Coleoptera, particularly to the Family Ripiphoridae. However (see Section 8.6), there is no molecular evidence for this, and the true affinity of the Strepsiptera is still unresolved.

The larvae of other Ripiphoridae are endoparasitoids of cockroaches or predators of other beetle larvae in rotting wood (the predatory species do not exhibit hypermetamorphosis).

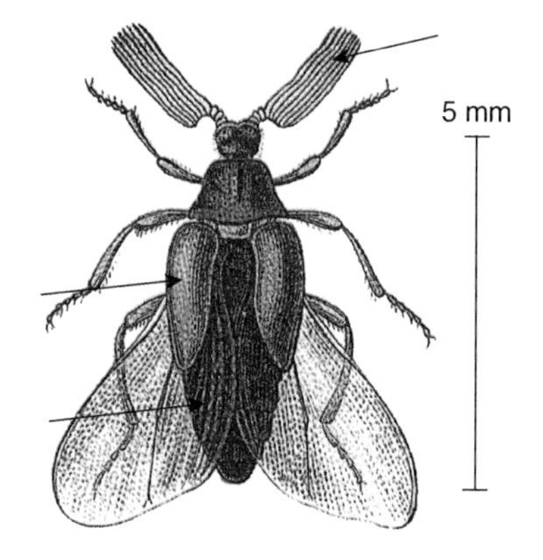

Fig. 12.42 Ripiphoridae (male *Ripidius pectinicornis*) (from Reitter 1908–1916).

Superfamily Tenebrionoidea

Family Alleculidae

The larvae bore in rotten wood, helping to break it down, or humus; the rather ovoid adults are frequent visitors to flowers. A few species are parasites of solitary bees. I mention the Family only because they are easily recognised by the spot character that the claws at the end of the tarsi are pectinate (comb-like).

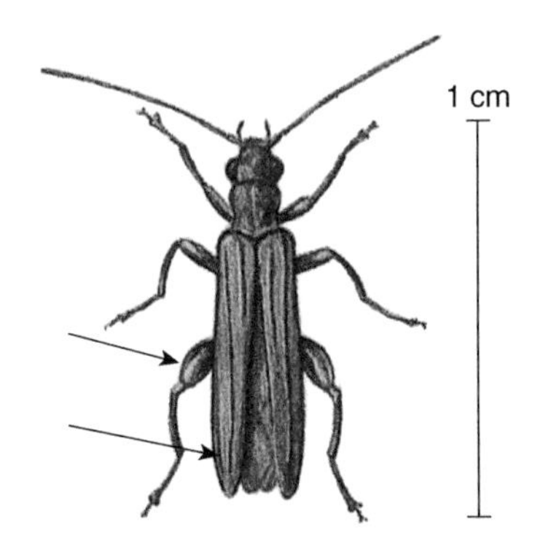

Fig. 12.43 Oedemeridae (male *Oedemera virescens*) (from Lyneborg 1968, with permission).

Family Oedemeridae

These look a bit like miniature Cerambycidae (see Section 12.3.12.1), except that the elytra are rather soft. The elytra are often metallic coloured and gape at the end of the abdomen; males often have swollen hind femora (as in Fig. 12.43). The larvae of most species bore wood but the adults frequently visit flowers to feed so, although the Family has no agricultural importance, you may well notice the adults and would like to know what they are. However, one species (*Nacerdes melanura*) bores marine timbers in the intertidal zone and can cause damage in docks to the wharves; hence its common name of 'wharf borer'. Like the Meloidae (see Section 12.3.11.1, Family Meloidae), oedemerids contain cantharidin.

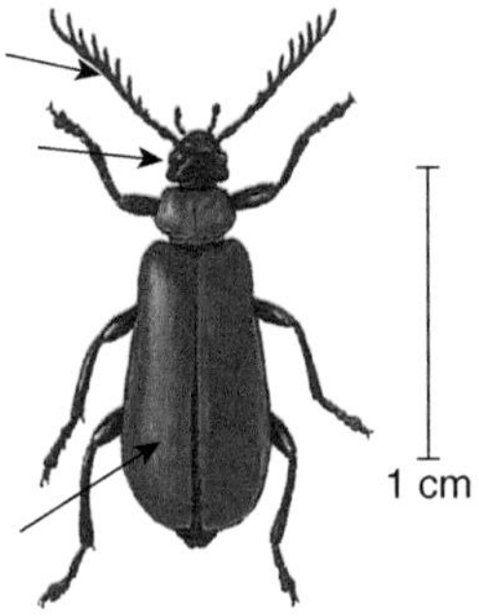

1 cm

Fig. 12.44 Pyrochroidae (*Pyrochroa coccinea*) (from Mandahl-Barth 1974, with permission).

Family Pyrochroidae

Both adults and the fungus-feeding larvae live under bark, but adults are often found visiting flowers. The adults are often black, and the Family is mentioned here because it includes the easily recognised and quite common red 'cardinal beetles' with black heads and long pectinate antennae (Fig. 12.44).

Family Tenebrionidae (darkling beetles)

This is a large Family, and in the tropics and subtropics the beetles are large as well! They can even be 16 mm long in the UK, for example the cosmopolitan *Tenebrio molitor* (flour beetle; Fig. 12.45a). The larvae (mealworms; Fig. 12.45b) are pests of stored flour, though they also attack whole grains. They have a quite hard and light chestnut-coloured cuticle, and are sold as bird food and also – stained in a variety of unnatural colours – to fishermen as bait. The length of the life cycle can be very variable depending on temperature and overwintering. Larvae can moult from nine to 20 times, and in the field there is just one generation a year. In continuous indoor rearing, the life cycle usually takes about 3

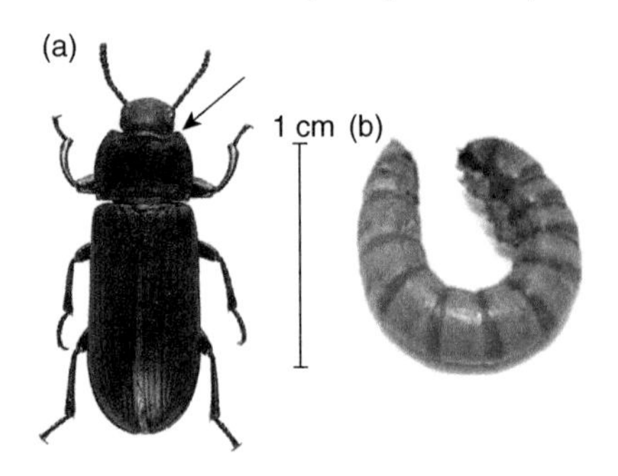

Fig. 12.45 Flour beetle (*Tenebrio molitor*): (a) adult (photo by Didier Descouens); (b) larva (mealworm) (courtesy of Lewis Ryan).

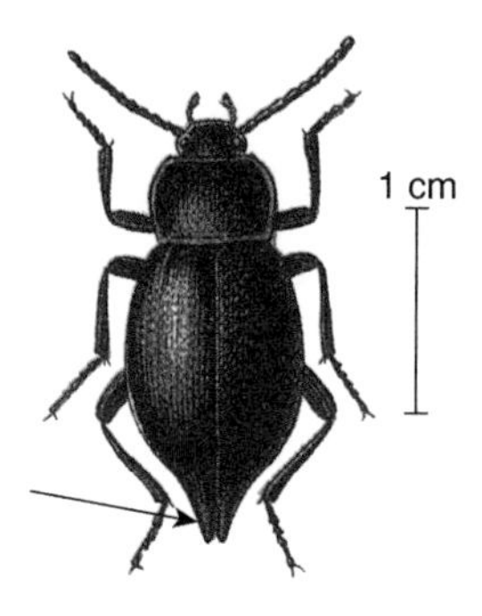

Fig. 12.46 A churchyard beetle, *Blaps mucronata* (from Reitter 1908–1916).

months. Adult Tenebrionidae are generally dark brown or black with clearly clubbed antennae and mostly (but not *T. molitor*) with no hind wings and the elytra unable to open. Another common largish species is the churchyard beetle, *Blaps mucronata* (Fig. 12.46), which scavenges on vegetable matter and has a very pointed tip to the abdomen. There are also a number of much smaller and more ovoid tenebrionids, which are cosmopolitan pests of flour and other stored products, especially *Tribolium confusum* (confused flour beetle) and the more reddish *T. castaneum* (red flour beetle; Fig. 12.47). Both have very similar biologies. They are pests of stored flour of wheat, maize etc., with *T. confusum* important in more temperate regions of Europe, North America and Australia etc., while *T. castaneum* is the flour beetle of warmer parts of these continents, Asia, Africa and South America. The insects live and oviposit in the produce and both adults and larvae feed on the flour. The insects are also found in stores of whole grain, feeding on the 'flour' created by breakage of the grain or damage by other insects. The adults are reddish-brown and about 3 mm long. The two species are hard to distinguish, and the best characters are that the antennal segments of *T. confusum* widen gradually towards the tip and the thorax has fairly straight sides in dorsal view, whereas in *T. castaneum* the apical three segments which form the club are abruptly wider than the preceding segments and the sides of the thorax are clearly curved. Glands on the abdomen and thorax release a pungent odour if the beetles are disturbed; this gives an undesirable taint to the flour. The creamy larvae have two short upwardly curved horns at the end of the abdomen. Pupation is in the flour, and the life cycle takes 7–12 weeks. Flour beetles will not develop or breed below 18°C.

2 mm

Fig. 12.47 Red flour beetle (*Tribolium castaneum*) (courtesy of Betty Greb, Agricultural Research Service, USDA).

12.3.11.2 Series Clavicornia = *Superfamily Cucujoidea*

The female tarsal formula is not 4–4–5 and the antennae are fairly obviously clubbed (though they are also clubbed in some Heteromera; see Section 12.3.11.1).

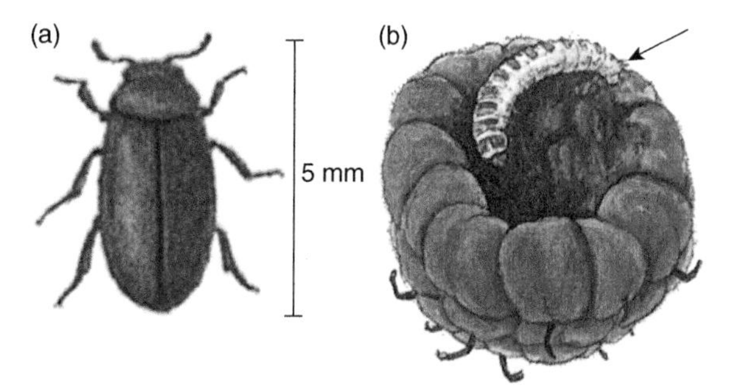

Fig. 12.48 Raspberry beetle (*Byturus tomentosus*): (a) adult and (b) larva feeding on raspberry (from Mandahl-Barth 1974, with permission).

Family Byturidae

This Family contains some fruit pests, particularly the raspberry beetle, *Byturus tomentosus* (Fig. 12.48). The adults spend most of the year inactive and sheltering under litter near the crop, and emerge to lay eggs on the receptacles of raspberry flowers. The larvae feed on the fruitlets of the berry, and tend to become obvious to consumers when the 8 mm long grubs float to the surface of the water in which the berries are being washed. They descend to the soil to pupate; the adults emerge in the spring and remain inactive until the oviposition cycle starts again. The insect occurs all over Europe.

Family Coccinellidae (ladybirds or ladybeetles (USA))

This is one of the most familiar Family of beetles, their abundance and appetite for aphids making them the best known of the 'gardener's friends'. Apart from the more elongate genera *Rhyzobius* (Fig. 12.50a) and *Coccidula*, they are rather round beetles. The elytra are often coloured red or yellow (warning colouration) with large spots (Fig. 12.49a), and the adults can produce a pungent smell in defence when they are attacked. However, there is lot of elytral colour variation within single species, including 'negatives' – that is black elytra with red spots! Much of the head is rather concealed under the pronotum, and the antennae are short and clearly clubbed. The tarsal formula looks like 3–3–3 but is actually 4–4–4, with a minute third segment. The tarsus can therefore be said to be 'pseudotrimerous' in analogy with the 'pseudotetramerous tarsus' of the Superfamily Chrysomeloidea (see Section 12.3.12). The larvae of many of the most important species that are aphid predators are black, with groups of hairs set on warts dorsally and laterally on the abdomen. Some of these warts may be yellow (Fig. 12.49b) and the pattern of the coloured warts aids in species recognition.

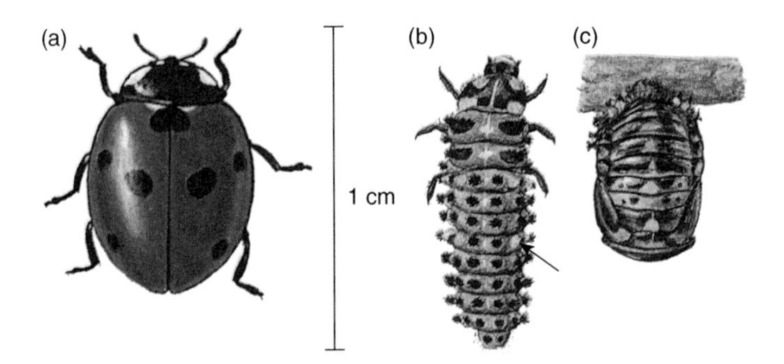

Fig. 12.49 Seven-spot ladybird (*Coccinella septempunctata*): (a) adult (from Mandahl-Barth 1974, with permission); (b) larva and (c) pupa (from Lyneborg 1968, with permission).

The best known species, such as *Coccinella septempunctata* (seven-spot ladybird; Fig. 12.49), *Adalia bipunctata* (two-spot ladybird) and *A. decempunctata* (ten-spot ladybird), are those that attack aphids. However, their importance in actually preventing aphid outbreaks on arable crops is often a little over-rated. The annual life cycle begins before aphids colonise annual crops, and so the first generation occurs on perennial trees and shrubs. Thus ladybirds may be important in controlling aphid outbreaks on fruit trees such as apples, but aphid populations have to reach quite large numbers on annual crops before the attractive odour from their honeydew reaches levels that result in the ladybirds migrating from the perennials. The next limiting factor is coccinellid behaviour; particularly the adults climb upwards whenever they reach a choice-point such

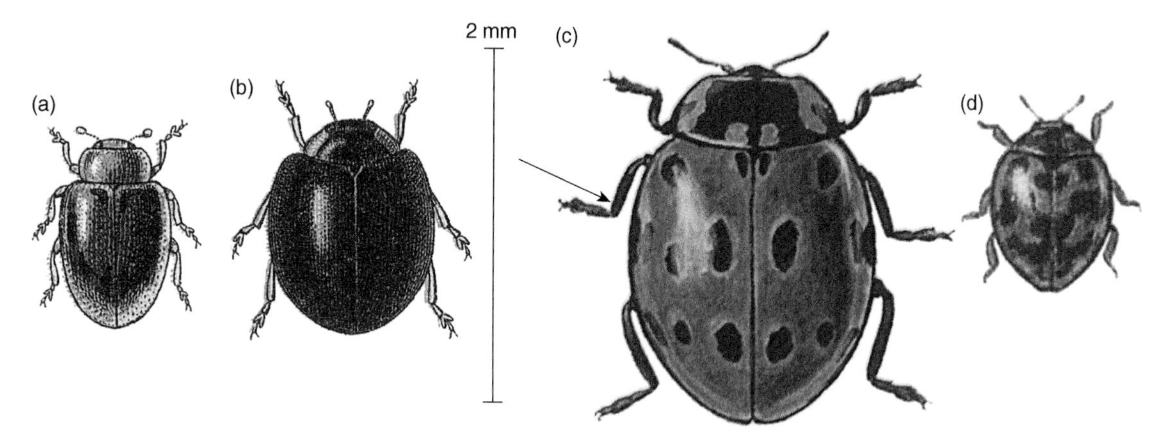

Fig. 12.50 (a) *Rhyzobius litura* and (b) *Exochomus flavipes* (from Reitter 1908–1916); (c) *Anatis ocellata* (from Mandahl-Barth 1974, with permission); (d) *Subcoccinella vigintiquattuorpunctata* (from Lyneborg 1968, with permission).

as where the petiole of a leaf joins the stem. They thus effectively clean the upper leaves of the plant of aphids, but often fail to encounter aphids on lower leaves. As a result they run out of food and depart from the plant while the lower leaves remain aphid infested. This may be stupid behaviour as far as the farmer is concerned, but evolutionarily it is quite sensible in that it prevents over-exploitation of the food resource, and the upper leaves can be recolonised by the aphids as food for other coccinellids. Female coccinellids seem to 'count' aphids in that they lay eggs near aphid colonies and in proportion to aphid abundance. An additional mechanism to aid the survival of the hatching larvae is that the first ones to hatch will cannibalise other eggs in the egg batch if aphids are not immediately to be found.

Pupation (Fig. 12.49c) is on the plant, and it is the emerged adults that overwinter. Many of these enter a state of arrested reproductive development (diapause) until the spring, and a common behaviour is an autumn migration to high altitudes where huge aggregations of overwintering ladybirds may form in the shelter of rocks, usually above the snow line. The traditional explanation for this behaviour is that the ladybirds are putting themselves into 'cold storage' to prevent attack and multiplication in their bodies of insect diseases.

Some species (e.g. in the genus *Exochomus* (Fig. 12.50b), and *Anatis ocellata*, the eyed ladybird (Fig. 12.50c), named from the light rings around its black spots) specialise on aphids in the Family Adelgidae on conifers, while *Cryptolaemus montrouzieri* (Fig. 12.51) predates mealybugs and is commercially available for release in glasshouses.

Fig. 12.51 *Cryptolaemus montrouzieri*: (a) adult and (b) larva (courtesy of Ward Stepman, BCP Certis).

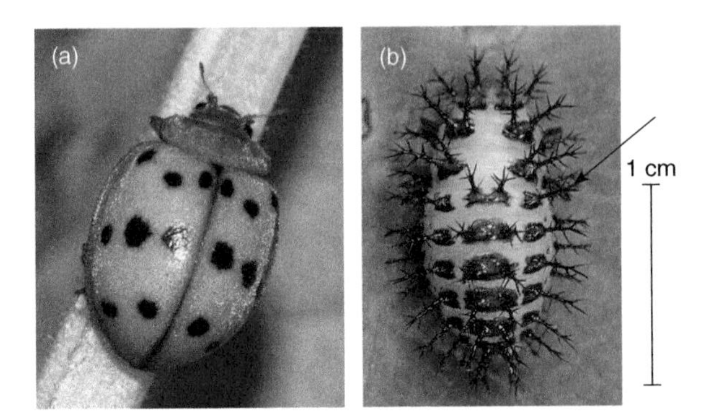

Fig. 12.52 Mexican bean beetle (*Epilachna varivestis*): (a) adult and (b) larva (courtesy of Tom Murray).

Adult coccinellid aphid predators may also take other food when aphids are scarce. Commonly, adults visit flowers to feed on pollen and nectar. However, it is important to be aware that not all Coccinellidae have a primarily predatory life-style. Some are herbivores. The genus *Scymnus* is one of rather small beetles, which are fungus feeders, as is the mildew feeder *Psyllobora vigintiduopunctata* (22-spot ladybird). The 24-spot ladybird, *Subcoccinella vigintiquattuorpunctata* (Fig. 12.50d), is a leaf feeder and the herbivorous genus *Epilachna* includes the serious legume defoliator *E. varivestis* (Mexican bean beetle; Fig. 12.52). The sides of the thorax and abdomen of the larva have stout branched spines, which remain evident in the pupa. The pest occurs in southern Europe, Africa, Asia and the USA and the yellow larvae strip the leaves, leaving only the main vein. The larvae pupate on the leaves. The adult is of variable reddish and brown colour with black spots like a typical coccinellid. The life cycle takes about 7 weeks.

Family Cryptophagidae (fungus beetles)

The antennae are quite long with an obvious club clearly divided into three segments, and the thorax and elytra have many short hairs and a rounded-off silhouette. The tarsal formula is 5–5–5, but can be 5–5–4 in males.

The beetles feed on fungi and fungal spores, often under loose flakes of bark and several species can become minor pests of grain and other stored foods provided they have become somewhat damp.

The pygmy mangold beetle (*Atomaria linearis*; Fig. 12.53), however, is a chewing herbivore on sugar beet and mangold seedlings in Europe. As with other pests of seedling sugar beet (e.g. beet leaf miner (see Section 10.4.2.2, Family Anthomyiidae) and Symphyla – relatives of millipedes) it has become a problem only with the changes to drilling to final stand and monogerm seed, both of which have removed the need for labour-intensive hand thinning, but at the same time has also removed their potential for compensating for early seedling losses. The adults appear in the spring to oviposit in the soil and bite into the roots and hypocotyls of the seedlings. The diagnostic symptom of little black pits results. Seedlings may be distorted but are often killed. The adults go on to feed on the still unopened leaves at the centre of young plants, and the holes then later become visible as the leaves expand. The larvae feed on the roots, and cause much less damage than the adults. Pupation is in the soil and there are two generations a year; the beetle overwinters in the soil as an adult.

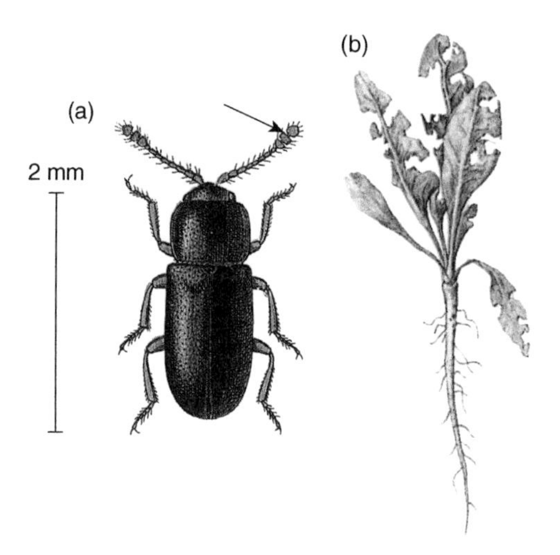

Fig. 12.53 Pygmy mangold beetle (*Atomaria linearis*): (a) adult (from Reitter 1908–1916); (b) damage to beet seedling (from Bayer 1968, with permission).

Family Cucujidae

This Family of scavengers has rather long antennae, and contains some pests of food stores. One of the most common and cosmopolitan pests of stored grain is the reddish-brown *Oryzaephilus surinamensis*, the saw-toothed grain beetle (Fig. 12.54). The common name derives from a feature of many members of the Family, that the side-margins of the thorax are toothed. Additionally, *O. surinamensis* has a pronotum and elytra which have strong longitudinal ridges. The larvae are free-living in the grain and attack grains that are already damaged, often by other stored products beetles or moths. The larvae feed preferentially on the germ, and pupation occurs among the grains. The pest also attacks many other stored foods, especially powdery ones like flour and sugar, and can chew its way through cardboard boxes and plastic bags. In warm temperatures the life cycle takes about 35 days, but the insect cannot breed below 19°C, and so in temperate countries it is easily controlled by cooling the stored product.

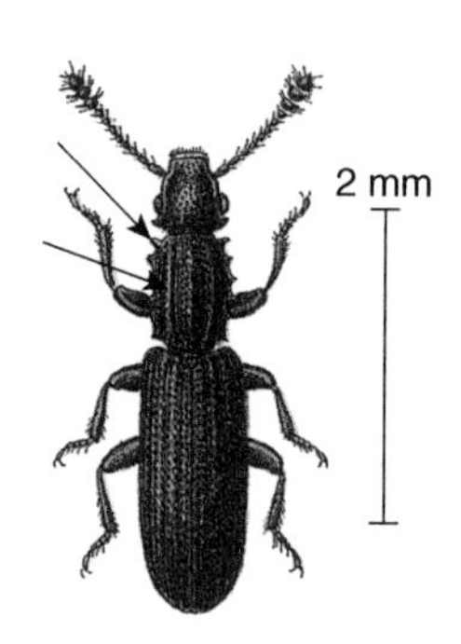

Fig. 12.54 Saw-toothed grain beetle (*Oryzaephilus surinamensis*) (from Reitter 1908–1916).

The larvae of the rust-red grain beetle (*Cryptolestes ferrugineus*) similarly prefer to feed on the germ of stored cereals throughout the warmer parts of the world, though they will also survive in cooler regions if the store is heated. The side margins of the thorax are expanded, but not obviously toothed.

Family Lathridiidae

This Family is similar in habits and appearance to the Cryptophagidae, except that the tarsi have only three segments. They may also be found in stored foods.

Family Mycetophagidae

These tiny hairy beetles survive wetting by living in the blotting-paper environment of the caps of mushrooms and toadstools. They are among the smallest insects (see Section 2.6.4).

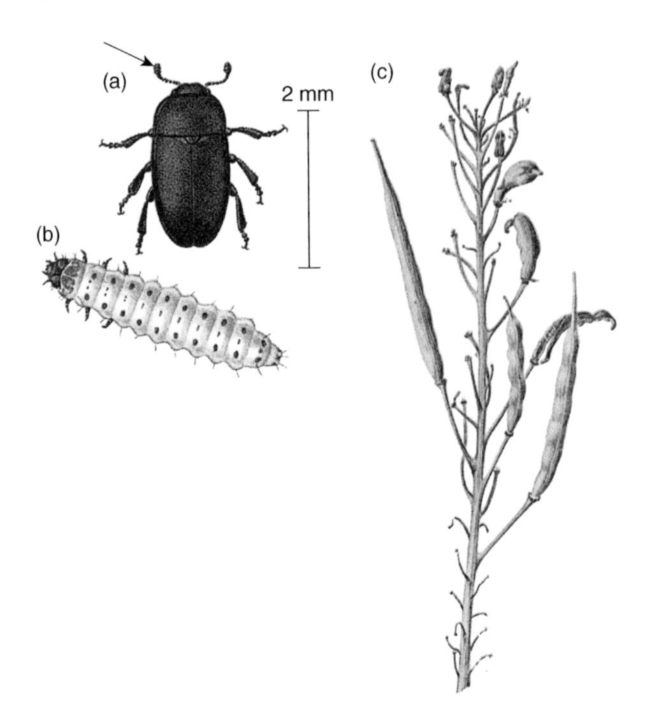

Fig. 12.55 Pollen beetle (*Meligethes aeneus*): (a) adult; (b) larva; (c) damage (pods have either failed to set or are malformed) (all from Bayer 1968, with permission).

Family Nitidulidae

These are mainly very small (<5 mm long) rounded beetles with a strongly developed three-segmented club at the end of the antennae. The tip of the abdomen often projects beyond the elytra and the tarsal formula is 5–5–5 or 4–4–4. Most species are scavengers in decaying organic matter or under tree bark, and the family also includes stored products pests such as the cosmopolitan *Carpophilus hemipterus*, which attacks mainly dried fruit, but also feeds on cotton bolls and maize cobs in the field. Both adults and larvae feed on the dried fruit and the produce is further spoilt by the accumulated frass. The elytra of the adults have diagnostic yellow patches and leave about a third of the abdomen exposed.

However, the most obvious and economically important nitidulids are the small black shiny pollen beetles (Fig. 12.55) of the genus *Meligethes* (especially *M. aeneus*). The whole life cycle is spent in flowers, where both adults and larvae feed on the pollen and anthers. Pods are aborted and seed production is greatly reduced. The larvae have just two instars before they drop to the ground to pupate, then emerging as adults which seek shelter until the following season. Little notice was taken of damage to flowers by *Meligethes* until the 1970s when oilseed rape became a major arable crop in Europe, and the beetles became abundant to the extent they badly affected seed yield.

12.3.12 Superfamily Chrysomeloidea

This Superfamily contains many of the most important beetle pests of crops, including the notorious Colorado beetle of potatoes. Although otherwise of very diverse appearance, the members of the Series have one thing is common – 'lobed feet' (Fig. 12.56). Firstly, the tarsi are 'pseudotetramerous' (i.e. the fourth segment is inconspicuous so

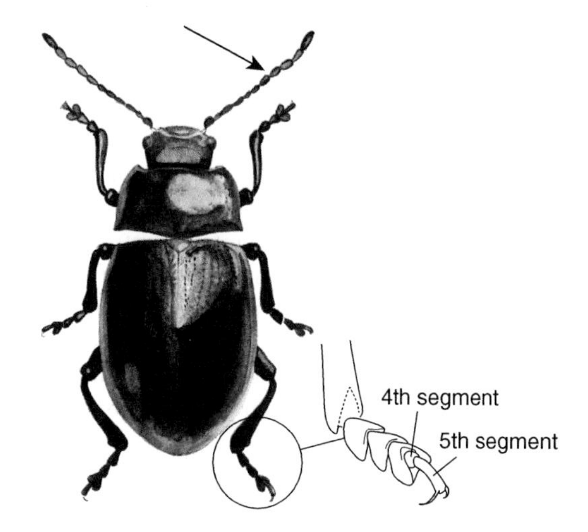

Fig. 12.56 The chrysomelid *Chrysolina fastuosa*, illustrating the pseudotetramerous tarsus of the Chrysomeloidea (beetle image courtesy of Heather Skidmore; inset by Gordon Riley, from Chinery 1993, with permission).

that the five-segmented tarsus appears only four-segmented). This character is hard to detect, and it is easier to see the sideways expansions of the first three tarsal segments (especially the third) which create the appearance I call 'lobed feet'. If a tarsus that looks four-segmented has these lobes, the beetle could be either in the Chrysomeloidea or the Curculionoidea, but the latter (see Section 12.3.13) mostly have clubbed antennae (unlike the filiform antennae of the Chrysomeloidea), and often a prominent snout (rostrum).

12.3.12.1 Family Cerambycidae (longhorn beetles)

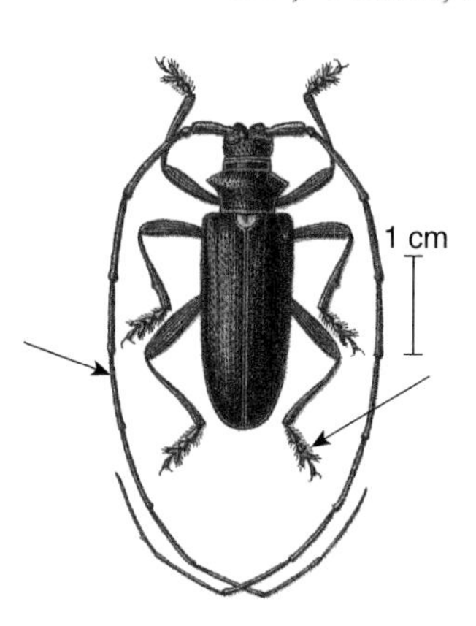

Fig. 12.57 *Monochamus sartor*, a cerambycid with exceptionally long antennae (from Reitter 1908–1916).

The common name derives from the characteristically long antennae, usually at least as long as the elytra, and often held swept backwards rather than projecting forwards (Fig. 12.57). The elytra are often broadest where they join the pronotum, so that the body shape of the beetles tapers somewhat towards the rear. They are often quite large; some tropical species can be 20 cm long. The fleshy stout larvae (Fig. 12.58) tunnel in the trunks of trees and may do so for many years because of the low nutrient content of wood. Commonly, frass holes are made at intervals through which the

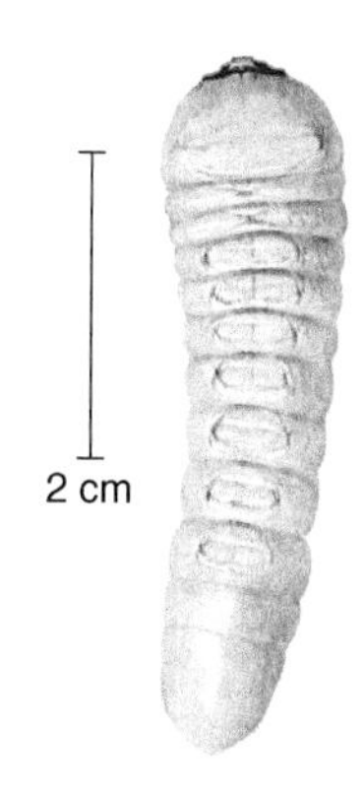

Fig. 12.58 Cerambycid larva (species not given) (from Zanetti 1977).

frass in the tunnel is expelled. Symbiotic bacteria, which can digest cellulose, have been found in the larvae of some species. The long large tunnels can make cerambycids serious pests of timber. Sometimes large tropical adult longhorns appear in temperate countries when imported timber infested with larvae has been used in building houses. I have also known one appear from a large wooden ornamental elephant, years after it was brought back as a souvenir from India.

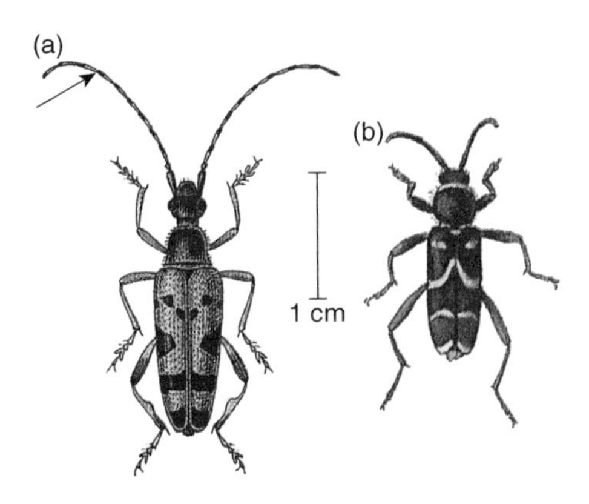

Fig. 12.59 (a) *Leptura maculata* (from Reitter 1908–1916); (b) the wasp beetle (*Clytus arietis*) (from Mandahl-Barth 1974, with permission).

Often encountered in Europe visiting flowers is *Leptura maculata* (Fig. 12.59a), with black spots on otherwise yellow elytra. Another species you may well meet is *Clytus arietis* (Fig. 12.59b), called the wasp beetle because of the yellow bars on black elytra. I recall a neighbour finding one and asking me 'Is this a baby wasp, because it hasn't got any wings?' I hope I need not elaborate on why this was a memorable **multiple** entomological howler.

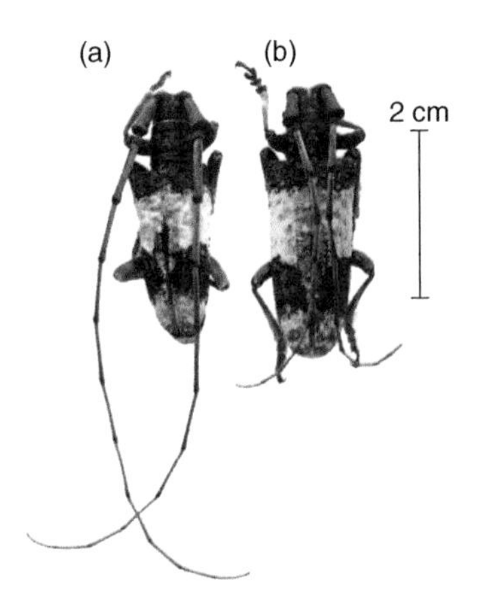

Fig. 12.60 White coffee borer (*Anthores leuconotus*): (a) male; (b) female (www.cerambycoidea.com).

The adult white coffee borer (*Anthores leuconotus*; Fig. 12.60) is actually light grey with a dark head and other markings (including near the tips of the elytra); it is about 3 cm long. It attacks particularly *arabica* coffee in the southern half of Africa, and attack is easily spotted from the wood shavings expelled from the burrows made by the feeding larvae. The younger larvae burrow in the bark and, if they ring the trunk, they can kill the tree; otherwise the foliage yellows and wilts. Later stages tunnel deeper into the wood, and the larval period can last for nearly 2 years. Pupation is in a large chamber carved in the wood by the larva.

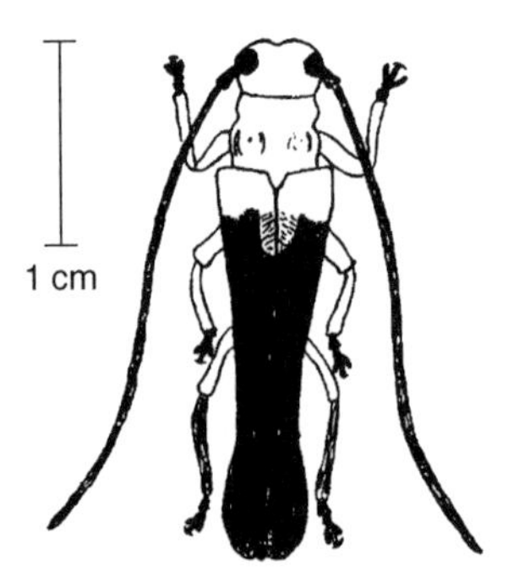

Fig. 12.61 Yellow-headed stem borer of coffee (*Dirphya nigricornis*) (from Hill 1983, with permission).

Another longhorn pest of mainly *arabica* coffee, occurring only in east Africa, is the yellow-headed stem borer, *Dirphya nigricornis* (Fig. 12.61). The eggs are laid under a little lid carved in the bark near the tip of a shoot, and the larvae then bore into the shoot and tunnel down towards thicker branches, making a line of holes for disposing of the frass to the outside at intervals. These lines of holes are diagnostic symptoms, as are the wilted shoot tips. Branches often break at one of these holes. Tunnelling often includes the main stem, and the pupae are often found in the stem near the ground. The adult has black antennae and a dark body, though the head, pronotum and the first part of the elytra are yellow or orange. The life cycle takes about a year.

Other pest longhorns are species of *Apriona*, which are widespread borers of mulberry, apple and fig in Asia, and *Batocera*, which has a similar Asiatic distribution but has mango as one of its main hosts in addition to apple and fig.

12.3.12.2 Family Chrysomelidae (leaf beetles)

That most species (especially the larvae) are leaf chewers, gives the Family its common name. The adults are usually rather round, with brightly coloured, shiny and often metallic-coloured elytra. Many are quite small (1 cm or less), but others are much larger. The largest UK species (>2 cm) is black with a violet sheen. It is the bloody-nosed beetle (*Timarcha tenebricosa*), the larvae of which feed on plants in the bedstraw family. The name 'bloody nosed' derives from the malodorous dark reddish fluid which oozes from the mouth and leg joints as a defence against attack. If you want to see 'lobed feet', you can't do better than the bloody nosed beetle – you won't need a hand lens! The beetles are flightless and the elytra are fused together.

Chrysomelid larvae are typical of the eruciform type (Table 12.1). They have a large wrinkled abdomen and are often dark in colour. The mouthparts are hard to see and the small thorax bears small legs. They can often look more like small slugs than beetle larvae and one species is a brilliant mimic of a bird dropping. There are many crop pests among chrysomelids.

Subfamily Bruchinae (seed beetles)

The antennae of bruchines (until recently regarded as a separate Family, the Bruchidae) are thickened towards the tip, but not in the form of a club. Very often the tip of the abdomen projects beyond the elytra.

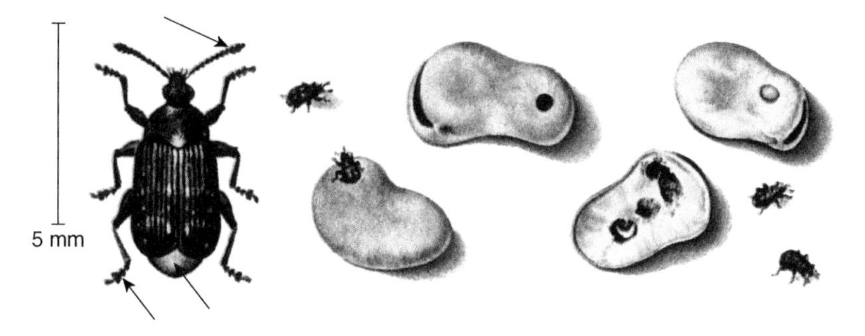

Fig. 12.62 Bruchinae (*Bruchus rufimanus*), beetle and damaged beans (from Bayer 1968, with permission).

Bruchinae are mainly pests of peas and beans (Fig. 12.62). The adults need to feed on the pollen and nectar of legume flower to mature their eggs, which are then laid on the pods. The hatching larvae tunnel through the pod wall to tunnel in and feed on the seeds within. Immediately on entry, the first part of the tunnel which of course increases in diameter as the larva grows, is a tight spiral. This may be to deter parasitoids using the entry hole to insert their ovipositor. Often a black bead of resinous material exuded by the pod wall shows externally where the larvae have entered, and a brown mark identifies the point of entry into the seed. Several larvae may tunnel in

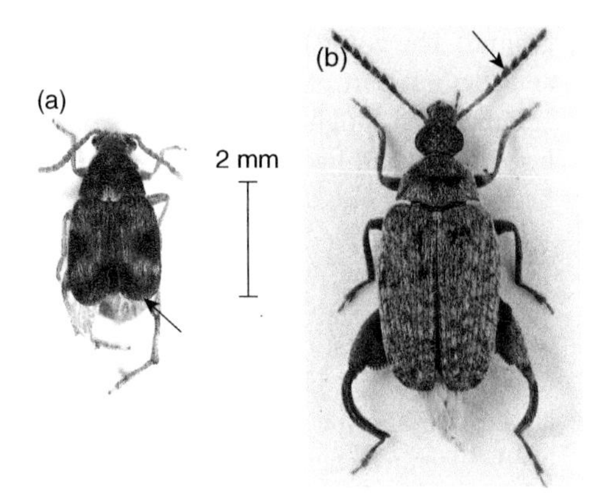

Fig. 12.63 (a) *Callosobruchus maculatus* (www.zin.ru); (b) groundnut seed beetle (*Caryedon serratus*) (www.mauritiusbeetles).

a single seed. The larvae eventually bore their way to near the surface of the seed, and pupate behind the 'window' so formed. Here they await the trigger of moisture before they emerge leaving an obvious round emergence hole. It is common for the adult to emerge from the pupa and remain behind the window till triggered by moisture to break through it. In wild plants the insects would eventually fall to the ground within the seed, but in crop legumes the infested seeds may well be taken into store. In some species that cannot develop in really dry seeds, there is no further spread of the infestation within the store after the emergence of the generation brought in at harvest.

Some species of Bruchinae, however, have adapted to life in store to the extent they no longer need adult food before oviposition, though fecundity is much lower than when adult food has been taken. Also, development to adult is not arrested by lack of moisture, so that successive generations can occur in a dry store once the first immigrants have arrived or infested seed has been brought in, and even with relatively low fecundities destructive populations can build up quickly. Several larvae may tunnel in a single seed, yet the tunnels never meet. The most important of these bruchine storage pests are in the genus *Callosobruchus*, with *C. maculatus* (Fig. 12.63a) perhaps being the best-known species.

These cosmopolitan *Callosobruchus* spp. usually infest cowpeas and other pulses by flying into the store and much more rarely enter pods in the field. The eggs are laid on the seeds, and the life cycle takes about a month. The adults are reddish-brown beetles, about 3 mm long, with darker markings particularly at the tips of the abdomen.

In other species, infestation normally starts in the field and is carried into store. *Acanthoscelides obtectus* is an important pest of beans and other pulses in southern Europe, Africa and the Americas from southern USA to most of South America. The adults are a blotchy grey and covered with yellowish hairs. There are about six generations a year.

Caryedon serratus (Fig. 12.63b), the groundnut seed beetle, attacks groundnut (and other legume) seeds in the field before being carried into the store, and is a serious problem in West Africa and parts of southeast Asia. Eggs are glued onto the pod wall or the seed in store soon after harvest, and then – like the other Bruchinae – the larvae eat out the cotyledons of the seeds. Pupation is in a thin cocoon on the outside of the seed or attached to the pod wall. The adult is rather oval and some 5 mm long with

Species	Distribution	Crop(s) attacked
Bruchidius atrolineatus	West Africa	Mainly field pest of cowpea, also in store
Bruchus atomarius	Europe and parts of Asia	Bean, lentil, pea
Bruchus dentipes	Widespread	Broad bean and other *Vicia* spp.
Bruchus lentis	Most warmer climates	Lentil only
Bruchus pisorum	Europe, parts of Asia, Canada	Peas in ripe pods
Bruchus rufipes	Europe, Asia, South Africa	Species of vetch
Callosobruchus chinensis	Cosmopolitan	Chickpea, cowpea, green gram
Callosobruchus theobromae	India	Pigeon pea

Table 12.2. Species of Bruchinae additional to those mentioned in the text.

distinctly serrate antennae. The hind femora are stout with a strong spine, and the life cycle takes 6–7 weeks under good conditions.

Table 12.2 lists some other important species of Bruchinae.

Subfamily *Alticinae (previously Halticinae)*

These are the flea beetles, so called because of their ability to jump. This ability lies in the enlarged hind femora (Figs 12.64a and 12.65a), which easily identify the Subfamily. The larvae of different flea beetle species feed on the roots or tunnel in the stem (Fig. 12.65b), but the root damage is not important compared with damage to the stem, but especially also with the feeding of the adults (see below in this Section). The larvae become full grown in a very short time and pupate in the soil. Most species have two or three generations a year in Europe before the emerging adults then return to shelter for the rest of the season and the winter (see below in this Section), but these pests occur on many different crops in many parts of the world, including the tropics.

Species of the genus *Phyllotreta* are serious pests of brassica seedlings, and have been known to make the crop only worth ploughing in. Examples of common species in Europe and elsewhere are *P. cruciferae* and *P. nemorum* (Fig. 12.64). The flea beetles spend most of the year sheltering outside the crop in sites like under fallen leaves in

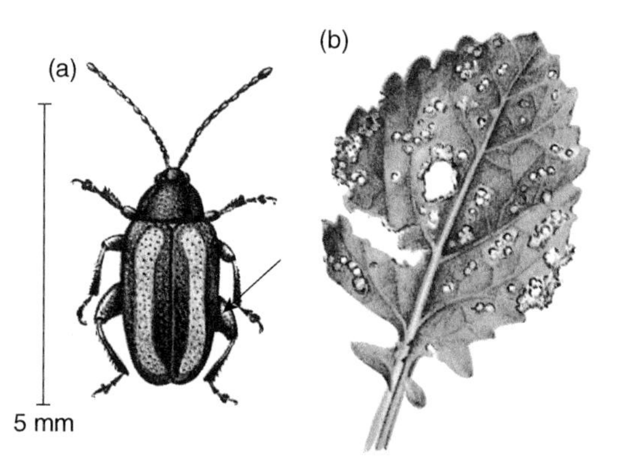

Fig. 12.64 *Phyllotreta nemorum*: (a) adult and (b) feeding damage done by the adults (from Bayer 1968, with permission).

hedges or at the edges of woodland. In early spring the adults fly to crops to lay their eggs. The adults feed on the cotyledons and leaves of young seedlings, biting out small areas of cells to leave almost invisible damage. The adults of most species are a uniform shiny black, but *P. nemorum* has two longitudinal yellow bands on the elytra; the larvae bore into the stems as well as making blister mines in leaves. Rather as with damage by capsid bugs (see Section 7.5.1.2, Family Miridae), growers who fail to spot early flea beetle damage soon notice it when the leaves expand and the damaged areas tear to form much larger holes with brown necrotic edges (Fig. 12.64b). By this time control measures are pointless, since the attack is usually over and the pest has left the crop. A non-insecticidal control measure used before DDT became available, and now resurrected by organic growers, is for two people to drag a sticky board over the crop to catch the beetles as they jump in response to the disturbance of the approaching board.

Chaetocnema concinna is the mangold flea beetle, the adults of which chew shot-holes in the leaves and cotyledons of mangold and sugar beet in Europe. Damage is not dissimilar from that of *Atomaria* (see Section 12.3.11.2, Family Cryptophagidae), and so this pest has also risen in prominence with the changes in management of the crop that leave less scope for compensation for seedling losses. By comparison, the second generation emerging in July feeds on the leaves of more mature plants, and really does insignificant damage.

Another genus of pests in Europe is *Psylliodes*, which at 3–4 mm long are quite large for flea beetles. *Psylliodes affinis* is the univoltine potato flea beetle; the adults feed on the leaves of potato and tomato while the larvae bore into the roots. *Psylliodes chrysocephala* (Fig. 12.65) is the cabbage stem flea beetle; here it is really only the larvae which do the damage, boring into stems and leaf veins, usually starting at the base of a petiole. When the stems are completely hollowed out, the plants die. There are two generations a year with considerable overlap so that adults, larvae and pupae may all be found at the same time. Other important genera of Alticinae are potato flea beetles in the genus *Epitrix* (some species even damage the tubers) in the USA and southern Europe and *Longitarsus*. *Longitarsus nigripennis* is a major pest of black pepper in India, whereas the main economic importance of the related *L. jacobaeae* is beneficial as a biological control agent of an important weed in North America, ragwort (*Senecio jacobaea*).

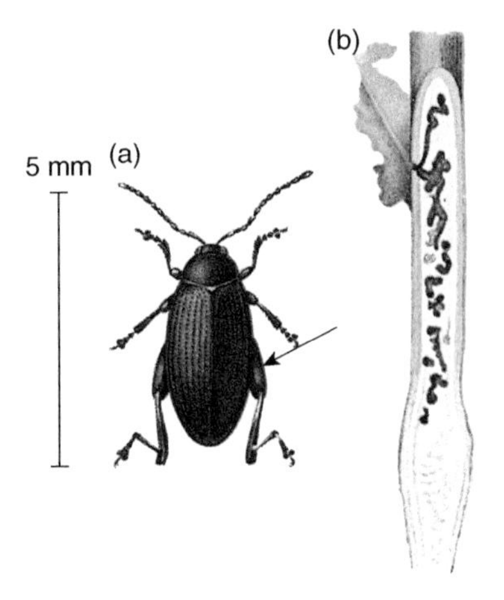

Fig. 12.65 Cabbage stem flea beetle (*Psylliodes chrysocephala*): (a) adult and (b) damage to oilseed rape stalk (from Bayer 1968, with permission).

Subfamily Chrysomelinae

Many of the defoliating chrysomelids are in this Subfamily. They tend to be host specific and so are known as pests in connection with particular crops. As for aphids, the name of the crop very often forms part of the common name. They usually have just one generation a year. Normally, both adults and larvae feed on the foliage, leaving black frass on the leaves, and pupation is in the soil. Most famous is the Colorado beetle *Leptinotarsa decemlineata*, a largish leaf beetle (*c.* 1.5 cm long) with bright yellow elytra between longitudinal black stripes (Fig. 12.66). The later instar larvae are pink

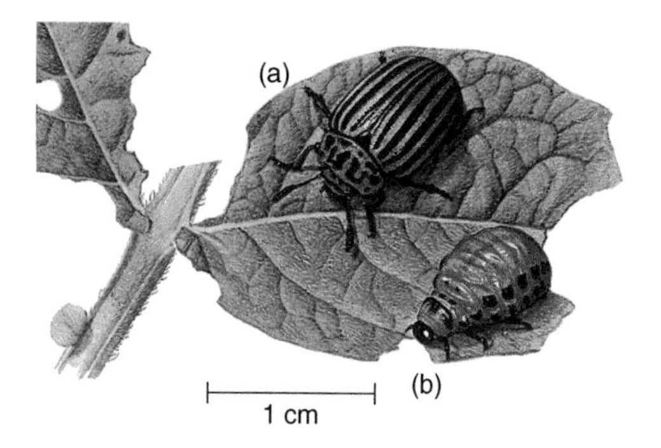

Fig. 12.66 Colorado beetle (*Leptinotarsa decemlineata*): (a) adult; (b) larva (from Zanetti 1977).

with lateral black spots on the abdominal segments. Larval development takes about 3 weeks and the whole life cycle about 5–6 weeks.

This pest of potatoes has been a causer of famines in its time. *Leptinotarsa decemlineata* was unknown until the entomologist Thomas Say discovered it on the eastern slopes of the American Rocky Mountains, where its food plant was buffalo-bur, a weed in the potato family. It then remained innocuous for several decades, until the pioneer settlers from Europe, planting potatoes for their sustenance, and the Colorado beetle met in the foothills of the Rockies. The survival and fecundity of the insect took off on potatoes, and it spread eastwards on potato crops at a rapid rate to reach the Atlantic seaboard by 1858. It soon transferred to Europe with the maritime trade, and has been a scourge of potato crops in southern Europe since then. Britain still (said he, thinking of climate change) has a climate that is probably unsuitable for the insect, for it has never established for any length of time, in spite of each year being brought in by holidaymakers from Spain who find the adult attractive and want to know what it is.

The Subfamily Chrysomelinae also contains the metallic blue mustard leaf beetle *Phaedon cochleariae*, which shot-holes the leaves of brassicas in Europe. It can occasionally be a serious pest on mustard crops, and is one of the few pests of consequence to watercress growers! It is most familiar to many entomologists as a research tool; it is the 'representative' beetle often cultured for teaching in universities and by industry for insecticide bioassay. Eggs are partly embedded in the foliage, and the larvae are yellowish with black spots on the abdomen and black legs. Other species of the genus *Phaedon* occur in the USA (*P. viridis*) and China and Japan (*P. brassicae*).

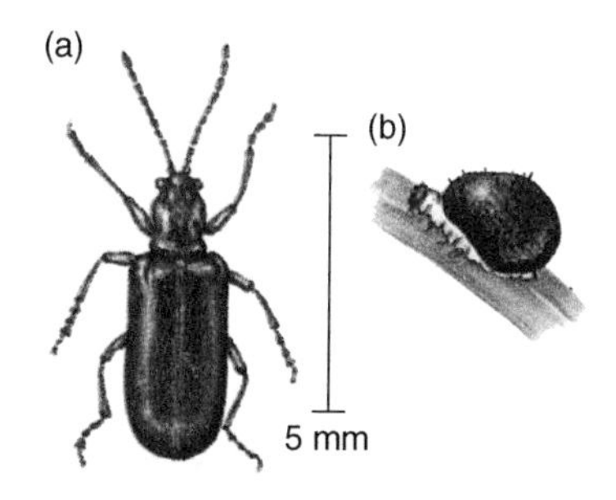

Fig. 12.67 Rice leaf beetle (*Oulema oryzae*): (a) adult and (b) larva (from Bayer 1968, with permission).

Subfamily Criocerinae

Here we find the black cereal leaf beetle, *Oulema melanopus*. This damages wheat in the USA; it also occurs in the UK but does not reach pest status. The related *O. oryzae* (Fig. 12.67), however, is a very serious problem on paddy rice in China and Japan. In this genus, both adults and larvae eat long strips out of the upper leaf epidermis, on which the eggs are laid. Severely damaged plants may be killed. The larvae camouflage themselves by covering themselves with their own frass (Fig. 12.67b). They thus appear almost black, but are actually yellow if the frass is removed. The larvae pupate on the leaf in a papery cocoon. Although the life cycle takes only about 6–8 weeks, the species is usually univoltine with the adults spending much of the year away from the crop, sheltering in leaf debris. The Subfamily also contains the asparagus beetle, *Crioceris asparagi* (Fig. 12.68). This pest is only of commercial interest to asparagus growers for it seems to be that relative rarity of a totally monophagous herbivore – it has never been recorded from any plant other than asparagus, though this includes both commercial and ornamental species. The adults are about 8 mm long, and more rectangular in outline than most Chrysomelinae. They are quite attractive; the head is black, the pronotum red and the elytra have large contrasting cream or yellow and black areas. The eggs are laid in a row on the feathery foliage on which the adults also feed, and the larvae are almost identical in green colouration to the host plant. They chew both the foliage and the spears. Pupation is in the soil or in hollow stems. There are two or three generations a year. The beetle is of European origin, but has also become established in North America.

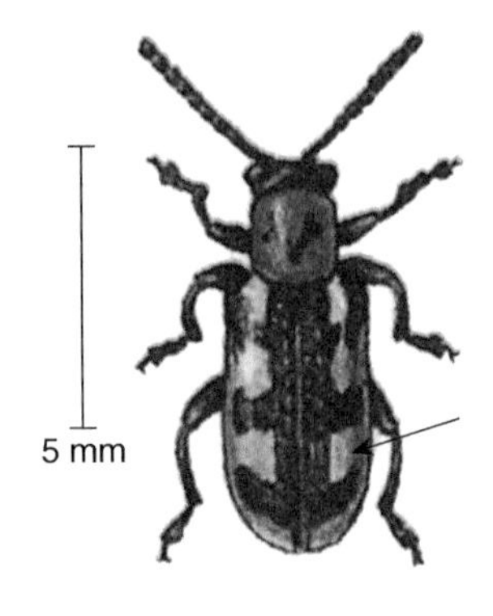

Fig. 12.68 Asparagus beetle (*Crioceris asparagi*) (from Lyneborg 1968, with permission).

Subfamily Galerucinae

In this Subfamily we find the economically important genus *Diabrotica*. *Diabrotica undecimnotata* and *D. balteata* (both known as the spotted cucumber beetle; the larva is known as the southern corn rootworm) are pests on cucurbits and maize in the USA and South America. The eggs are laid underground, and the larvae burrow into the roots, stems and cotyledons of young plants. Pupation is in the soil. The adults feed on the leaves and transmit bacterial wilt of cucurbits, a serious problem. The adult beetle is about 7 mm long, and is a yellowish-green with a black head and 12 large spots on the elytra (Fig. 12.69). The life cycle takes about 8 weeks and *D. undecimpunctata* has one generation a year; in contrast, *D. balteata* may go through as many as six generations. Other important *Diabrotica* species are *D. virgifera* (western corn rootworm), which spread into Europe in the 1990s, and *D. barberi* (northern corn rootworm).

Fig. 12.69 Spotted cucumber beetle, *Diabrotica* sp. (courtesy of Insect Sciences Museum of California).

Subfamily Hispinae

This has pest species associated with rice. Attacked leaves shrivel and die. The adults are recognisable by the spines on the thorax and abdomen (Fig. 12.70). Eggs are inserted into slits made in the rice leaf, and the larvae then burrow into the leaf and feed there, causing blotch mines. Pupation is within the mine. The adults also feed on the leaves, making white feeding scars. The life cycle takes 3–4 weeks. *Dicladispa armigera* (paddy hispid) is a serious problem of the rice crop in India and southeast Asia, whereas *Trichispa sericea* (rice hispid) is important in rice nurseries rather than after transplanting, and occurs in Africa.

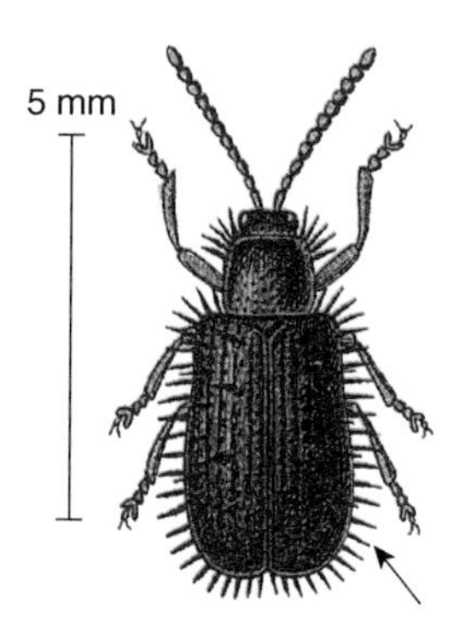

Fig. 12.70 Hispinae (*Dicladispa testacea*) (from Reitter 1908–1916).

Subfamily Cassidinae

These are the tortoise beetles, mentioned because they are not uncommon and are striking in appearance. The adults are often found on thistles, and you need to turn the beetle over to see the head, abdomen and legs etc. since the whole dorsal surface of the insect (the elytra and thoracic shield) hides the living beetle and forms a dome which, at the edges, seals flat against the leaf surface (Fig. 12.71a). Most are well-camouflaged by the green dome (e.g. *Cassida viridis*), but some other tortoise beetles are partly red, yellow or brown. Larvae of Cassidinae are also easy to recognise. Large spiny projections ring the whole body (Fig. 12.71b) and, for camouflage, the larvae impale their faeces and cast skins on these projections, especially on the two long posterior ones. *Cassida nebulosa* attacks Chenopodiaceae such as beet and spinach, and sweet potato (Convolvulaceae). The damage caused results in holes in the leaves (often clear windows with an intact epidermis) rather than in removal of large areas of leaf lamina. Pupae are attached to the leaves.

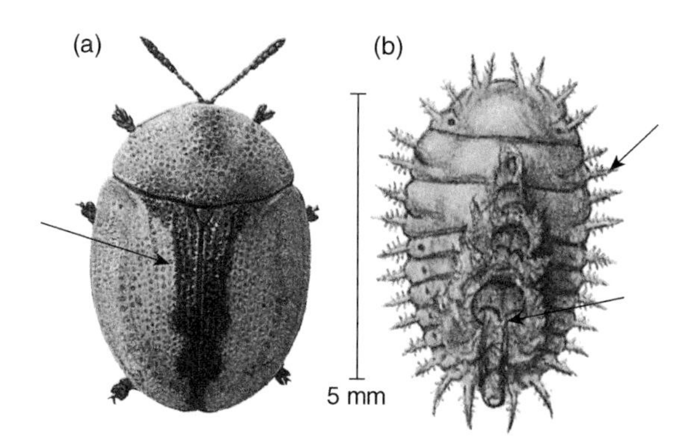

Fig. 12.71 (a) Adult *Cassida sanguinolenta* (from Zanetti 1977); (b) larva of *C. nebulosa* (from Lyneborg 1968, with permission).

Subfamily Clytrinae

Lastly in the Chrysomelidae I mention *Clytra quadripunctata* purely because of its bizarre life history. The larvae feed, not on plant leaves, but scavenge food materials in the mounds of wood ant nests. The most remarkable part of the life cycle is the way the larvae get to the ant nest in the first place. The female beetle climbs up the stem of a tree and then along a branch overhanging an ant nest. From there she drops eggs onto the mound, and the hatching larvae find themselves where they need to be.

Thus you only find the larvae in wood ants' mounds that have an overhanging branch from an adjacent tree!

12.3.13 Superfamily Curculionoidea

In contrast with the Chrysomeloidea, the antennae of this Superfamily broaden out at the end to terminate in a noticeable club. The larvae are usually apodous. Although not very easy to see in the Subfamily Scolytinae, the mouthparts in all the Subfamilies are carried at the end of a more-or-less elongated snout or 'rostrum'. Although this makes the head of many curculionoids look not unlike that of an elephant, it is important to remember than in the beetles the mouthparts are at the tip and not the base of the rostrum! The larvae (Fig. 12.5e) are nearly always white, curved and apodous (no legs), and therefore live in enclosed sites such as seeds, wood and the soil next to roots. All the species mentioned in this book are now in one Family, the Curculionidae, although until recently the Subfamilies Scolytinae, Platypodinae and Apioninae were treated as the Families Scolytidae, Platypodidae and Apionidae outside the Family Curculionidae.

12.3.13.1 Family Curculionidae

A number of taxa with clubbed and mostly geniculate antennae attached part-way along the rostrum.

Subfamily Scolytinae (bark beetles)

These bark-boring beetles are small and dark coloured. The pronotum tends to hide the head when viewed from above and the clearly clubbed but short antennae are geniculate (elbowed), though the elbow is often concealed in a dorsal view (Fig. 12.72). A rostrum is not obvious. It is common for the elytral edges to be raised up to form a sort of shovel for clearing wood dust when the adult tunnels. The tunnels that the insects make produce attractive patterns on the inside of bark when this is peeled away leaving the other half of the pattern on the branch or trunk (Fig. 12.73). Indeed, the tunnels follow the cambium, which provides far better nutrition than either the bark or the wood on either side. There is a central tunnel excavated by the

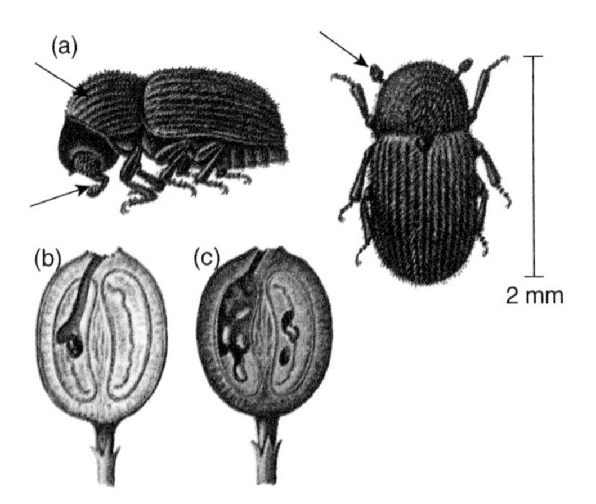

Fig. 12.72 Coffee berry borer (*Hypothenemus hampei*): (a) adult; (b) borer and feeding gallery in coffee berry; (c) injury caused by larvae (all from Bayer 1968, with permission).

adult female (with the male helping to clear the wood dust). Mating occurs when the tunnel has been started and the female then progresses the tunnel, laying eggs at intervals. The hatching larvae then bore away from the central tunnel, forming radiating branches. These branches of course widen as the larva grows, and pupation occurs at the end of the branch tunnel. A hole, outside and above where the pupa formed, shows where the adult emerged. Different species make different patterns of galleries. Some species bore into the wood rather than in the cambium. Many scolytines obtain much of their nutrition by grazing fungi on the walls of the tunnels; this is especially important for the wood-boring species, often called 'ambrosia beetles'. Spores of the appropriate fungus are carried by the mother and the tunnel is inoculated as oviposition proceeds.

Fig. 12.73 A typical bark beetle gallery pattern (from Zanetti 1977).

The genus *Scolytus* is easily spotted by the toothed front tibiae. The elm bark beetle (*S. scolytus*; Fig. 12.74) is the species which entered Britain from other parts of Europe in the 1920s to spread the infection of Dutch elm disease fungus in the UK. A major epidemic in the 1970s led to the death of most English elms (*Ulmus procera*) and a permanent change in the UK agricultural landscape.

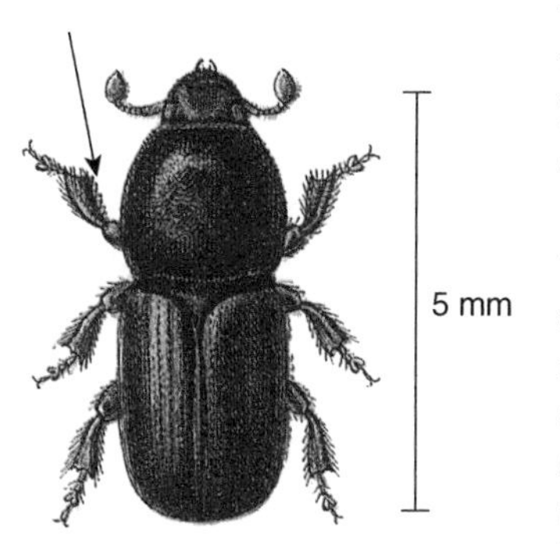

Fig. 12.74 Elm bark beetle (*Scolytus scolytus*) (from Reitter 1908–1916).

Scolytinae can tunnel so extensively as to kill trees, but usually any trees that die were already sickly and prone to attack. Healthy trees tend to be resistant to attack, and healthy conifers exude copious resin when damaged, embalming the adult scolytines and forming what are known as 'pitch tubes'. Thus the genus *Ips* and the western bark beetle *Dendroctonus* perhaps often act beneficially in removing sickly trees to allow new seedlings in the naturally regenerating forests of North America. A striking demonstration of this phenomenon occurred at the beauty spot of Lake Tahoe in the north of California. This became a focus for entrepreneur developers who saw there would be a demand for holiday homes near the edge of the lake, with the coniferous wooded setting and the shade so provided as a main attraction. Many buildings were put up cheek by jowl; indeed at times trees were pulled away from houses by ropes and winches to allow balconies to be built, after which the tension on the tree was slowly released till the return of the tree to its previous angle was arrested by contact with the balcony. The roads and other changes to the drainage of the site resulted in waterlogging and trees started dying, accelerated by the attention of bark beetles. This caused great consternation to those who had bought houses because of the woodland setting and resulted in great trouble for the developers.

Fig. 12.75 Tea shot-hole borer (*Euwallacia fornicatus*) (reproduced with permission of CSIRO, Australia).

Euwallacea fornicatus is the tea shot-hole borer (Fig. 12.75). It is dark brown to black, and about 4–5 mm long. Unlike many other scolytines, this insect does not specialise on unhealthy trees, but perfectly healthy ones. Much of the severity of the damage relates to the weakening of large branches of the bush that then break off as tea-pickers move through the plantation. Eggs are laid near branch junctions, and there the larvae tunnel in and, as just mentioned, weaken the junction. This borer is one of the ambrosia beetles (see earlier in this Section) which feed mainly on a fungus infection of their galleries introduced at oviposition from spores carried by the female; the fungus stains the wood in a diagnostic way. The fungus may eventually kill the bush/tree as happens with Dutch elm disease (see earlier in this Section). The sex ratio is heavily biased at 10 females to one male; the former fly short distances, but the males are wingless. The wingless males never leave the galleries in which they and the females of the new generation mate before the females disperse. The life cycle takes 30–35 days. The insect is found in the tropics from Madagascar, across Southeast Asia to Papua New Guinea.

Xyleborus dispar is the shot-hole borer of plums, apple and pear in Europe; though local in occurrence it can then be quite important, tunnelling into the branches and trunk and inoculating its galleries with ambrosia fungus.

A serious pest of both *robusta* and *arabica* coffee is the coffee berry borer, *Hypothenemus hampei* (Fig. 12.72), which is not a typical scolytine bark beetle at all in that it bores not in bark but in coffee berries. As with *E. fornicatus*, the sex ratio is heavily biased in favour of females at 13 : 1, and the males are wingless. The beetles are small black insects, 1.2–1.8 mm long. The adult female cuts holes in the maturing coffee bean and oviposits up to a dozen eggs in each of these chambers. Parthenogenesis can occur, whereby unfertilised eggs develop; it is thought that this is the result of a bacterial infection (*Wolbachia*). The holes cut by the female are seen near the tip of the green berries, and are diagnostic. So is the blue-green stain which appears on the surface of infected berries. Again the males do not leave the berries, but fertilise the females there before they disperse to new berries on the same tree, or to new trees. One tree may sustain three to five generations in a season, and there can be 100 insects in one berry. The life cycle takes between 25 and 45 days, and the pest is found world-wide wherever coffee is grown.

Subfamily Platypodinae

Like the Scolytinae, these are 'ambrosia beetles' whose larvae cultivate fungus gardens as their sole source of nutrition in their tunnels. However, Platypodinae do this in the xylem and not in the phloem and bark as do the scolytines. They only attack dying or recently dead trees. They may thus cause economic damage by tunnelling in standing dead timber or felled logs before milling. The Subfamily is mostly found in the tropics. Adults (Fig. 12.76) have distinctive parallel-sided and blunt-ended

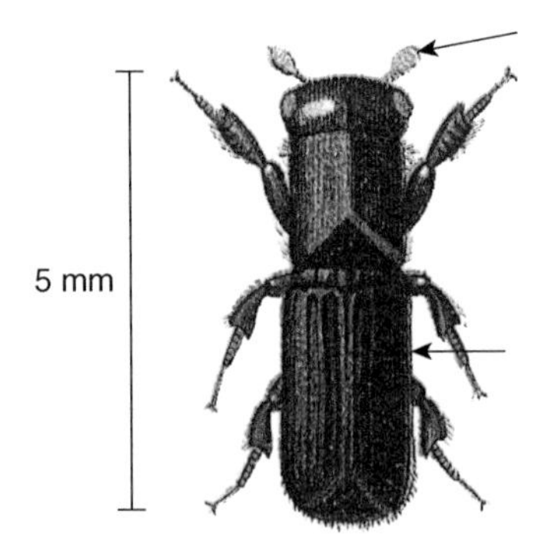

Fig. 12.76 Platypodidae (*Platypus cylindrus*) (from Reitter 1908–1916).

elytra and the length of the first tarsal segment exceeds the combined length of the remaining ones.

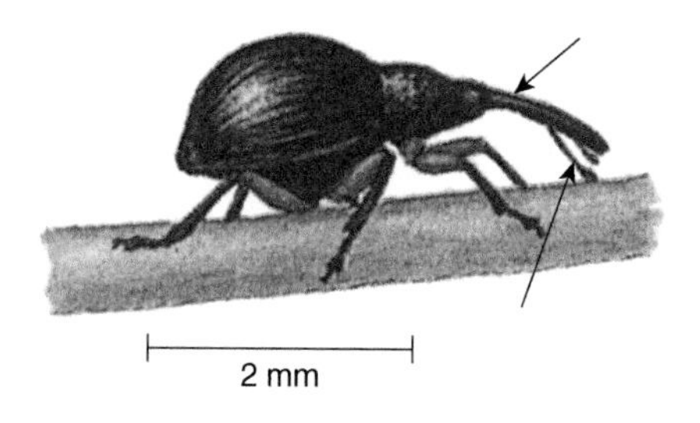

Fig. 12.77 *Apion apricans* (from Lyneborg 1968, with permission).

Subfamily Apioninae (seed weevils)

These are small weevils with an obviously elongated rostrum (Fig. 12.77); in contrast to the true weevils (see Section 12.3.13.1, Other Subfamilies), the antennae are straight and are only occasionally geniculate (elbowed). The elytra are narrower at the front than the back, giving the beetle a pear-shaped appearance. The trochanters are elongated compared with those of the true weevils (Section 12.3.13.1, Other Subfamilies). Although the elytra may be hairy, hairs are not flattened to coloured scales as found in the true weevils. Several species of the genus *Apion* are crop pests. The larvae of *Apion trifolii* (the clover seed weevil) attack the ovules and seeds of clover (particularly red clover) while the adults make holes in leaves and florets, which in a heavy attack can appear shredded. Clover is attacked after the adults have spent the first part of the spring feeding on the leaves of a variety of unrelated plants. The larvae pupate in the clover flower heads and emerge after only a week, so that a second generation is possible. Adults of the bean flower weevil (*A. vorax*) do minor damage to the foliage of field beans, but the damage they cause by transmitting *Broad bean stain virus* can be more significant.

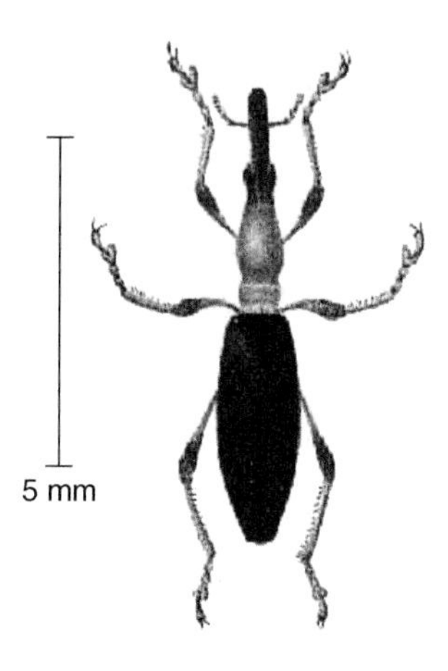

Fig. 12.78 Sweet potato weevil (*Cylas formicarius*) (reproduced with permission of CSIRO, Australia).

Two species of *Cylas* are pests of sweet potato. Both are serious pests of the crop with very similar biologies. Eggs are laid singly on the stem or actually into a tuber. The larvae tunnel inside both organs to feed, pupating there after a few weeks. Usually, their tunnels in the tubers become secondarily infected with bacterial rots, and so rotting will continue after harvest and storage. The adult is a slender black weevil, 6–8 mm in length. *Cylas formicarius* (Fig. 12.78) is the sweet potato weevil found both in Africa and India, Southeast Asia and the Pacific rim, whereas *C. puncticollis* is known as the African sweet potato weevil on account of its distribution being limited to tropical Africa, where both *Cylas* species infest the crop together.

Other Subfamilies (true weevils)

These have a wide antennal groove running forward from the point of insertion of the geniculate antennae. Very many are also coloured, either uniformly or in patterns, but magnification will show that the colours derive from flattened hairs like the scales of the Lepidoptera (Fig. 12.79). Often the antennal groove can be picked out because it is shiny and devoid of scales.

The Subfamilies in this group can be divided into long- and short-nosed weevils. Long-nosed weevils have an obviously extended rostrum (remember that the mouthparts are at the very tip), which may be extremely long (Fig. 12.80b). By contrast, the short-nosed weevils have a rostrum which is hardly prolonged at all, and may not be

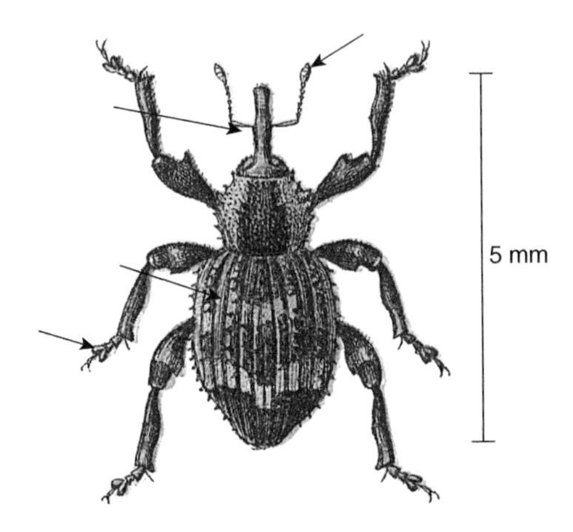

Fig. 12.79 A long-nosed weevil (from Reitter 1908–1916).

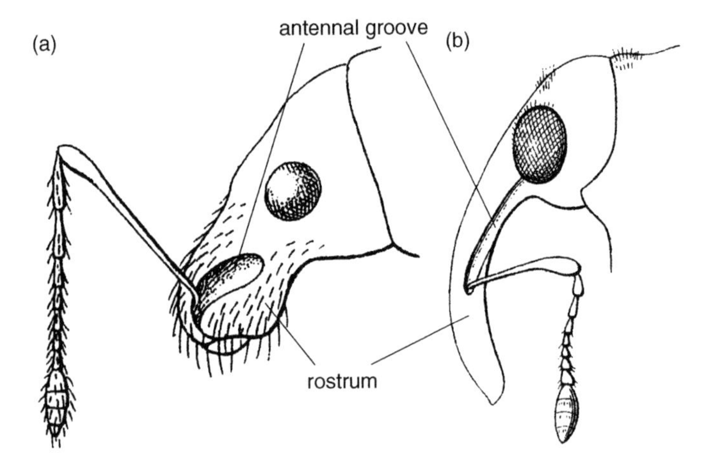

Fig. 12.80 Side view of head of (a) a short-nosed and (b) a long-nosed weevil (from Reitter 1908–1916).

detectable at all in dorsal view. However, viewed from the side, the antennal groove is obvious and gives away the Subfamily (Fig. 12.80a).

Larvae are apodous and often curved, they are usually creamy white with a light brown head capsule (Fig. 12.5e). That the larvae cannot move and thus develop adjacent to roots or within plant structures such as stems, petioles and fruit gives ample scope for species to be plant pests or pests of stored products, although leaf feeding by the adult can often contribute significantly to the damage caused by weevils.

Many larger weevils do not fly; the wings may be atrophied and the forewings (i.e. the hard elytra) fused along the mid-line.

Subfamily Entiminae These are the short-nosed weevils, also called broad-nosed weevils. Figure 12.81a shows the damage caused by the light grey-brown adults of *Sitona lineatus* (pea and bean weevil), named 'lineatus' from the longitudinal light stripes on the pronotum and elytra. The rather even and regular notching at the edges of the lower leaves of pea and bean plants (Fig. 12.81b) is probably more familiar to

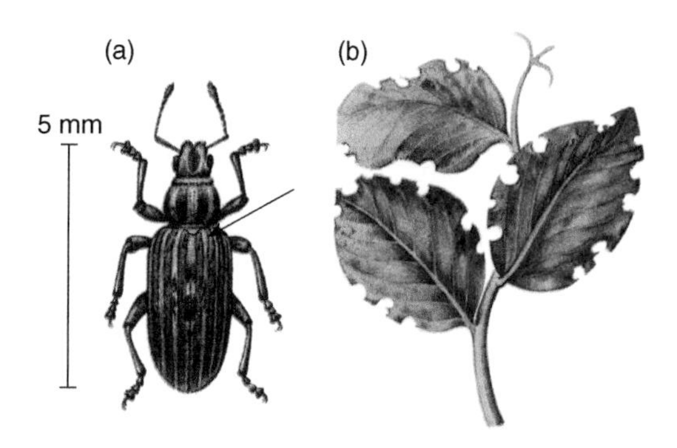

Fig. 12.81 Pea and bean weevil (*Sitona lineatus*): (a) adult and (b) notches on bean leaf (from Bayer 1968, with permission).

most gardeners than the adult weevil. However, this damage is more cosmetic than yield reducing and the real damage is done by the larvae feeding on and destroying the root nodules. Such damage also occurs to red clover grown as a forage crop, and economic losses are often caused. The insect also transmits two viruses, *Broad bean true mosaic virus* and *Broad bean stain virus*. However, *Apion vorax* (see Section 12.3.13.1, Subfamily Apioninae) is a more important vector of the latter virus. There is only one generation a year. The larvae pupate in the soil at a depth of about 5 cm and the emerging adults usually find the crop has been harvested and fly out of the field to seek any fresh young host foliage, before spending much of the rest of the year sheltering in leaf litter outside the crop fields.

Shiny cereal weevils (*Nematocerus* spp.) attack nearly all cereals, especially maize, barley and wheat in East Africa. Like *Sitona*, the adults notch the leaves, but with *Nematocerus* severe defoliation can result. Eggs are laid on the leaves, but the hatchlings drop to the ground and burrow into the soil where they feed on the roots, stem and germinating seeds. Pupation is in a cell in the soil. The adults are flightless; as with many other flightless weevils the elytra are fused.

Weevils in the genus *Phyllobius* are common in Europe on tree foliage or as flower visitors, and are easily recognised by the slightly powdery bright emerald-green scales which cover it completely – the weevils are of medium size. There is one generation a year. *Phyllobius pyri* is a minor pest on apple and pear foliage as adults, but it is the feeding by its larvae on the roots of grass swards which is more serious.

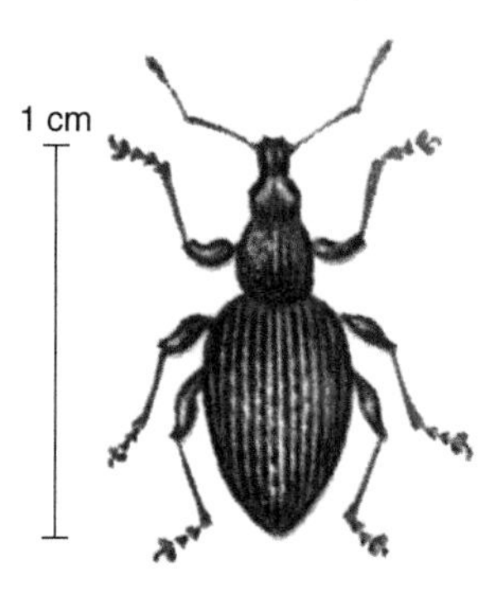

Fig. 12.82 Vine weevil (*Otiorynchus sulcatus*) (from Bayer 1968, with permission).

A group of wingless brown/black largish weevils in the genus *Otiorhynchus* are major problems in a variety of crops in temperate regions. All species have a particularly long antennal scape (second segment), and some have ventral teeth on the front femora. The adults are nocturnal feeders on leaves, leaving characteristic notches in the margins. Being weevils, the larvae are of course apodous and are white with a brown head capsule. They live and feed in the soil.

A serious pest of ornamentals, soft fruit, vines and forestry seedlings is the vine weevil (*O. sulcatus*; Fig. 12.82). Once again, although the adults are in evidence as leaf chewers, it is the larvae at the roots that do the real damage, which can be very serious. Damage frequently occurs to potted plants in nurseries. The larvae pupate and so overwinter in the soil, and there is one generation a year. Persistent soil insecticides gave good control until they were banned and ceased to be available, and the insect has really only developed as a serious commercial pest since that time. Fortunately, the nematodes *Steinernema* and *Heterorhabditis* are commercially available and can be used effectively as a biological control agent in the soil provided the latter is warm and kept moist. The clay-coloured weevil (*O. singularis*) is a pest of apple and raspberry. Here the larvae do little damage, and it is the feeding on the buds and flush growth by the adults that is the problem.

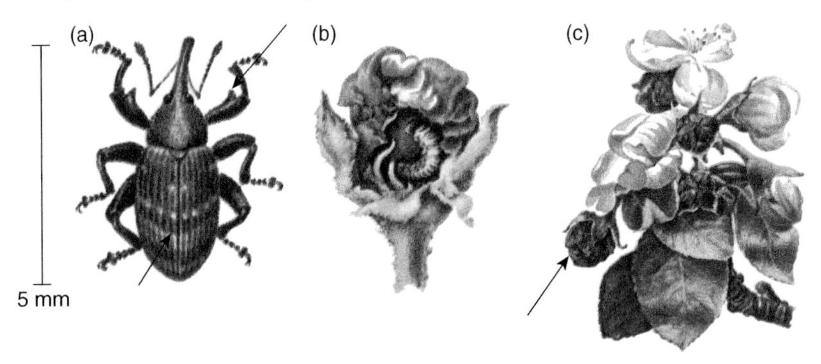

Fig. 12.83 Apple blossom weevil (*Anthonomus pomorum*): (a) adult; (b) larva feeding in blossom; (c) flower truss with normal and 'capped' blossoms (all from Bayer 1968, with permission).

Subfamily Curculioninae This is the first Subfamily of long-nosed weevils I shall mention. Apple blossom weevil (*Anthonomus pomorum*) is a small brownish beetle (Fig. 12. 83a) with a pale chevron on the elytra made by sloping bars on each side, and a large tooth on the front femur. The adults fly to apple trees when the flowers are still closed and at the 'pink bud' stage. They lay eggs in the flower and the larvae feed on and destroy the reproductive structures (Fig. 12.83b). However, they also bite and sever the base of the petals, so that attacked flowers do not open when the others do and remain as what are known as 'capped blossoms' with the dead petals turning brown (Fig. 12.83c). The larvae have a short development time and pupate in the flower, with the adult that emerges cutting a round hole in the cap in order to escape. It is a very brief life-cycle, hardly longer than the life of a flower and for the rest of the year the adults first aestivate (summer inactivity) and join this onto hibernation to spend nearly the entire year hiding in leaf litter close to the orchard. Similar biology applies to the closely related *A. piri*, the larvae of which feed inside and kill fruit buds on apple and pear, beginning in February; the insect can be a serious pest of pear on mainland Europe.

In the same genus, and again with that tooth on the front femur, *A. grandis* is one of the few insects to have had a song composed about it – the 'Boll weevil song' of Burl Ives. This rather large black shiny (i.e. no scales) weevil (Fig. 12.84) is indeed one of the major pests in the Americas of one of the most important world crops – cotton. In a way, the life-cycle is not dissimilar to *A. pomorum* in the long inactive adult life in litter outside the cotton fields after the crop has been harvested. However, should other cotton plants still be available in the area, several generations a year are possible. The adults feed on the flower buds ('squares') and flowers (Fig, 12. 84b) and the females make punctures in the squares and bolls into which they lay their eggs. The larvae (Fig. 12.84c) then tunnel into the tissues where they later pupate; the squares yellow and die and most, together with attacked young bolls, are shed. Larger attacked bolls remain, but fail to develop normally and the lint is stained and rotten. The life cycle takes 3 weeks and there can be seven to ten generations a year.

Subfamily Rhynchophorinae *Cosmopolites sordidus* (banana weevil) is a large (10–16 mm long) black long-nosed weevil (Fig. 12.85a) attacking both banana and plantain in nearly all banana growing countries in the world. The adults feed on decaying plant material and lay their eggs singly in cavities they chew in the pseudostem or in roots near the soil surface. They are active at night. The larvae (Fig. 12.85b) tunnel into the pseudostem and from there into the banana rhizome and the roots, but fairly shallowly in the soil.

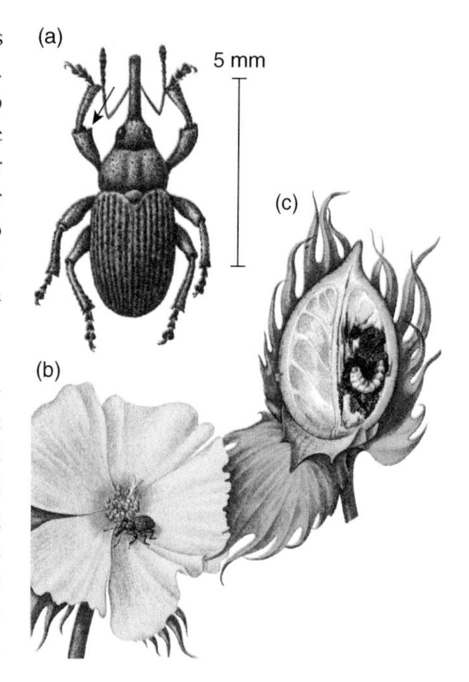

Fig. 12.84 Boll weevil (*Anthonomus grandis*): (a) adult; (b) adult on cotton flower; (c) infested boll (all from Bayer 1968, with permission).

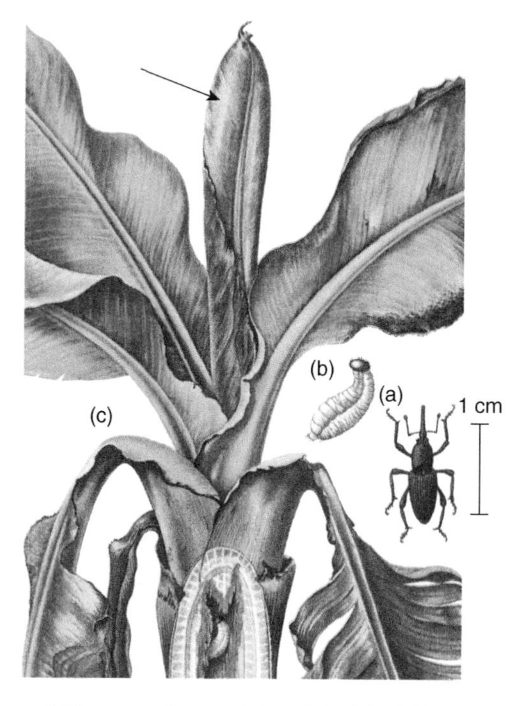

Fig. 12.85 Banana weevil (*Cosmopolites sordidus*): (a) adult; (b) larva; (c) damage with dying heart (all from Bayer 1968, with permission).

Rotting fungi then move into the tunnels and magnify the problem. The damage is so destructive that first the young leaves (Fig. 12.85c) and then whole plants wilt; the plants may finally fall over and die. Pupation is in the plant tissue, and the life cycle takes 3–5 weeks, though egg hatch may sometimes be delayed by as much as 5 weeks. The adults can survive for 6 months without food. Farmers try to trap the arriving beetles (which usually walk; they rarely fly), attracted by a volatile from dead dry banana leaves, by setting up pieces of banana stem which they later collect and destroy.

Probably the most serious pest of sisal is *Scyphophorus acupunctatus* (sisal weevil). It occurs widely in the tropics, especially East Africa, and also in the southern USA, the West Indies and the north of South America. Eggs are laid where damage and rotting has made the plant vulnerable; the weevil may even chew on the spike to start local rotting for successful oviposition. Over 3–8 weeks the larvae tunnel into the base of the flower spike and the plant usually dies eventually, especially if a young plant. The weevil is therefore particularly serious in nurseries and new plantings. To pupate, the larvae uses plant fragments to make a cocoon. The life cycle takes 7–14 weeks, with four or five generations a year. The adults are small (10–15 mm) black weevils that themselves feed at the base of the spike and remain in the area; dispersal is limited and slow.

Three species of small and mainly brown weevils in the genus *Sitophilus* are cosmopolitan pests of stored products. The females bite a small hole in the grain, insert an egg, and then seal the hole with a gelatinous secretion. Larvae feed within the grain/seed, first feeding on the living germ area and then eating the storage endosperm and then pupating. Pupation is in the grain and the neat round exit holes where the emerged adults have left the now hollow grain are obvious black circles. The minimum generation time in optimum conditions is around 30 days. Populations can grow rapidly and the respiration of the beetles heats the stored product, which not only kills it but drives water vapour to other areas where it condenses to wet the grain. This in turn allows moulds to grow so that the damage caused by the weevil extends further than its feeding.

Sitophilus oryzae (rice weevil) and *S. zeamais* (maize weevil), respectively, attack stored rice and maize. They are both very similar, with large round punctures on the thorax and elytra and four large lighter patches on the elytra. However, whereas *S. zeamais* is 3.5–4.0 mm long, *S. oryzae* is smaller at 2–3 mm. Unlike these two species, *S. granarius* (grain weevil; Fig. 12.86), which attacks wheat is wingless and cannot fly. It is again about 2–3 mm long, but has no light patches on the elytra, and the punctures (especially so on the thorax) are elongated rather than round.

Fig. 12.86 Grain weevils, *Sitophilus granarius* (from Zanetti 1977).

Subfamily Ceutorhynchinae Three species of *Ceutorhynchus*, a genus of largish grey long-nosed weevils, are pests of cabbage

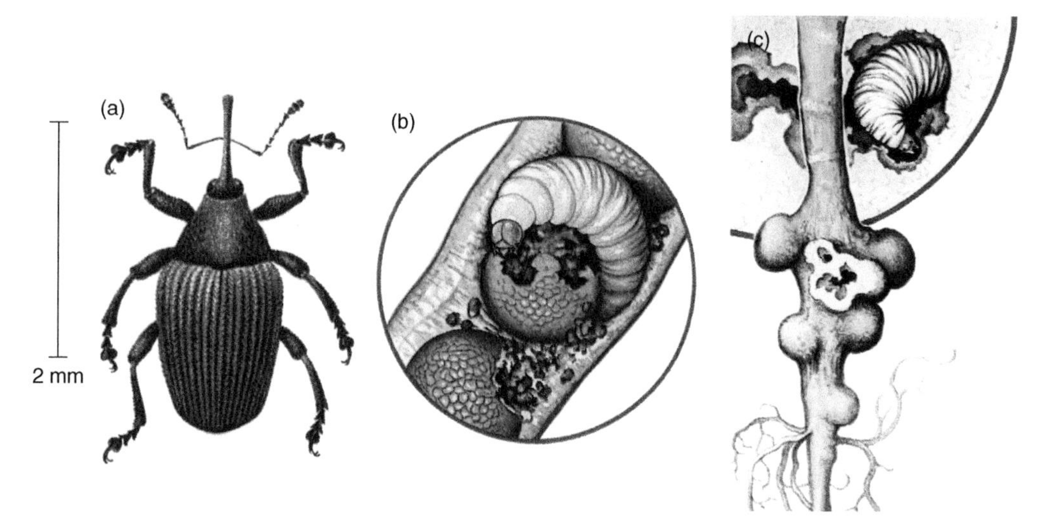

Fig. 12.87 (a) Cabbage stem weevil (*Ceutorhynchus obstrictus*); (b) larvae feeding in the oilseed rape pod; (c) turnip gall weevil (*C. pleurostigma*) galls on root of turnip (all from Bayer 1968, with permission).

and related plants, especially oilseed rape. The common names of the cabbage seed weevil (*C. obstrictus*; Fig. 12.87a) and cabbage stem weevil (*C. pallidactylus*), are indicative of the part of the plant damaged; both species are distributed in Europe and North America. The cabbage seed weevil is really of economic importance only in oilseed rape. The eggs are laid on the pods, and the larvae tunnel through the pod wall to feed on the developing seeds within (Fig. 12.87b). Pupation occurs in the soil and there are two generations a year. Stems of young plants and the base of the petioles of older plants are tunnelled by the larvae of cabbage stem weevil, and this can be a problem in most brassicas. Affected stems and leaves wilt and die. Before this, the symptom to look for at the base of petioles is a sort of silvering where air in the larval tunnel has replaced dense plant tissue. Pupation is again in the soil, but there is usually only one generation a year. The third species is the turnip gall weevil, *C. pleurostigma*. Here the adults, which are distinguishable by a spine on the front femur, emerge in May and feed on foliage, damage which is not important. Eggs are not laid on the roots till late summer, and the larvae make round galls on the roots of cruciferous crops, especially turnip (Fig. 12.87c). Damaged roots will be rejected for market. The species is restricted to Europe; there may be a second generation in some years.

Subfamily Hyperinae The alfalfa weevil *Hypera postica* is a serious pest of forage legumes in Europe, Asia and North America. The adult is only about 3 mm long with clearly defined light-coloured longitudinal stripes and bifid scales. The female lays up to 40 eggs in a cavity she makes in buds or stems in the spring. The larvae initially feed in the cavity, but after 3–4 days they move to the shoot tips and onto the leaflets; they also destroy flower buds. Pupae are formed in silk cocoons in the soil or on the foliage. After about 3 weeks the adults emerge and feed on the foliage before seeking shelter for hibernation. There is one generation a year.

13 Class Arachnida

13.1 Introduction

OK, this is supposed to be a book about entomology. So what is another Class of the Phylum Arthropoda other than the Class Insecta doing here? The answer is that, for applied entomologists rather than purists, some of the Arachnida (particularly the spiders and mites) are 'honorary insects' because they co-occur with insects in crop ecosystems as beneficial and pest organisms, which have to be recognised as such when making pest management decisions.

However, the Class is much broader than those members that occur in crops. Ticks live on animals and are thus also relevant for this book as pests of livestock in agriculture. But the Class has some Subclasses which will not be mentioned again after this brief introduction. These Subclasses include the Scorpionidea (scorpions, though agricultural workers in the tropics need to look out for them!), the Pseudoscorpionidea (common tiny predators that have scorpion-like pedipalps and live under bark and in leaf litter) and the Opiolones (harvestmen).

In this Chapter, we shall concern ourselves with only two Subclasses, the Araneida (spiders) and the Acarina (mites and ticks). The basic life history in these groups is rather simple. There is no major metamorphosis, and the 'nymphs' hatching from the eggs are not much different from miniature adults, and usually share the same habitat. Like insects there are a number of instars with moulting until the adult stage is reached.

13.2 Subclass Araneida (spiders)

Spiders (Fig. 13.1a) are immediately different from insects in having only two divisions to the body. The anterior division is the ***prosoma*** (=front body) and the posterior one is the ***opisthosoma*** (=back body). In relation to the insect division into head, thorax and abdomen, the division between the prosoma and the opisthosoma is part way down the abdomen. The prosoma bears a battery of single-lens eyes and the mouthparts. The latter are composed of six-segmented ***pedipalps*** and two-segmented ***chelicerae*** (the poison jaws).

Also on the prosoma are the legs, noticeably different from the insects in being eight instead of six. Also each leg has seven segments instead of the five (coxa, trochanter, femur, tibia and tarsus) of the insects. Yet the same names for the segments are used, and the number made up by having three trochanters – the first, the second and the third. Spiders have no antennae, and the first pair of legs often substitute rather than being used for walking. All spiders are apterous.

Handbook of Agricultural Entomology, First Edition. H. F. van Emden.

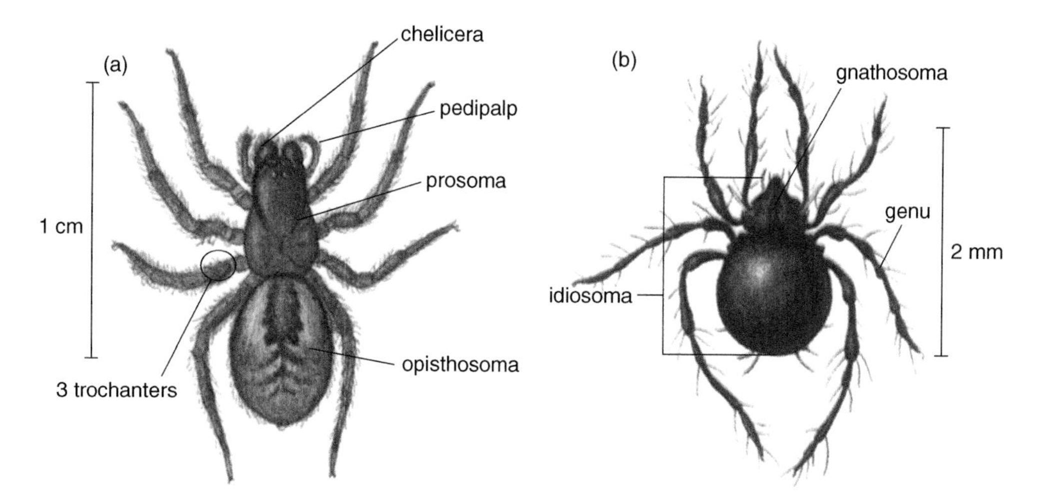

Fig. 13.1 External features of (a) a spider (from Lyneborg 1968, with permission) and (b) a mite (dorsal views) (from Mandahl-Barth 1974, with permission).

The ventral surface of the anterior part of the opisthosoma has the genital structures and openings as well as paired openings leading to the lung books, cavities which contain gill-like leaves at which the gas exchange of respiration occurs between the air in the cavities and the body fluid. At the posterior end of the opisthosoma are the spinnerets, a group of short barrels from which silk is produced to construct the familiar spiders' webs. The silk is also used by spiders (including non-web-spinners) for aerial dispersal. Young spiders throw out strands of silk which act as a parachute in reverse (paragliding?), so that spiders are lifted up by upward thermal currents, sometimes to a considerable height (up to 5000 m), and side-winds can then displace them a long way.

13.2.1 Family Araneidae

This Family includes the web-spinning spiders, which trap prey on the sticky web and then firmly immobilise it by spinning silk around it. A paralysing venom is injected from the poison jaws, which are then used to suck out the body contents of the victim. Often pest insects such as winged aphids are trapped, but so are beneficial insects such as small hover flies. It is, however, unlikely that any significant pest control or damage to biological control results.

Spider silk is incredibly thin yet strong, in fact even stronger than steel. Still nothing synthetic has been produced that can adequately replace silk from spiders for the fine cross-hairs of gun sights. A typical width of spider silk is only 0.003 mm.

In many species, the females care for the young, sharing food with them and carrying them around.

13.2.2 Family Lycosidae

These are the free-living hunting spiders, which forage for prey on the ground or by climbing plants. They are found in most crops in most countries, and are ubiquitous generalist predators *par excellence*.

They have been much studied in European cereal crops as predators of aphids, though perhaps partly because it is much easier to sample predators like spiders that blunder into pitfall traps than it is to sample plants for predators such as ladybirds and hover fly larvae. Many workers now feel their importance in temperate cereal crops has been overestimated, but there seems no doubt that they are an important component of the guild of predators in tropical crops, particularly rice.

Unfortunately, spiders are highly sensitive to most insecticides, and therefore tend to be absent in heavily sprayed crops.

13.3 Subclass Acarina (mites and ticks)

The mites (Fig. 13.1b) and ticks have just the one body region (the ***idiosoma***), though often the anterior end with the mouthparts is much narrower and projecting (the ***gnathosoma***). However, there is no cuticular demarcation of the gnathosoma from the idiosoma. As in spiders, there are no antennae; once again the first leg substitutes. The mouthparts are also segmented pedipalps (with up to five segments in the Acarina), but the chelicerae are three- and not two-segmented. The bottom two segments are fused to a cone and the third ones form the projecting stylets with which the food substrate is pierced to allow the juices to be sucked up (Fig. 13.5a).

The walking legs have up to six segments, with a segment called the ***genu*** (knee) between the femur and tibia (Fig. 13.1b), but they are not uniformly eight in number. Four pairs may be the commonest pattern but, as will be described later, one group of mites has only two pairs, and in most Families of mites the first instar (e.g. Fig. 13.6c) has only three pairs (like an insect). This first instar is known as the 'larva', and the subsequent eight-legged instars as 'nymphs'. The successive nymphal stages are called the protonymph (i.e. 'first nymph' but of course the **second** instar), deutonymph ('second nymph' but **third** instar) and tritonymph ('third nymph' but **fourth** instar). Like spiders, mites and ticks are apterous, but mites similarly disperse to great heights and long distances on silken threads.

It is when we consider the gas exchange mechanism for respiration that the Acarina spring a surprise. In all aspects considered in the above paragraphs, mites share characters with the spiders, but in respect to respiration they resemble the Class Insecta in breathing air through spiracles and circulating it in the body in tracheae.

Identification of mites to species is very difficult, involving very high magnification to look at the number, position and direction of the setae and, in some circumstances, the interference colours they can show under special lighting through the microscope.

13.3.1 Order Mesostigmata

Meaning 'middle spiracles', this Order is recognised by the single pair of spiracles situated laterally and close to the coxae in the middle of the idiosoma. There are no ocelli, and the pedipalps are five-segmented.

13.3.1.1 Family Phytoseiidae

These are rather oval mites found actively moving on plants. The last segment of the pedipalps has a slim two-pronged projection (the ***apotele***). The Family is of economic importance as predators of pest mites, and several species are reared commercially for release in glasshouses and orchards. Such release of predatory mites began in the 1960s with the red *Phytoseiulus persimilis* (Fig. 13.2) in all-year-round chrysanthemum and cucumber houses to control the glasshouse red spider mite (*Tetranychus urticae*

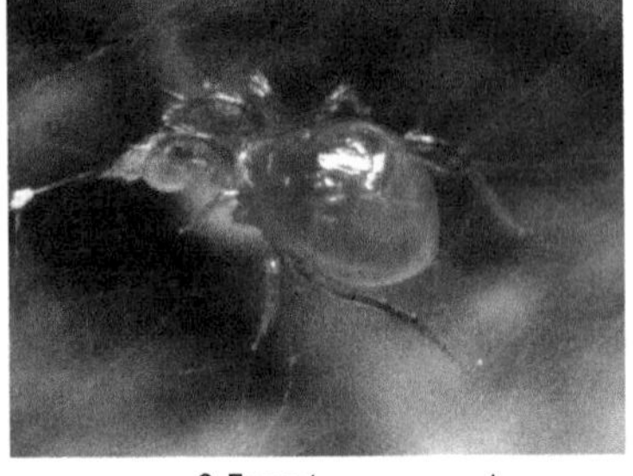

Fig. 13.2 *Phytoseiulus persimilis* (courtesy of Ward Stepman, BCP Certis).

and *T. cinnabarinus*). The release was part of an integrated biological control package for glasshouses, and the predatory mites were so effective that growers actually devoted some glasshouse space to rearing the pest mite for release in the commercial houses to stop the predator dying out! The predator hitches lifts on people, and so quickly invaded these pest-rearing houses. It then had to be controlled with an insecticide chosen because it did not kill the pest – one of the quirkiest stories of biological control! A second phytoseiid predator, *Amblyseius californicus*, is also commercially available for the control of *Tetranychus* species in glasshouses. *Typhlodromus pyri* (Fig. 13.3) is a pale phytoseiid applied outdoors at the rate of 25,000 per hectare to control fruit tree red spider mite (*Panonychus ulmi*). The predator is unfortunately very susceptible to chemical pesticides.

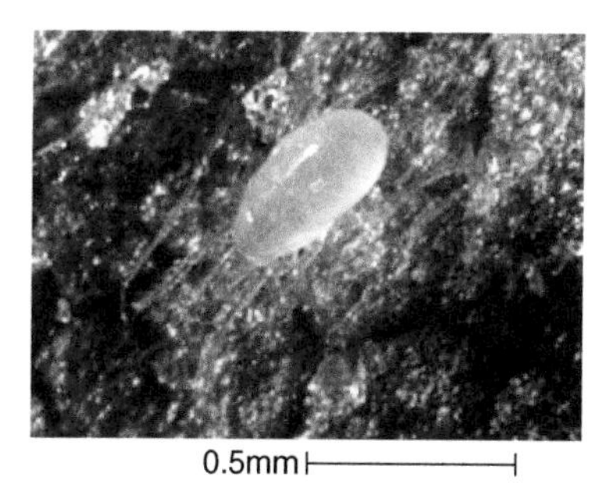

Fig. 13.3 *Typhlodromus pyri* (courtesy of Syngenta Bioline Ltd.).

13.3.1.2 Family Varroidae

Varroa destructor is an external parasite of honey bees; it is reddish-brown in colour and transverse in shape, about 1.6 mm long but up to 2 mm wide (Fig. 13.4). The mite transmits 'varroatosis', a particularly lethal form of the disease called 'deformed wing virus'. This is an important contributory factor to a syndrome known as 'colony collapse disorder'. This problem has caused enormous economic losses of entire bee colonies. The mite has long been known as a parasite of the Asian honey bee (*Apis cerana*), but jumped to the European honey bee (*A. mellifera*) in the Philippines in the early 1960s, since when it has spread to this species across most of the world, reaching the UK in 1992. At time of writing, Australia still seems free of the mite, though it reached the neighbouring New Zealand in 2000.

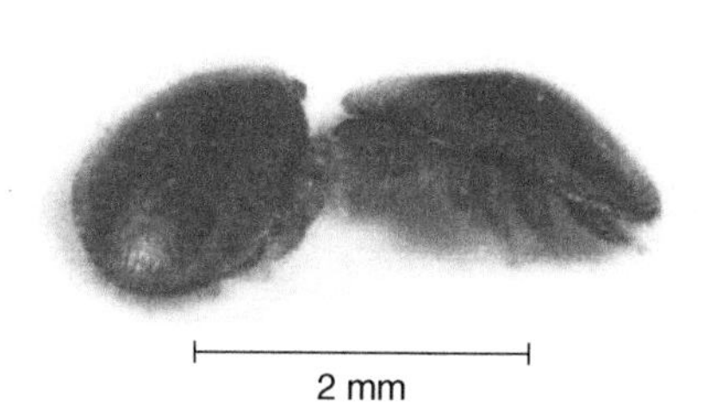

Fig. 13.4 Varroa mite (courtesy of Rothamsted Research).

The mites board adult bees when these visit flowers, and make open wounds to suck the haemolymph; at this point they inject the virus directly into the body of the bee. In the hive, the mite lays several eggs on a larva in a brood cell, and these eggs hatch just before the pupated bee emerges. The mites then spread to other cells and other adult bees.

13.3.2 Order Ixodida (ticks)

These are the largest mites (Fig. 13.5), and are blood-sucking external parasites of vertebrates and birds. When fully engorged with blood they can be more than 25 mm long. The gnathosoma has backwardly pointing barbs for anchorage in the skin of the host. Ocelli may be present or absent.

13.3.2.1 Family Ixodidae (hard ticks)

The name 'hard ticks' refers to a large semirigid plate (the **scutum**) which covers much or all of the body dorsally. Several are pests of vertebrates such as sheep (the sheep

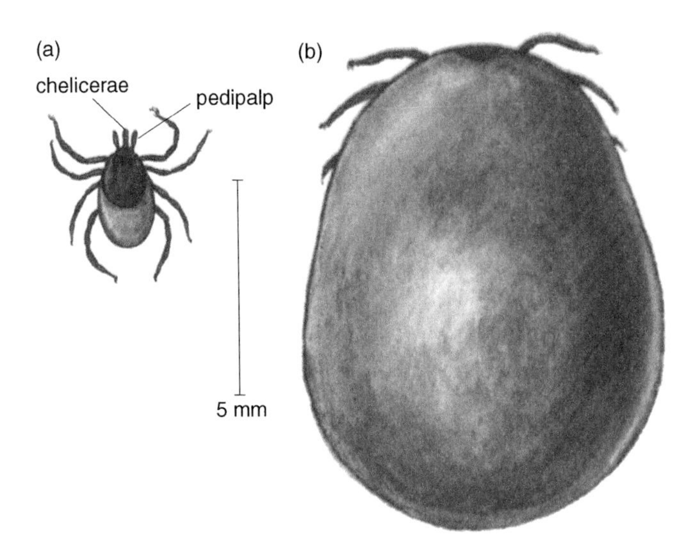

Fig. 13.5 The sheep tick, *Ixodes ricinus*: (a) before and (b) after feeding (from Mandahl-Barth 1974, with permission).

tick is *Ixodes ricinus*), cattle and man. They may transmit disease organisms such as rickettsia and those causing Lyme disease.

13.3.2.2 Family Argasidae (soft ticks)

The scutum is usually absent or indistinct. The argasids parasitise bats and birds, for example the fowl tick (*Argas persicus*).

13.3.3 Order Prostigmata

As 'prostigmata' suggests, the one pair of spiracles is sited well forwards, even as far as between the chelicerae. However, spiracles may sometimes be absent in this Order. The number of pedipalp segments varies from three to five. Ocelli may be present or absent. The tibia and tarsus are often narrower than the genu and other preceding segments, and together look rather as if the genu ends in a curved claw. This Order contains most of the plant-feeding mite pests.

13.3.3.1 Family Tetranychidae (spider mites)

These are mites with round bodies and long thin legs ending in the tibiotarsal claw referred to above. The mites look like miniature spiders and are reddish or greenish in colour; they may have a pair of bright red pigment spots anteriorly on the idiosoma. Silk glands serve as spinnerets, and are present on the pedipalps; silk strands and webbing are commonly produced.

Tetranychids rival aphids in their speed of population increase in spite of breeding by sexual reproduction. Eggs are laid in or on the food plant, and the hatching mites go through the typical acarine sequence of six-legged larva, protonymph, deutonymph, tritonymph, adult. However, some individuals (i.e. it is not a species characteristic) moult direct from the deutonymph to adult.

The fruit tree red spider mite (*Panonychus ulmi*; Fig. 13.6) attacks most fruit, particularly apple and nearly all soft fruit. It is distributed widely throughout Europe and the USA, as well as parts of Asia, temperate parts of Africa and Australasia. It has many

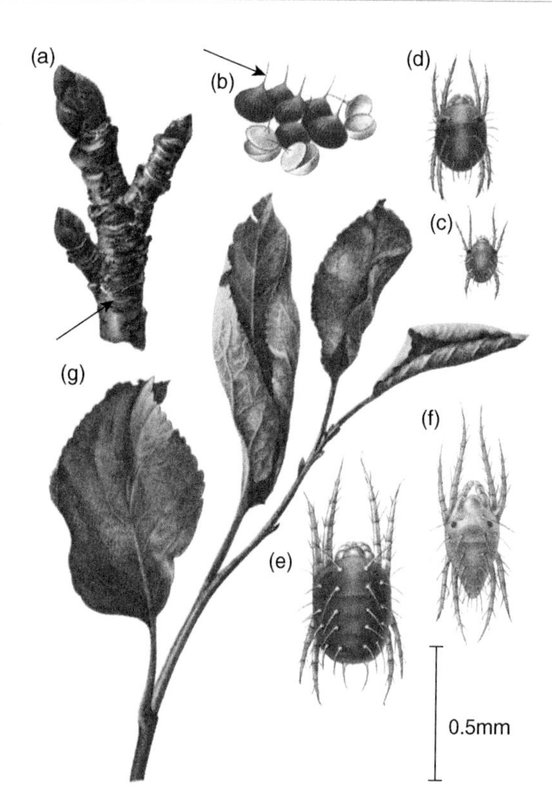

Fig. 13.6 Fruit tree red spider mite, *Panonychus ulmi*: (a) winter eggs on twig; (b) eggs enlarged; (c) 'larva'; (d) nymph; (e) adult female; (f) adult male; (g) leaf damage (all from Bayer 1968, with permission).

wild hosts including hawthorn, and young mites produce silken threads and are dispersed by the wind for long distances. The stylet-like mouthparts suck out the contents of individual cells which therefore lose their green colour leading to 'silvering' as the cell contents become replaced by air. As damage increases, the leaves begin to look mottled. Eventually 'bronzing' occurs when the leaves have lost virtually all their liquid contents. It may then even be possible to rub leaves into powder with the fingers.

Such destructive damage occurred in European apple orchards after World War 2, when DDT was introduced. In unsprayed orchards there are many species of predator that keep the mite in check. DDT, however, wiped out the predators not already killed from the 1920s onwards by highly toxic winter washes such as tar oil. These coat small insects and eggs and suffocate them. The red spider mite on apples is a classic example of a 'man-made pest'. Previously, it had hardly been noticed. Controlling it now meant major changes and reductions in the spray schedule for apples, but the mite rapidly developed tolerance to any pesticide applied. Nowadays, the predatory mite *T. pyri* (see Section 13.3.1.1) is an important component of the control strategy, and decisions on whether or not to spray are based on monitoring populations of both the pest and the predator.

The mite selectively survived the suffocating winter washes that killed its predators because its red winter eggs are laid in growth rings and other crevices; they breathe through films of liquid with a snorkel tube. This is an adaptation to surviving the films of water and lichens that run down the twigs after rain and cover the growth rings. At the time of maximum red spider mite problems, there were so many winter eggs that it looked as if someone had marked the growth rings on the twigs with red paint! The eggs are slow to develop in the spring, and do not hatch till May in the UK (i.e. after flowering). The young mites move onto the foliage and become adult to lay their own eggs in what may be less than 4 weeks, enabling about five generations a year in the UK. With each female laying up to 90 eggs on the leaves, increase can be very rapid. These summer eggs are quite unlike the winter eggs. They are smaller, orange rather than red and without a snorkel.

As mentioned earlier, there is a large guild of predators that prevent the mite building up in unsprayed orchards. These include Heteroptera, Neuroptera, Diptera, Coleoptera and of course other Acarina. A particularly important bug was the mirid *Blepharidopterus angulatus*, the black-kneed capsid (after its diagnostic feature!). This bug overwinters as a (snorkel-less) egg, and so was particularly hard hit by the winter washes. If spraying in an orchard is discontinued, it takes about 2 years for the guild of predators to re-assemble.

Fig. 13.7 Red citrus mite, *Panonychus citri*, with damage to citrus leaves on left. (Note: the mites are normally much redder than illustrated) (from Bayer 1968, with permission).

Panonychus citri (Fig. 13.7), the red citrus mite, causes silvering and necrosis of citrus leaves. Young twigs may die back and fruit drop can occur on attacked branches. The bright red eggs have a long snorkel, from the tip of which threads radiate to the leaf surface like guy ropes (Fig. 13.7). All active stages of the mite are red with strikingly white setae, though the artist of Fig. 13.7 seems to have used atypically pale specimens! The pest occurs in all the warmer parts of the world where citrus is grown, but is especially a problem in California and Florida.

The glasshouse red spider mite, often called *Tetranychus telarius*, is actually two species, *T. urticae* and *T. cinnabarinus*. *Tetranychus urticae* is also found on outdoor plants (see below in this Section); it is greener (Fig. 13.8) than *T. cinnabarinus* and

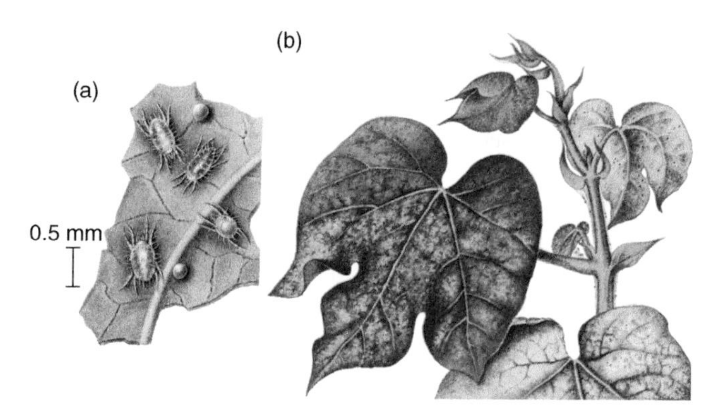

Fig. 13.8 (a) Two-spotted mite, *Tetranychus urticae* and (b) damage to cotton leaves (from Bayer 1968, with permission).

lays white eggs. As day length shortens to less than 14 hours, the mites become quite bright red and hibernate. By contrast, *T. cinnabarinus* is dark reddish all the time and lays eggs which are red or at least pink. It does not hibernate, but breeds continuously in the warmth of a heated greenhouse.

Both species predominate on the old leaves of plants, and the symptoms of attack (Fig. 13.8b) are often confused with leaf senescence. The mites move over the leaves on silk scaffolding produced from the pedipalps; as the mites make the leaf brittle with the removal of most of the cell contents, the silk serves to preserve the integrity of the leaf (rather like historic regimental flags in cathedrals are kept in silk bags to stop them disintegrating further). As the infestation progresses, damage moves up the plant and eventually – when the plant has really deteriorated – the mites assemble on the highest point, usually a flower or the top of a supporting cane. Here they hope to catch the wind or a passing horticulturist to disperse. Thus, like their predator *Phytoseiulus,* mentioned earlier, they are easily carried to another greenhouse on gloves, old canes etc.

It was mentioned earlier that *T. urticae* also occurs on outdoor crops. Across most of the world, particularly in the tropics and subtropics, it is better known by its American name of the two-spotted mite (after the paired red pigment spots characteristic of several tetranychid species). It can feed on hundreds of plant species and is a major and ubiquitous pest of most vegetables, especially of tomatoes, peppers and beans, but is also a pest of maize, cotton and hops as well as of many ornamentals.

The coffee red mite *Oligonychus coffeae* attacks both tea and coffee. On tea it is known as the red tea mite, but it is only an occasional local pest on that crop and tends to attack a group of bushes rather than spreading through the crop. The bright red eggs are laid on upper leaf surfaces near main veins, and change to orange just before they hatch. The deutonymph moults directly to the adult. In the same genus, *O. pratensis* is the Banks grass mite, a pest of maize and sorghum in Africa, the USA and Central and South America.

The genus *Bryobia* contains several pest species. The front legs are very long and held out forwards like antennae. *Bryobia praetiosa* is the polyphagous clover mite of clover, grasses and ornamental flowers; other species are pests of fruit, for example *B. cristata* (grass–pear bryobia mite) and *B. rubrioculus* (apple and pear bryobia mite).

13.3.3.2 Family Tenuipalpidae (false spider mites)

Brevipalpus phoenicis is one of several species of *Brevipalpus* which are known as 'red crevice mites' and which, as a group, occur throughout the world and can occasionally cause serious damage to ornamental trees, to many orchard crops and to tea and coffee because of toxic saliva. *Brevipalpus phoenicis* is especially found on the crops in the above list. Symptoms of feeding show as spotting of the foliage; fruit and yield loss is especially serious on plants suffering from some environmental stress, when plants may even be killed. Males are found only rarely, and reproduction is mainly by parthenogenenesis. The mites are orange or red and the eggs are oval and bright red; they are inserted in bark crevices or glued onto young leaves. Females lay just one egg per day, and it is the females that overwinter. The young mites are also red, and seem to have bursts of feeding activity before quite long periods of inactivity. The life cycle takes about 6 weeks.

13.3.3.3 Family Tarsonemidae

These are small semitransparent brownish or cream mites shaped like little beer kegs. The hind legs are reduced so that at first glance only three pairs of legs are obvious in dorsal view; in that view the most obvious clue to the existence of hind legs is a long bristle originating on the tarsus.

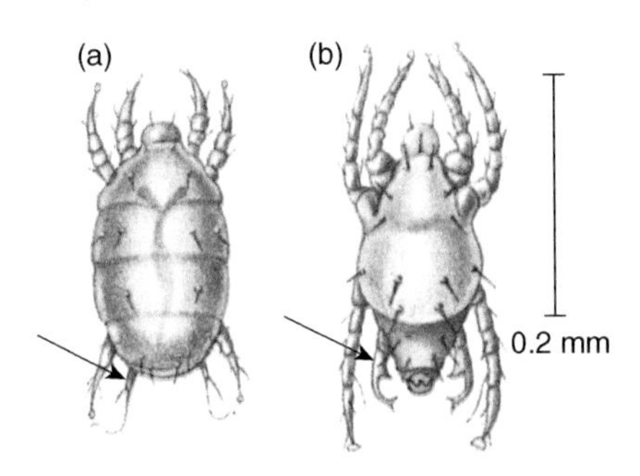

Fig. 13.9 Broad mite (*Polyphagotarsonemus latus*): (a) female and (b) male (from Bayer 1968, with permission).

The broad mite (*Polyphagotarsonemus latus*; Fig. 13.9) is, as its name suggests, a polyphagous species. In temperate regions it is a glasshouse pest of numerous ornamentals and vegetables such as cucumber and tomato. In the tropics and subtropics it is mainly a pest of tea (where it is known as the yellow tea mite) and cotton, but it is also a minor pest on very many other crops, including coffee, potato, tomato, citrus, beans and peppers. Colonies of mites occur on the undersides of leaves while the latter are still young, but such leaves are vacated when they mature. Attacked plants are stunted and the leaves of new growth are cupped or show other distortions, and corky brown areas may develop between the main veins on the underside of the leaves. The eggs are flattened underneath, and the upper surface has several rows of white tubercles. The minute white larvae stay near the egg-shells from which they emerged and after 2–4 days they moult into a quiescent nymphal stage, which only lasts a couple of days before the moult to adult. The females have very reduced hind legs. Males, which are the most active interplant movers, may carry quiescent female nymphs in a T-shaped pouch on their back at the tip of the abdomen (the hind legs are adapted for picking them up), but male nymphs are rarely moved. However, these can migrate quite actively themselves when they emerge.

The cyclamen mite (*Phytonemus pallidus*) has a subspecies *P. pallidus fragariae* (strawberry mite) which is an important field pest of strawberries. Otherwise the species is a glasshouse pest of ornamentals, particularly cyclamen and African violet. It occurs in Europe, Asia and North America. Adult females overwinter in the crowns of the plants and lay their white eggs in the spring. These eggs are quite large in that they are about half as long as the female mite! The growing points are attacked, the young leaves become deformed and the plants are stunted. The life cycle can take as little as 1–2 weeks, with continuous breeding possible in warm temperatures. Biology is not unlike that of the broad mite, with female nymphs being carried around by the males, but the quiescent stage takes the form of the nymph remaining in the unshed larval cuticle. The mites normally only move from plant to plant if the leaves are in contact, but they also disperse along strawberry runners.

The bulb scale mite (*Steneotarsonemus laticeps*) is a particular problem on forced narcissus bulbs. The mites are found in groups feeding between the scales at the neck of the bulb. Infested stored bulbs dry out and show brown necrotic scars if cut across below the neck. The emerging leaves of infested bulbs are weak, distorted, and with an unusually bright green colour. At this time the mites may spread onto the leaves, the bases of which appear to be dusted with fine white powder.

13.3.3.4 Family Eriophyidae ('gall mites', but there are also free-living species)

These are of a most unusual appearance for mites; in fact when I first saw them I thought they were something quite different! First of all they only have two pairs of legs, and then the body is annulated at very close spacing rather like an earthworm (Fig. 13.10a). They are very small – only up to 0.2 mm long and resemble flakes of ground coconut to the naked eye.

The blackcurrant gall mite (*Cecidophyopsis ribis*) is a generally distributed and major pest of blackcurrants, particularly as it transmits the *Reversion virus*. This disease causes flowering, fruit set and therefore the yield of bushes to decline. Another symptom is that the leaf shape and venation revert to the wild-type ancestral appearance, with fewer main veins and fewer serrations at the leaf edges. The life cycle is the key to the pest status of *C. ribis*. Mites enter new buds at the end of flowering in May (see the following paragraph) and there is a burst of egg production in early autumn, which results in some 4000 individuals per bud (I did say they were small mites!). A second burst of egg production follows in late winter, raising numbers to perhaps as many as 36,000 in a bud. Many of the buds swell to produce the 'big bud'

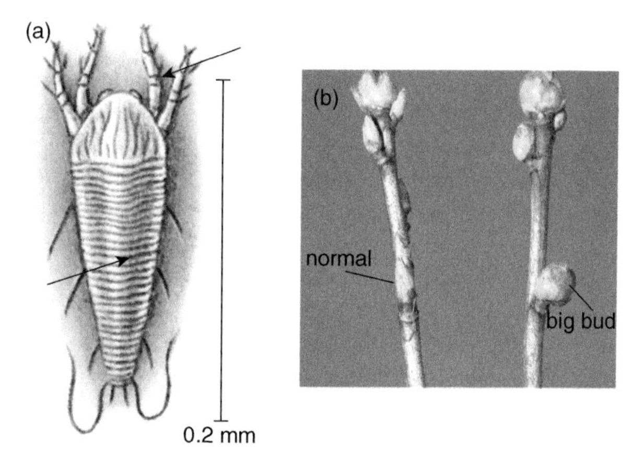

Fig. 13.10 (a) An eriophyid mite, *Calepitrimerus vitis* (from Bayer 1968, with permission). (b) Normal and 'big' blackcurrant bud (attack by *Cecidophyopsis ribis*) (courtesy of Rémi Coutin, INRA).

symptom (Fig. 13.10b). I cannot vouch for the truth of the story, but it is said that when big buds first appeared in a Scandinavian blackcurrant breeder's plants, he thought he had bred a super-vigorous variety which many growers in Europe then purchased and so rapidly spread the problem. Whether true or not, it is a nice story.

In early spring, mites swarm from the overcrowded buds onto the leaves, where they attach themselves with their terminal sucker and stand upright with their legs in the air, especially on the highest part of the leaf (usually the centre). However, I have seen illustrations that suggest numbers may be so high as to form an almost continuous fringe of little mites along the entire leaf edge. Why do the mites do this? They hope to grasp the legs of a passing pollinator such as a honey bee, which would enable them to disperse to another bush or even another plantation. This dispersal behaviour of hitching a lift on another animal is called phoresy, though wind also disperses many individuals. The peak of this swarming on the leaves occurs during flowering, which of course precludes the use of pesticides to control them because of the danger to pollinating insects. Towards the end of flowering, the remaining mites begin to move to enter new buds, and by the time the fruit begin to swell the leaves are again deserted.

Not all infested buds swell to form big buds, and a bush with more than 12% big buds cannot be saved and is better grubbed out and burnt.

There are two important eriophyid pests of citrus. One is the citrus red mite (*Phyllocoptruta oleivora*) – not to be confused with the red citrus mite *Panonychus citri* (see Section 13.3.3.1). The other is the citrus bud mite (*Aceria sheldoni*).

Of these, *Phyllocoptruta oleivora* is a common pest of citrus in many countries. The small white eggs are laid in depressions in the leaves or on the surface of the fruit. Affected fruit become discoloured, with lemons becoming silvery and oranges and grapefruit russet-coloured. On grapefruit this peel symptom is known as 'shark skin'. Leaves and young shoots may also show damage.

Aceria sheldoni is especially serious on lemon and grapefruit in the Mediterranean, the USA, South America, Australia and Africa. The eggs are laid in the buds, but the mites (which are yellowish or pinkish in colour) inhabit the buds, developing blossoms and their calyces. Nearly all parts of the plant, that is the twigs, flowers and the fruits become distorted; the fruits can look quite bizarre, and multiple buds are especially diagnostic.

There are many other eriophyid crop pests, for example *Eriophyes mangiferae* is the mango bud mite of the near East, south-east Asia and USA, and *Phytoptus pyri* is the pear leaf blister mite of Europe, India and Canada.

13.3.4 Order Astigmata

As the name suggests, no spiracles or tracheae are present. Ocelli are usually absent, and the two-segmented pedipalps are not obvious. Although some Eriophyidae in the Order Mesostigmata also have no spiracles, the unusual appearance of eriophyids prevents any confusion.

13.3.4.1 Family Acaridae

Acarus siro (flour mite) is a widespread pest of stored products in the milled or otherwise fragmented form, such as flour and oatmeal. The size and colour of the mites make them just about invisible in such substrates and their presence might pass unnoticed, were it not for the darker excreta and the unpleasant taint added by both the excreta and the cast skins.

13.3.4.2 Family Sarcoptidae

These are spherical mites with short stout legs, and which are well known as skin parasites of mammals and birds. The itch mite (*Sarcoptes scabiei*) causes scabies in humans and a form of mange in domestic and farm animals.

13.3.5 Order Cryptostigmata (beetle mites)

These mites, with their heavily armoured body, dark colour and winged expansions (***pteromorphs***) laterally at the front of the idiosoma, are easily recognised (Fig. 13.11), though of little economic importance. Most are herbivorous in leaf litter. However, in the Family Mycobatidae, the cherry beetle mite (*Humerobates rostrolamellatus*) is often mistaken for a pest of cherry since it clusters on split cherries and plums. Otherwise it lives harmlessly, but often in large numbers, on the tree bark feeding mainly on surface colonisers such as lichens and algae.

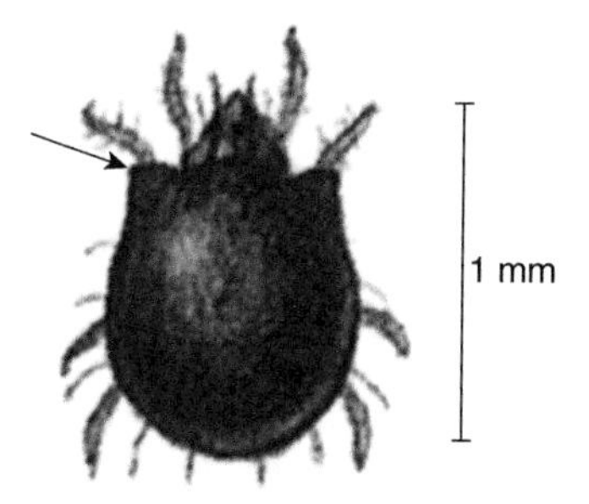

Fig. 13.11 A beetle mite, *Oribata setosa* (from Lyneborg 1968, with permission).

Bibliography

At the end of each reference, letters in brackets indicate: A, agricultural entomology; C, pest control/management; E, general entomology; I, source of illustrations. References that are solely a source of illustrations are in a smaller typeface.

There are of course many more similar texts; here I have selected ones that are currently recommended by colleagues in different parts of the world. Additionally, many groups of pests and beneficial insects (locusts and grasshoppers, thrips, aphids, stem-boring Lepidoptera, ladybirds and mites come immediately to mind) have their own dedicated textbooks. Although pest control/management is not the subject of the present book, I thought it helpful to include this topic in the bibliography.

Alford, D.V. (1999). *A Textbook of Agricultural Entomology*. Blackwell Publishing Ltd., Oxford, UK, 314 pp. (A,I)

Artigas, J.N. (1994). *Entomología Económica: Insectos de Interés Agrícola, Forestal, Médico y Veterinario* (in 2 volumes). Ediciones Universidad de Concepción. Concepción, Chile, 2069 pp. (A)

Atwal, A.S. (1976). *Agricultural Pests of India and South East Asia*. Kalyani Publications, Ludhiana, India, 498 pp. (A)

Bailey, P.T. (ed.) (2007). *Pests of Field Crops and Pastures: Identification and Control*. CSIRO Publishing, Collingwood, Australia, 528 pp. (A,C)

Bayer (1968). *Pflanzenschutz Compendium, volume 2*. Farbenfabriken Bayer, Leverkusen, Germany, pp. 281–511. (I)

Becheyne, J. (ed. Von Hajek, C.M.F.) (1956). *Beetles*. Thames and Hudson, London, UK, 158 pp. (I)

Bennett, W.G., Owens, J.M., Corrigan, R.M. and Truman, L.C. (1988). *Truman's Scientific Guide to Pest Control Operations*. Purdue University, Duluth, USA, 495 pp. (C)

Blackman, R.L. and Eastop, V.F. (2000). *Aphids on the World's Crops: an Identification and Information Guide* (2nd edn). John Wiley & Sons, Ltd., Chichester, UK, 466 pp. (I)

Borror, D.J., Triplehorn, C.A. and Johnson, N.F. (1999). *An Introduction to the Study of Insects* (6th edn). Saunders College Publishing, Philadelphia, USA, 874 pp. (E)

Capinera, J.L. (2001). *Handbook of Vegetable Pests*. Academic Press, San Diego, USA, 727 pp. (A)

Caswell, G.H. (1962). *Agricultural Entomology in the Tropics*. Edward Arnold, London, UK, 152 pp. (A)

Chapman, R.F. (1971). *The Insects. Structure and Function* (2nd edn). English Universities Press, London, UK, 819 pp. (I)

Chapman, R.F. (1998). *The Insects. Structure and Function* (4th edn). Cambridge University Press, Cambridge, UK, 770 pp. (E)

Chinery, M. (1986). *Insects of Britain and Western Europe*. Collins, London, UK, 318 pp. (E,I)

Chinery, M. (1993). *Insects of Britain and Northern Europe*. Collins, London, UK, 320 pp. (I)

Colyer, C.N. and Hammond, C.O. (1951). *Flies of the British Isles*. Frederick Warne, London, UK, 383 pp. (I)

Comstock, J.H. (1940). *Introduction to Entomology* (9th edn). Comstock Publishing Co., Ithaca, USA, 1064 pp. (I)

Handbook of Agricultural Entomology, First Edition. H. F. van Emden.

Cranshaw, W. (2004). *Garden Insects of North America*. Princeton University Press, Princeton, USA, 656 pp. (A)

Dent, D. (2000). *Insect Pest Management* (2nd edn). CABI, Wallingford, UK, 410 pp. (C)

Dipper, F.A. and Powell, A. (1984). *Field Guide to the Water Life of Britain*. Reader's Digest Association, London, UK, 336 pp. (I)

Edwards, F.W., Oldroyd, H. and Smart, J. (1939). *British Blood-Sucking Flies*. British Museum (Natural History), London, UK, 156 pp. + 45 plates. (I)

van Emden, H.F. and Service, M.W. (2004). *Pest and Vector Control*. Cambridge University Press, Cambridge, UK, 3439 pp. (C)

Folsom, J.W. and Wardle, R.A. (1934). *Entomology with Special Reference to its Ecological Aspects* (4th edn). Blakiston, Philadelphia, USA, 605 pp. (I)

Forsyth, J. (1966). *Agricultural Insects of Ghana*. Ghana Universities Press, Accra, Ghana, 163 pp. (A)

Gallo, D., Nakano, O., Neto, S.S. *et al.* (2002). *Entomologia Agrícola*. FEALQ, Piracicaba, Brazil, 920 pp. (A)

Gratwick, M. (ed) (1992). *Crop Pests in the UK: Collected Edition of MAFF Leaflets*. Chapman and Hall, London, UK, 490 pp. (A)

Hill, D.S. (1974). *Synoptic Catalogue of Insect and Mite Pests of Agricultural and Horticultural Crops*. University of Hong Kong, Hong Kong, 150 pp. (I)

Hill, D.S. (1983). *Agricultural Insect Pests of the Tropics and their Control* (2nd edn). Cambridge University Press, Cambridge, UK, 746 pp. (A)

Hill, D.S. (1987). *Agricultural Insect Pests of Temperate Regions and their Control*. Cambridge University Press, Cambridge, UK, 659 pp. (A)

Hill, D.S. (1994). *Agricultural Entomology*. Timber Press, Portland, USA, 635 pp. (A)

Hoffmann, G.H. and Schmutterer, H. (1983). *Parasitäre Krankheiten und Schädlinge an landwirtschaftlichen Kulturpflanzen*. Ulmer, Stuttgart, Germany, 488 pp. (A)

Horowitz, A.R. and Ishaaya, I. (2004). *Insect Pest Management: Field and Protected Crops*. Springer, New York, USA, 365 pp. (C)

Kogan, M. and Jepson, P. (2007). *Perspectives in Ecological Theory and Integrated Pest Management*. Cambridge University Press, Cambridge, UK, 570 pp. (C)

Koul, O., Dhaliwal, G.S. and Cuperus, G.W. (2004). *Integrated Pest Management: Potential, Constraints and Challenges*. CABI, Wallingford, UK, 329 pp. (C)

Kranz, J., Schmutterer, H. and Koch, W. (eds) (1978). *Diseases, Pests and Weeds in Tropical Crops*. John Wiley & Sons, Inc., New York, 704 pp. (A)

Le Pelley, R.H. (1959). *Agricultural Insects of East Africa*. Chiswick Press, London, UK, 307 pp. (A)

Lewis, T. and Taylor, L.R. (1967). *Introduction to Experimental Ecology*. Academic Press, London, UK, 401 pp. (I)

Lyneborg, L. (ed. Darlington, A.) (1968). *Field and Meadow Life*. Blandford Press, London, UK, 164 pp. (I)

Mandahl-Barth, G. (ed. Clegg, J.) (1973). *Pond and Stream Life* (3rd edn). Blandford Press, London, UK, 108 pp. (I)

Mandahl-Barth, G. (ed. Darlington, A.) (1974). *Woodland Life* (2nd edn). Blandford Press, London, UK, 179 pp. (I)

Masutti, L. and Zangheri, S. (2001). *Entomologia Generale e Applicata*. CEDAM, Padua, Italy, 978 pp. + 32 plates. (A,E)

Metcalf, C.L. and Flint, W.F. (revised by Metcalf. R.L.) (1962). *Destructive and Useful Insects: their Habits and Control*. McGraw-Hill, New York, USA, 1087 pp. (A,C,I)

Nair, M.R.G.K. (1986). *Insects and Mites of Crops in India*. Indian Council of Agricultural Research, New Delhi, India, 408 pp. (A)

Neuenschwander, P., Borgemeister, C. and Langewald, J. (eds) (2003). *Biological Control in IPM Systems in Africa*. CABI, Wallingford, UK, 413pp. (C)

Norris, R.F., Edward, P., Caswell-Chen, E.P. and Kogan, M. (2002). *Concepts in Integrated Pest Management*. Prentice Hall, Lebanon, USA, 586 pp. (C)

Parra, J.R.P., Botelho, P.S.M., Corrêa-Ferreira, B.S. and Bento, J.M.S. (eds) (2002). *Controle Biológico no Brasil. Parasitoides e Predadores*. Ediciones Manole, San Paulo, Brazil, 609 pp. (C)

Pearson, E.O. (1958). *The Insect Pests of Cotton in Tropical Africa*. Commonwealth Institute of Entomology, London, UK, 355 pp. (I)

Pedigo, L.P. and Rice, M. (2009). *Entomology and Pest Management* (6th edn), Prentice Hall, Lebanon, USA, 784 pp. (E,C)

Radcliffe, E.B., Hutchinson, W.D. and Cancelad, R.E. (2009). *Integrated Pest Management: Concepts, Tactics, Strategies and Case Studies*. Cambridge University Press, Cambridge, UK, 529 pp. (C)

Rappaport, R. (1992). *Tropical Agricultural Handbooks: Controlling Crops Pests and Diseases*. Macmillan, London, UK, 106 pp. (C)

Reitter, E. (1908–1916). *Fauna Germanica. Die Käfer des Deutschen Reiches* (in 5 volumes). Luiz, Stuttgart, Germany, 1667 pp. + 158 plates. (I)

Richards, O.W. and Davies R.G. (eds) (1977). *Imms' General Textbook of Entomology* (10th edn in 2 volumes). Chapman and Hall, London, UK, 1354 pp. (E,I)

Ross, H.H., Ross, C.A. and Ross, J.R.P. (1982). *A Textbook of Entomology* (4th edn). John Wiley & Sons Inc, New York, USA, 696 pp. (E,I)

Scott, R.R. (ed.) (1984). *New Zealand Pest and Beneficial Insects*. Lincoln University College of Agriculture, Canterbury, New Zealand, 373 pp. (A)

Spuler, A. (1910). *Die Schmetterlinge Europas, volume 3*. Schweizebartsche Verlagbuchandlung, Stuttgart, Germany, 95 plates. (I)

Thacker, J.R.M. (2002). *An Introduction to Arthropod Pest Control*. Cambridge University Press, Cambridge, UK, 360 pp. (C)

Toro, H., Chiappa, E. and Tobar, C. (2009). *Biología de Insectos*. Ediciones Universidad de Valparaiso, Valparaiso, Chile, 249 pp. (E)

Tremblay, E. (1994–2003). *Entomologia Applicata* (in 7 parts for different taxa, but not completed; the Hymenoptera and much of the Coleoptera are missing). Liquori Editore, Naples, Italy, 1675 pp. (A)

Uvarov, B. (1966). *Grasshoppers and Locusts: a Handbook of General Acridodology, volume 2*. Cambridge University Press, Cambridge, UK, 481 pp. (I)

Vasantharaj David, B. (2003). *Elements of Economic Entomology*. Popular Book Depot, Coimbatore, India, 562 pp. (A,C)

Vasantharaj David, B. and Ananthakrishnan, T.N. (2006). *General and Applied Entomology*. Tata McGraw-Hill, New Delhi, India, 1184 pp. (A,E)

Wagner, M.R., Cobbinah, J.R. and Bosu, P.P. (2007). *Forest Entomology in West Tropical Africa: Forest Insects of Ghana*. Springer Science and Business Media, Houten, The Netherlands, 244 pp. (A)

Walter, G.H. (2003). *Insect Pest Management and Ecological Research*. Cambridge University Press, New York, USA, 387 pp. (C)

Zahradník, A. and Chvála, M. (ed. Whalley, P.) (1989). *Insects. A Comprehensive Illustrated Guide to the Insects of Britain and Europe*. Hamlyn, London, UK, 508 pp. (I)

Zanetti, A. (1977). *The World of Insects*. Sampson Low, Maidenhead, UK, 256 pp. (I)

Index

Notes:
Page numbers in **bold** refer to figures, tables and boxes
The varying breadth of the host ranges of different insects makes indexing for individual crops impractical. For example, pests with tomato included in their host range may also be referred to as attacking vegetables, solanaceous crops or glasshouse crops.

Handbook of Agricultural Entomology, First Edition. H. F. van Emden.

Printed and bound by CPI Group (UK) Ltd, Croydon, CR0 4YY
16/04/2025
14658459-0003